Traveling from New Spain to Mexico

TRAVELING FROM NEW SPAIN TO MEXICO

MAPPING PRACTICES OF NINETEENTH-CENTURY MEXICO

Magali M. Carrera

DUKE UNIVERSITY PRESS
Durham & London 2011

Printed in the United States of America on acid-free paper ♾ Designed by Jennifer Hill. Typeset in Minion Pro by Tseng Information Systems, Inc.

Library of Congress Cataloging-in-Publication Data appear on the last printed page of this book.

To Alan

i carry your heart with me (i carry it in
my heart) i am never without it

e. e. cummings

CONTENTS

ILLUSTRATIONS

PREFACE

Maps depict real and imagined spaces. They approximate the boundaries and physical features of a space within the lines of longitude and latitude. At the same time, maps may also visualize the content of a physical space, locating wondrous places, curious sights, and strange people. Both of these mapping practices, geodetic and cultural, beguile and persuade the viewer to believe that truth is inherent in the map. And we want to believe because the map image offers a quick, succinct way to understand our physical and cultural relationship to other places—near and far.

Mapping practices depict real and imagined places, as well. This is especially manifest in sixteenth- and seventeenth-century representations of New Spain, viceregal Mexico. In early modern European world atlases, the landmass of New Spain is depicted as a territory located within the coordinates of longitude and latitude, as in Nicolas Sanson's *L'Amerique Septentrionale* (referring to the northern regions of the Americas; 1674) (figure 5, chapter 1, p. 29). At the same time, prints as well as the frontispieces and decorative cartouches of atlas pages repeatedly mapped the Americas allegorically as bare-breasted women wearing short, feathered skirts and headpieces, as seen in Jan van der Straet's 1638 engraving *America* (figure 4, chapter 1, p. 28). Mapping strategies such as Sanson's and van der Straet's visualize the physical boundaries as well as cultural content of New Spain as locatable, knowable, open, and accessible. Subsequently, late eighteenth-century travelers merged these earlier visualizations with supposed scientific principles to fabricate elaborate discourses about New Spain as a space full of awe-inspiring landscapes and abundant resources or, conversely, a deteriorated place, with degenerate flora, fauna, and people. As a result, from the sixteenth century to the early nineteenth, New Spain, which had been an unknown and, thus, an invisible physical and cultural space, became visualized as a place through maps and images.

Traveling from New Spain to Mexico: Mapping Practices of Nineteenth-Century Mexico elucidates the complex and diverse visual practices through which Mexico came to locate and define its national space. At the time of its independence from Spain in 1821, Mexico was well formulated as a geographic and cultural place but not as a political reality. Antonio García Cubas (1832–1912), a highly respected nineteenth-century geographer and cartographer, provided Mexico with critical mapped images of itself—both real and imagined—in the late 1850s. The magnificent *Carta general de la República Mexicana* (General Map of the Mexican Republic) from his *Atlas geográfico, estadístico, é histórico de la República Mexicana* (1858) (figure 46, chapter 5, p. 152) depicts the space of Mexico in the graticule, grid lines, of longitude and latitude; and at the same time, its title block illustrations imagine Mexico as a cultural and physical place. García Cubas's later works continue this fabrication of Mexico, culminating with the *Atlas pintoresco* [picturesque] *é historico de los Estados Unidos Mexicanos* (1885), whose thirteen pages, such as the *Carta agrícola*, which exhibits Mexico's immense agricultural wealth (figure 76, chapter 6, p. 213), present diverse Mexicos to export to the international stage as well as to inspire citizens.

In my original conceptualization of this project, I believed that I would undertake a rather straightforward iconographic and contextual analysis of Antonio García Cubas's major geographic and cartographic projects, which have received little scholarly study. However, my subsequent research made it clear that his atlas and map projects have complex and convoluted origins, and, at the same time, are highly interconnected to other nineteenth-century texts and images. The study of García Cubas's works requires a holistic assessment of how a newly independent Mexico came to map itself within the broader history of mapping. Consequently, García Cubas's works may be placed at a visual nexus: they are unequivocally related to the mapping practices of earlier centuries and, at the same time, embedded in emerging conditions and practices of nineteenth-century visuality, which were affected by new technologies and processes of visual reproduction (lithography, daguerreotype, photography) and methods of display (albums, museums, exhibitions, world fairs). By connecting old and emerging visual strategies, Antonio García Cubas fabricated credible and inspiring nationalist visual narratives for a rising sovereign nation, his beloved *patria*.

Consisting of an introduction and seven chapters, *Traveling from New Spain to Mexico* examines Mexico's program to map itself and define its national space across a palimpsest of earlier mapping practices as well as through an emerging nineteenth-century visual culture. The introduction establishes the theoretical and methodological foundations of this study. It begins with an overview of recent scholarly research on the history of Mexico's nineteenth-century art and cartography. I also delineate the current theoretical perspectives of the history of cartography and visual culture studies that inform my research.

The next two chapters summarize the 300-year history of European practices of constructing a space called New Spain, and then Mexico, from the sixteenth century to the middle of the nineteenth. These mappings were numerous and not sequential in their evolution; they demonstrate that New Spain, and ultimately Mexico, gained its form, content, and meaning through the cumulative effects of disparate kinds of images. Such mappings made visible to Europeans a territory that was initially conceived of as invisible, and they inaugurated an assortment of speculative narratives about the territory's meaning. In order to distinguish non-Spanish mappings from those of Spain and avoid overwhelming the reader with detailed information, I have divided this overview. Chapter 1 focuses on examples of French, Dutch, and English imagery through which New Spain's contents became envisaged by way of mobile and circulating images and objects that were subsumed into global paradigms. In chapter 2, I review Spain's particular contribution to the fabrication of New Spain. The Spanish crown's initial and paramount interest in mapping lay in using maps as way-finding devices to mark how to reach and then return from the Americas. Spanish mappings, in contrast to those from non-Spanish-speaking cultures, also focused on administrative objectives. Further, by the mid-eighteenth century, it is possible to locate within these mapping practices a formative but distinctively criollo—a Spaniard born in New Spain—cultural identity across texts, images, and objects. By the end of the eighteenth century, a small but expanding inventory of images—touting a unique geography, culture, and history—would feed into incipient nationalism and form a foundation for visualizing Mexican independence.

Chapter 3 examines the impact that an influx of foreign travelers had upon Mexico during the first decades of the nineteenth century. Their travel narratives were produced in the form of texts as well as lithographs, daguerreotypes, and stereographs. The result of these travelers' productions, however, was quite distinct from those of previous centuries because through travel writing and travel illustration the desultory themes and ideas of non-Spanish and Spain mappings began to cohere into extended narratives about Mexican national character and identity. Likewise, an emerging but disparate inventory of mobile images would begin to be aggregated into visual fabrications of Mexico—not New Spain—that supported these narratives.

In 1835, the historian and statesman Carlos María de Bustamante, posed three questions in an essay: "¿Quienes somos? ¿de donde venimos? ¿para donde caminamos?" (Who are we [Mexicans]? Where do we come from? To where are we going?). The next three chapters trace Mexico's search for answers to these questions through its visual culture over the following seven decades. Chapter 4 investigates how Mexican independence required both new and reconfigured allegorical imagery to destabilize and replace those fabricated by Europeans: seminude females were replaced by gowned, female figures of liberty. At the same time, while such renovated imagery gave form to the idea of the nation, it could not manifest the Mexican state, the entity that administers the nation. As a result, the inventory of geographic and cultural imagery originating in previous mapping practices and travel writing was recontextualized and expanded by state institutions, such as the national museum and art academy, into exhibitions that allowed Mexican citizens to locate themselves within certain historical and geographic coordinates.

Antonio García Cubas also proposed answers to Bustamante's questions through his mapping projects. Chapter 5 closely examines his works undertaken between 1857 and 1880, which were produced in concordance with and, at the same time, resistance to earlier mapping traditions and travel narratives. I also review the production and content of García Cubas's geographical educational materials during these decades.

Under the authoritarian regime of Porfirio Díaz (1876–1911), Mexican spaces required reformed mappings in the 1880s. Chapter 6 examines some of these new mappings as manifest in contemporary histories, and focuses on García Cubas's *Atlas pintoresco é historico* (1885), a synthesis and culmination of his earlier works. As the nation of Mexico unfolds within this atlas, I analyze its thirteen pages, identifying its thematic and image sources. By the end of the nineteenth century, García Cubas's cartographic projects had unified and framed the fragments of real and imagined Mexico that had emerged in the earliest mappings of New Spain and were expanded upon and mobilized in nineteenth-century travel writing and illustration. In this chapter, I also interrogate the silences within García Cubas's geography and cartography. Specifically, I elucidate Mexican women's absence across the spaces of the maps and atlases.

On 15 September 1910, as part of the Centenario de la Independencia de México (celebrating the one hundredth anniversary of Mexico's independence), three parades were planned to march through the streets of Mexico City. Within these parades, floats from each state illustrated its natural and commercial resources; floats from business companies displayed Mexican industrial production; and over 1,000 heavily costumed individuals re-enacted important events from Mexican history. This study concludes, in chapter 7, by analyzing these moving pageants as they, figuratively and literally, performed and displayed the nation of Mexico as it traveled from being New Spain to becoming an independent nation. Here, the real and imagined Mexico envisioned in Antonio García Cubas's maps and atlases came to life.

ACKNOWLEDGMENTS

This book is the result of diverse research opportunities, scholarly interactions, and personal support, which I wish to acknowledge. In spring 2003, I presented an exploratory paper on García Cubas's *Atlas pintoresco é historico* at the Twentieth International Congress of the History of Cartography at Harvard University. Exchanges with the national and international scholars attending this conference opened my eyes to the breadth and excitement of an expanding field. I also recognized that pursuing further study would require development of my knowledge of the field of cartographic history and theory. Fortunately, the following summer, I had the privilege of participating in the National Endowment for the Humanities Summer Institute, Reading Popular Cartography, at the Newberry Library's Hermon Dunlap Smith Center for the History of Cartography. The institute, under the superb leadership of Dr. James Akerman, was excellent and provided the background I needed to conceptualize a book project. The Newberry's librarians offered support and direction for my study as well. I gratefully acknowledge Jim's mentorship and ongoing support of this project. He pointed me to the right resources, the right scholars, and reminded me to stay close to the cartographic product and avoid getting lost in theory.

In summer 2005, I received a NEH Summer Stipend, which allowed me to focus on completing most of the research for the project in libraries and archives in Mexico, as well as the John Carter Brown Library, and Brown University's Hay Church Library and Rockefeller Library collections. Again, the help of the librarians in these institutions was critical. I also worked on eighteenth-century maps at the Museo Naval in Madrid, where I received the generous assistance of the collection's curators. Further, I have worked extensively at the Library of Congress's Map and Geography Division, whose librarians have had infinite patience with my multiple requests and introduced me to new and important material. I am grateful to Anthony Mullan, LOC librarian, for kindly sharing his wealth of knowledge about cartography in the Americas.

In addition, through participation in various scholarly conferences, I have benefited from discussions with excellent scholars of the geography and cartography of the Americas, including Jordana Dym, Karl Offen, Michael Edney, Ernesto Capello, Ricardo Padrón, Christina Connett, and John Hébert. I especially want to express my gratitude to Raymond Craib, who readily responded to my inquiries, generously shared materials, and read sections of manuscript drafts.

Institutional support has also been critical for completing my project. The University of Massachusetts, Dartmouth, provided research, travel, and publication subvention funds, as well as a sabbatical leave that allowed me to finish the first draft of the manuscript. At the same, university colleagues have listened to my ongoing fascination, bordering on obsession, with maps and mapping. I thank the librarian Linda Zieper, who is always helpful in locating research material for me. Harvey Goldman assisted me with photography in Mexico. Allison Cywin, curator of the Visual Resources Center, offered extraordinary assistance in the preparation of the images for my manuscript. Her "can-do" attitude is always reassuring. I thank Charlene Ryder, secretary to the Department of Art History, who has always been there for me with her quick and accurate assistance as well as her friendship. Dr. Michael Taylor, chairperson of the Department of Art History, has supported my projects and career in innumerable ways. While he claims to be a traditional art historian, he is not; he always asks incisive questions that make me think and rethink my research premises. Dr. Jen Riley, professor of English and Women's Students, is always available for intellectual discussion and personal encouragement.

My profound appreciation goes to Valerie Millholland, senior editor at

Duke University Press, who oversaw this project from prospectus to book. Anonymous readers offered excellent comments on the manuscript. I also express my sincerest gratitude to Miriam Angress, associate editor, who guided the preparation of the images for production; Sonya Manes, copyeditor, who brought consistency and clarity to the manuscript; and Mark Mastromarino, assistant managing editor, who guided the book to its final publication.

As with my previous book, enduring friendships and my family have sustained me through this project. Janet Freedman and Andrew Peppard provide continuing and rock-steady friendship. As we have raised our families together, Marion Wainer has availed me of her lively friendship, wisdom in child rearing, and thoughtful intellectual perspectives; Henry Wainer continually encourages and celebrates my successes. For my mother, Jo Carrera, all words of gratitude are inadequate to express my appreciation of her unswerving support and love. Alan Heureux, my companion in life, generously provides me with the emotional support to undertake my research and writing; I hope I can live up to the exuberant pride he always shows for my work. I thank Arin and Ana Heureux, my children, who support me without question and whose unfolding lives as remarkable young adults I observe with awe.

Despite all of the best suggestions and efforts of these colleagues, readers, editors, friends, and family, I remain responsible for all the weaknesses, omissions, and errors of this book.

Introduction

RESEARCH AND THEORETICAL PERSPECTIVES

> A man sets out to draw the world. As the years go by, he peoples a space with images of provinces, kingdoms, mountains, bays, ships, islands, fishes, rooms, instruments, stars, horses, and individuals. A short time before he dies, he discovers that the patient labyrinth of lines traces the lineaments of his own face.
>
> Jorge Luis Borges

Like Borges's character in the epigraph above, Antonio García Cubas produced maps and atlases of Mexico during the second half of the nineteenth century that did not so much locate land, people, or places in the lines of longitude and latitude as sketch the face of the nation. But exactly how and why did Mexico come to map a national identity for the first time in the middle of the nineteenth century? As I will elucidate in the following chapters, by the time it gained independence in 1821, Mexico, as a geographic space, was embedded in mapping traditions and practices that focused on locating New Spain as a bounded cartographic space and, at the same time, identifying its cultural content. For example, the "America" page of the 1638 *Theatrum Orbis Terrarum, sive, Atlas Novus* (Theater of the World, or a New Atlas), by Cornelius and Joan Blaeu, depicts a map of the American continent set in lines of longitude and latitude surrounded by small images of the supposed types of indigenous peoples who inhabit the landmass (figure 3, chapter 1, p. 27). The history of the mapping of New Spain and, subsequently, the formation of independent Mexico's national identity in the nineteenth century were formed through the intersection of spatial location with cultural location.

More than thirty years after independence from Spain, Antonio García Cubas created Mexico's *Carta general la República Mexicana* (1856). Addressing both geographic and cultural description of the Republic, this self-mapping project shaped a spatial stage for a refashioned cultural content of the Mexican nation. Over the next forty years, García Cubas's major geographic cartographic works—the *Atlas geográfico, estadístico é histórico de la República Mexicana* (1858) and, especially, the *Atlas pintoresco é historico de los Estados Unidos Mexicanos* (1885)—filled this stage with verbal and visual descriptions of Mexico's physical geography as well as its human and historical geographies. García Cubas's concepts of maps and their associated imagery were derived from earlier mapping and visual traditions and most influenced by the imagery developed in the late eighteenth century and first sixty years of the nineteenth.

Recognizing these contextual complexities of García Cubas's projects, I situate this study within recent scholarship of the academic fields that are the major focus of this work: the history of art, history of cartography, and visual culture studies of nineteenth-century Mexico. This overview is followed by a discussion of the theoretical perspectives that connect these fields and informs my analyses.

Interest in the art of nineteenth-century Mexico has proliferated in the last decades in the form of scholarly writings as well as museum exhibitions and catalogues. Scholars such as Esther Acevedo, Jaime Cuadriello, and Fausto Ramírez have published comprehensive analyses of the visual arts of late-viceregal and independent Mexico in their numerous books and articles. Their research and that of other academics appear in important exhibition catalogues sponsored by the Instituto Nacional de Bellas Artes and the Museo Nacional de Artes in Mexico City. These include *Nación de imágenes: La litografía mexicana del siglo XIX* (Nation of Images: Mexican Lithography of the Nineteenth Century; 1994) and the three-volume *Los pinceles de la historia* (Brush strokes or Sketches of History) series (1999, 2000, 2003), which provide excellent visual arts inventories and scholarly essays that delineate the development of the nation of Mexico through its visual history. In addition, the Fomento Cultural Banamex produced important exhibitions and catalogues—such as *Viajeros [Travelers] europeos del siglo XIX en México* (1996), *Pintura y vida cotidiana en México 1650–1950* (Painting and Daily Life in Mexico, 1650–1950; 1999) by Gustavo Curiel, et al., and *México: The Projects of a Nation 1821–1888* (2001)—which examine continuity and change in the visual art of Mexico.

Scholars have also undertaken detailed studies of the institutions and documentation of nineteenth-century art. For example, Eduardo Báez

Macías's *Guía [Guide] del archivo de la Antigua academia de San Carlos* reprints extensive documents that record the development of the national art school of New Spain, and Ida Rodríguez Prampolini's *La crítica de arte en México en el siglo XIX* reproduces the art and exhibition critiques from nineteenth-century Mexican newspapers and journals. Further, Esther Acevedo and Stacie Widdifield edited a multivolume series, entitled *Hacia otra historia del arte en México* (*Toward Another History of Art in Mexico*), which contains excellent scholarly articles that analyze the construction of Mexico through its artistic production from the late-viceregal period to the 1920s.

Stacie Widdifield has also written various articles and an influential book, *The Embodiment of the National in Late Nineteenth-Century Mexican Painting* (1996). Other scholars—such as Barbara Tenenbaum, Mauricio Tenorio-Trillo, and Rebecca Earle—continue to take up the topic of Mexico's visual art of the nineteenth century as part of larger historical studies. In addition, diverse doctoral dissertations coming from universities in the United States and Mexico—such as those of Raymond Hernández-Durán (*Reframing Viceregal Painting in Nineteenth-Century Mexico*; 2005), Christina Bueno (*Excavating Identity*; 2004), and Luis Granados (*Cosmopolitan Indians and Mesoamerican Barrios in Bourbon Mexico City*; 2008)—indicate that nineteenth-century Mexican visual art is of growing interest to young scholars. Overall, this scholarly literature on nineteenth-century Mexican art provides a broad understanding of the production and documentation of the visual arts. For the most part, this international art-history scholarship remains focused on stylistic, iconographic, and historical analyses with some broad reference to theoretical writings, such as those on nation building or visual culture.

Recent studies of the history of cartography of Mexico constitute a more limited bibliography.[1] Elías Trabulse's 1983 *Cartografía mexicana: Tesoros de la nación siglos XVI a XIX* (Mexican Cartography: National Treasures from the Sixteenth to Nineteenth Centuries) offers a useful survey of the general cartography of Mexico with a few examples from the nineteenth century. Héctor Mendoza Vargas's *México a través de los mapas* (Mexico Through Maps; 2000)—containing essays by various scholars focused on maps, mapping history, and mapmakers between 1520 and 1920—provides important examples of current cartographic research. His recent anthology, coedited with Carla Lois, *Historias de la Cartografía de Iberomérica: Nuevos caminos, viejos problemas* (Histories of Iberoamerican Cartography: New Roads, Old Problems), brings together current cartographic and geographic research. Luisa Martín Merás's 1993 *Cartografía marítima hispana: La imagen de América* (Hispanic Maritime Cartography: The Image of America) provides an excellent overview of Spanish maritime mapping of the Americas. In addition to these more synoptic works, several studies of specific periods of Mexico's cartographic history have been published during the past twenty years. For example, Barbara Mundy's *The Mapping of New Spain* (1996) surveys the early mapping of New Spain and examines the production of maps during the colonial period. Richard Kagan reviews Spanish and indigenous mapmaking of Mexico and Peru as part of his larger study of cities in *Urban Images of the Hispanic World 1493–1793* (2000). *La Gran Linea: Mapping the United States–Mexico Boundary, 1849–1857* (2001) by Paula Rebert investigates the mapping of the border between Mexico and the United States after Mexico lost half of its territory to the United States. Ricardo Padrón's book *The Spacious Word: Cartography, Literature and Empire in Early Modern Spain* (2004) obviously addresses mapmaking in Spain, but the author also brings cartography of New Spain into his analysis. Finally, Raymond Craib's articles and his book *Cartographic Mexico: A History of State Fixations and Fugitive Landscapes* (2004) have led the way to a more comprehensive understanding of the history of Mexican cartography in the context of the growth of the nation. Overall, this growing bibliography continues to work on the recovery of a history of cartography of Mexico.

It is both curious and problematic that within the scholarly discourse on the visual art or the cartography of nineteenth-century Mexico, a connection between mapping and visual culture has not been a critical thread.[2] In fact, a review of all the essays in the exhibition catalogues on nation building mentioned above reveals that no articles address how maps and mapping practices impact nationalist art. Nor do cartographic studies consider maps in relation to art production. This oversight is particularly noteworthy because, as visual artifacts, maps share the same cultural context as other visual arts. This is to say that through self-mapping, Mexico would begin to imagine the content and meaning of a place previously known as New Spain and formulate a space called Mexico. The visual arts—painting, sculpture, architecture, and, especially, prints and photography—produced, reproduced, and circulated components of this imagined space.

This separation of map production from nationalist art production is the result of disciplinary boundaries as well as a pervasive notion in scholarly as well as popular writing of the existence of a cartographic ideal. This ideal links maps categorically to a scientifically measurable, that is, empirical, territory, and ignores their relevance to the field of visual art.[3] This cartographic ideal assumes that "maps necessarily refer to a physical landscape and their history is the history of their ever-lasting accuracy and comprehensiveness."[4] Such a view also "presupposes some foundational object against which the distortions and interpretations can be measured."[5] As Matthew Edney, a leading historian of cartography, has cogently argued, this empiricist paradigm "comprises a set of generally unstated and unexamined presumptions about the nature of maps. . . . Couched as 'common sense,' these presumptions are thoroughly intertwined in a potent web of belief and conviction which continues to permeate academic and professional writing."[6] Some of the common assumptions Edney points out include the belief that maps distill, condense, simplify, or concentrate reality; are necessarily rooted in direct experience and observation; are made to be actively and functionally used; and contain meaning that is stable but latent until a map reader retrieves it.[7] Consequently, both scholarly and layperson assessments of mapping traditions insist that although displaying varying levels of craftsmanship, maps portray a geographic reality.

These presumptions form a metanarrative that informs and structures much of the historiography of Western cartography. Concomitantly, these assumptions structure the evaluative criteria and assessment of all maps, mapmaking, and map use. For example, the indigenous mapping traditions of ancient and colonial Mexico, as seen in the map of the indigenous community of Amoltepec (figure 12, chapter 2, p. 45), were viewed consistently from the conquest through to the twentieth century as inept and flawed because they did not follow the requirements of the cartographic ideal: they lacked the universal, or standard, configuration of European-produced maps (i.e., the use of latitude and longitude or the rhumb lines formed from compass readings). In such an evaluation, indigenous maps could only be deemed as defective because they did not represent a physical territory accurately, distorted reality, and, accordingly, could be best assessed as representative of some early stage in the progressive development of cartography.

This interlocking of maps with empirical exactness informed much of the scholarship of cartography into the 1980s and resulted in formalistic studies focused on the identification of authorship, provenance, and assessment of accuracy. In the late 1980s, the critical writings of J. B. Harley, David Woodward, and other scholars of the history of cartography assessed these restrictive associations of maps and questioned their common presumptions. Harley sought to expand the scholarship of cartographic history, writing that "it is a major error to conflate the history of maps with the history of measurement . . . maps are perspectives on the world at the time of their making."[8] Harley and Woodward called for more comprehensive, theoretically informed analyses of maps and mapping along with conventional formal and stylistic study. As coeditors of the multivolume epic *The History of Cartography* (1987–), Harley and Woodward redefined maps, moving away from the strictly empiricist assessment paradigm toward a comprehensive contextual approach. The working definition that guides *The History of Cartography* is "Maps are graphic representation of things, concepts, conditions, processes, or events in the human world."[9] This reconceptualization invigorated the history of cartography scholarship resulting in critical analyses that probe not only the formal and historical elements or empirical accuracy of a map but the social, economic, and political framework of its making.

Using Harley's and Woodward's reformed analytical approach, the Amoltepec map was in fact quite accurate and did represent a reality; however, their makers' underlying principles of what constituted accurateness and reality did not parallel those of the adjudicating and presuming Western world. Ironically, within this cartographic ideal evaluative system, Antonio García Cubas's 1858 *Carta general la República Mexicana* was judged as an exact and truthful map (figure 46, chapter 5, p. 152); however, reconsidered in this broader, renovated perspective, the national wholeness the *Carta general* envisions was in fact far from the political, social, and economic reality of mid-century Mexico.

Scholars from the early 1990s to the present have extended this contextual redirection of cartographic scholarship away from earlier assumptions and toward more theoretically informed analyses.[10] Along with these studies on specific periods or national histories of cartography, the scholarly analyses of maps and mapping continue to expand theoretical inquiries. This recent analysis is well assessed in *A History of Spaces: Cartographic Reason, Mapping and Geo-coded World* (2004) by John Pickles. Inspired by Michele Foucault's call to write a history of space, Pickles contends that "Mapping technologies and practices have been crucial to the emergence of modern 'views of the world,' Enlightenment sensibilities and contemporary modernities."[11] Basing his theoretical views in the works of Harley and Woodward along with those of numerous social theorists, Pickles looks at the strategic efficacy of mapping practices, explaining that "Cartographic institutions and practices have coded, decoded and recoded planetary, national and social spaces. . . . Maps and mappings precede the territory they 'represent.' [As a result,] territories are produced by the overlaying of inscriptions we call mappings."[12]

Pickles's theoretical inquiry further integrates the history of mapmaking with the emergence of the modern project, which he defines as the "age of world-as-picture" and the "world-as-exhibition."

> The world-as-picture and as-exhibition was, in part, produced by technologies and practices of representation, including cycles of mapping, each of which left their residual impress on contemporary ways of seeing: the geometric experiments of perspective; the exploratory portolan [sea] charts and the deep cultural fascination with boundaries (coastlines) that gave rise to them; the parcelling of land in the regional and national cadastres [taxation maps]; the national topographic mapping programmes; the emergence of the globe as cultural icon; and the more recent remote remapping of all aspects of social life.[13]

In placing maps and mapping into these visual constructs, Pickles further draws on Walter Benjamin's theoretical analyses of the visual display and exhibition in nineteenth-century Paris through which

> representation entered fully into the commodity relation by its production of an economy of display in which the spaces of the city were restructured as spaces of visual display and mass consumption. The visual, informational and the exotic were commodified for bourgeois consumption through ur-forms of a new visual and global imaginary: the national exhibition (Crystal Palace), the panorama, the plate-glass window, and the shopping arcade in which the work of people, places and goods were gathered for display and consumption. . . . Maps and mapping have been at the heart of this economy of display and demarcation.[14]

Thus, Pickles situates maps and mappings in the sociopolitical and historical contexts emphasized by Harley and Woodward but also pushes further to make the critical point that the geopolitics of map representations is tied to the circulation of commodities and systems associated with visual culture.

In identifying issues of systems and circulation, Pickles identifies three constructs that underlie the production and functioning of maps: the cartographic gaze, cartographic imagination, and cartographic bricolage. The cartographic gaze refers to controlled ways of seeing based in the representation of three-dimensionality. It is based in the understanding of vision within unilinear geometric perspective articulated by Leon Battista Alberti, a fifteenth-century polymath, which re-created the one-to-one relationship between viewer and object. In Alberti's structure, the "mediating point would have been the transparent pane of glass through which the Renaissance artist saw the object and onto which, as the glass turned into canvas, he was to paint it."[15] The cartographic gaze also bases its power in map projection, a geometric system, which projects the earth's landmasses onto a flat surface. This system allows this depiction of the spherical earth through the use of meridians, parallel straight lines, and the parallels of longitude placed at right angles to the meridians, preserving the shape of landmasses but not their area. Both Albertian perspective and map projection assume a "Cartesian commitment to vision as the privileged source of 'direct' information about the world" and "dominated by a commitment to modeling a God's-eye view."[16] Here, "seeing is believing" and, thus, the viewer of the map is duped into believing that mapped vision is equated with omniscience.

Cartographic imagination is a mode of reasoning that utilizes the certitude of this cartographic gaze to create, analyze, and structure the world; it has "influenced the very structure and content of language and thought itself."[17] Finally, underlying and implementing the cartographic gaze and imagination is what Pickles identifies as cartographic bricolage, the cumulative processes of map fabrication. Maps, in their technical power to depict a seamless unity and wholeness, conceal the fact that they are created from fractured, contradictory, and mobile elements.[18] As a result, maps may be understood as assemblages, collages, and montages of fragments of earlier maps; debris from previous misrepresentation of spaces and shards of current and past visual images. Simultaneously, bricolage erases the traces of its amalgamation practice. Pickles suggests that

> If *bricolage* can serve us not only to describe the origins of modern mapping practices but also as a general metaphor for all mapping practices,

> it also posses a challenge to modernity and linear histories of development, progress and evolution in techniques of mapping and in the flux of representational styles.[19]

Studying maps and the practices of mapping requires constant awareness of the fragmented appropriating processes that underlie their production.

While theoretical conceptualization of maps and mapping can be perplexing in their terminology and descriptions of complex practices, one overriding observation is clear: for all their appearance of clarity and wholeness, maps are messy affairs that cannot be assumed to represent an empirical reality or represent a single perspective. Rather than static, inert, and transparent images, maps may best be envisaged as palimpsests—the result of endless overwriting by previous social constructions, economic milieus, and historical contexts—that are never complete.[20] Viewers of maps comprehend such palimpsests in different ways at different times. This designation allows the map its presence as a physical object but also recognizes the importance of map fabrication processes as well as map reception.

The theoretical perspectives outlined above inform this study as it investigates mapping practices for a space known as New Spain, then Mexico. Such practices locate this landmass spatially, culturally, and conceptually. Mapping practices are manifest as diverse visual images, including maps. For example, the Blaeus' *Theatrum* page maps New Spain as a landmass set in the lines of latitude and longitude; it also maps its content through ethnographic-like figures. In this way, this study heeds Harley's call to "interrogate maps as actions rather than as impassive descriptions" through interrogating the visual complexities and webs that produce mappings—not just maps. It is not so much the maps as the broader mapping practices underlying maps that are the focus of this study. Therefore, I consider mapping in an expansive constructive sense as synthesizing diverse data gathering and practices, resulting in displays of this information in text, images, or cartography through which New Spain / Mexico was fabricated into a seemingly stable and coherent entity. As a result, my study does not focus on the accuracy of Antonio García Cubas's or others' maps. Instead, I emphasize the use of mapping practices to fabricate visually a coherent entity ultimately called Mexico.

Understanding that the significance of a map is not located in its mathematical measurement, I focus on the visual dynamics of mapping. Such analysis leads to the recognition that maps and mapping are firmly intertwined in a society's visual culture. As a result, this study further locates maps and mappings in historically located seeing, that is, in the theoretical perspectives of the recent scholarship of nineteenth-century visual culture.

VISUAL CULTURE

In 1904, Antonio García Cubas published his memoir, *El libro de mis recuerdos* (The Book of My Recollections). At first glance, it appears to be an odd collection of descriptions of the places and people of Mexico rather than a typical first-person narrative. As a geographer, however, it makes sense that García Cubas frames his heavily illustrated memoir as a personal travel narrative through time and space, that is, through important locales in Mexico and numerous historical events and topics.

Published when he was seventy-two years old, the cartographer's memoir repeatedly references a theme that underpins all of his cartographic and geographic work: García Cubas's work is situated in intense seeing. This is evident not only in his magnificent maps but also in his verbose statistical studies, because in every example he is picturing a Mexico for a reader and viewer. For example, García Cubas concludes his opening pages of the introduction with a revealing statement:

> Opening the *Libro de mis recuerdos* raises the veil of the past and there appears in the *escena* [stage] a society that by its customs differs essentially from the present one. In that [scene] the moral element shone more and in this the material element stands out: basing itself in both characteristics and the nation will be great.[21]

The cartographer uses the phrase "raising the veil" to activate the metaphor of a theater curtain and stage, thereby locating a bounded or delimited space for producing and consuming imagery.[22] This *theatrum* approach may be seen as referencing traditional cartographic practices found in European atlases of the seventeenth and eighteenth centuries.[23] García Cubas, however, does not use the theater metaphor to reveal a single, complete scene as found in atlas formats. Instead, he positions the viewer as an eyewitness to stationary travel; that is, before the viewer's eyes, the memoir's texts and illustrations unfold as a journey through space and time that illuminates Mexico's moral character and material possibilities.

Further, the audience García Cubas addresses, of course, includes Mexicans and thus sets up a self-reviewing scene or display in which Mexicans view Mexican places, people, and history in search of portents of a grand future for the country. Later in this memoir, he reminds the reader that unlike many narrators, who with malicious intention make the reader look through glasses that poorly accommodate one's vision, deforming and exaggerating facts, "I want you to observe the pictures that I offer you with your natural vision."[24] Here, García Cubas firmly believes that his production of seeing reveals the truth that is inherent in the image that he presents and that understanding the nation requires looking or self-reflection in texts and images.

His use of the seemingly simple metaphors of stage and theater as well as looking glasses links his mappings to broad nineteenth-century discourses on seeing and viewing associated with the deployment of new visual conditions and practices. These innovations included new methods of visual reproduction that appeared in Europe, the United States, and Mexico. These included lithography; photography; photolithography; and formats of visual display, albums, exhibitions, museums, and world's fairs.[25]

These conditions and practices were both manifested in and integrated through an emerging Mexican geographic and cartographic awareness appearing in the three decades after 1821 that utilized visuality to locate, activate, and regulate nationalist culture. García Cubas's introduction points to the need to examine the visual conditions and practices that affected how Mexico, as a new nation, could and would display itself to others *and* to itself through diverse mappings. This focus on seeing and sight firmly situates García Cubas's works within nineteenth-century international visual culture and his construction of the spaces of Mexico within that visual history.

The term *visual culture* needs clarification here as it appears frequently in recent book titles found in visual arts bibliographies with diverse usages.[26] Within these writings, Vanessa Schwartz and Jeannene Przyblyski, authors of *The Nineteenth-Century Visual Culture Reader*, provide a cogent and functional definition and explanation of the term.

> Visual culture can be defined first by its objects of study, which are examined not for their aesthetic value per se but for their meaning as modes of making images and defining visual experience in particular historical contexts. Visual culture has a particular investment in vision as a

> historically specific experience, mediated by new technologies and the individual and social formation they enable. Moreover, it identifies and underscores the status of the visual as a sensory experience that is itself conditioned by a historical understanding of physiology, optics, and cognitive science.[27]

Thus, with its emphasis on vision and visuality, the study of visual culture calls for interdisciplinary analysis of inventories of visual objects, institutions, processes, and historically located observers. Seeing and sight are part of complex cultural systems, not neutral phenomena and visual culture analysis shifts "away from things viewed towards the process of seeing," and focuses on the "social formation of the visual field."[28] Through this perspective, works of art have their own specificity—neither autonomous entities nor mere reflections of social and political processes; concomitantly, analysis of the visual materials does not rely on the assumption of inherent qualities but their performance in social contexts. Here, Pickles's constructs of cartographic gaze, cartographic imagination, and bricolage may be seen as subsets of visual culture. Placing "visual objects, image production, and reception at the center of a historically based inquiry has also suggested a reorganization of historical periods which can be sorted as 'scopic regimes,' with distinct patterns of regulating . . . [the] uses of particular objects, technologies, and ways of seeing."[29]

These visual culture perspectives allow the location of both the history of cartography and the history of art within a common scopic regime of the nineteenth century not rooted solely in style, iconography, or medium but in modes of visual production and consumption. Working within this framing of visual culture, García Cubas's work may be dislodged from its place in the restrictive and separated fields of knowledge of the history of cartography and the history of art, as well as nineteenth-century history, delimited by the dates 1800–99. His maps and mappings can be conceptualized through analyses that connect his work to these fields of knowledge at once through the identification of broader, threaded nineteenth-century discourses on the power of image viewing to constitute and reflect meaning, the impact of new technologies of seeing and reproduction, and the power of visual display. García Cubas's maps and atlases derive from and expand upon a history of seeing New Spain, as he attempts to see and fabricate Mexico.

Tracing these scopic threads will take us back to the eighteenth century to look not for origins but at New Spain's emphatic interest in new ways of using seeing and new reasons to see. These viceregal interests in seeing and sight are renovated and become dynamic practices in the nineteenth-century culture.[30] The concept of a nineteenth-century scopic regime as distinctive patterns regulating the uses of particular objects, technologies, and ways of seeing, is examined by Jonathan Crary in *Techniques of the Observer: On Vision and Modernity in the Nineteenth Century*. Crary proposes that a "rupture" occurred in European visual history between the eighteenth century, and an early nineteenth-century understanding of seeing, sight, and optics. Increased information about the physiology of the eye and expansion of optical science brought the recognition that sight was not initiated in the physical world, that is, the object viewed was not the source of sight. Instead, "Knowledge of a phenomenal world begins with the excited conditions of the retina and develops according to the constitution of the organ [eye/brain]."[31] Perception was increasingly analyzed as a function of physiology rather than as a function of the mechanics of light and optical transmission. Seeing was understood as occurring because of the physiology of the viewer and not by the physical presence of the viewer *at* a site of seeing. As a result, the eye and the viewed object—person, landscape, and so on—no longer needed to be in a static one-to-one time and space relationship. With this renovated understanding of the physiology of the eye came the recognition that the viewer and viewing could be manipulated by imaging technologies, such as the stereoscope, kaleidoscope, zootrope, phenakistiscope, and the camera. Consequently, Crary argues,

> What takes place from around 1810 to 1840 is an uprooting of vision from the stable and fixed relations incarnated in the camera obscura. If the camera obscura, as a concept, subsisted as an object ground of visual truth, a variety of discourses and practices—in philosophy, science, and in procedures of social normalization—tend to abolish the foundations of that ground in the early nineteenth century. In a sense, what occurs is a new valuation of visual experience: it is given an unprecedented mobility and exchangeability, abstracted from any founding site or referent.[32]

In other words, the earlier understanding of unified and homogeneous vision within unilinear geometric perspective articulated by Alberti was no longer the only way to understand and reproduce seeing. There emerged a modern and heterogeneous system of seeing or vision—that is, a scopic regime—that was distinct from classical, meaning Renaissance, modes of vision.[33]

Along with the shifts in the understanding of seeing described above, nineteenth-century visual culture was also driven by the explosive production and circulation of images made possible by new technologies such as lithography, chromolithography, daguerreotype, and photography. These practices, discussed more thoroughly in chapter 4, promoted a renovated understanding of the viewer by recognizing that perception is fundamentally durational and that cognition is an active processing of aggregating information. Thus, the viewer has enormous potential for consumption—and distortion—of visual information and proliferating visual images required better control or, at least, direction of the viewer's attention and consumption.[34]

Crary asserts that this interest in attentiveness is inherent in modernization and capitalism.

> Modernization is a process by which capitalism uproots and makes mobile that which is grounded, clears away or obliterates that which impedes circulation, and makes exchangeable what is singular. . . . Modernization becomes a ceaseless and self-perpetuating creation of new needs, new consumptions and new production. Far from being exterior to this process, the observer as human subject is completely immanent to it. Over the course of the nineteenth century, an observer increasingly had to function within disjunct and defamiliarized urban spaces, the perceptual and temporal dislocation of railroad travel, telegraphy, industrial production, and flows of typographic and visual information.[35]

Effectively, in the visual field, modernity is a continual process of dematerialization of bodies, objects, places, and relations through creation of multiple copies that are interchangeable. The exchangeability of forms is a condition for the rematerialization of bodies, objects, places, and relations into new hierarchies and institutions to promote the conditions under which the viewer could be activated as an attentive perceiver, that is, an observer-consumer.[36] As a result, specific methodologies such as exhibitions, and institutions such as museums, appeared to control and direct the seeing of this nineteenth-century viewer.[37] Creating an attentive observer—one who practices guided vision, complying with observing rules, codes, and conventions, to consume these divergent, proliferating, mobile, and circulating images and imagery—becomes an important objective and result of nineteenth-century visual culture in Europe and well as in Mexico.

Although critics find valid weaknesses and overstatements in Crary's theories, his writings have stimulated broader scholarly considerations of the changes that mark nineteenth-century visual culture.[38] This theoretical work on visual culture has begun to look beyond Continental Europe for sources and application of its concepts. In her important book *Vision, Race, and Modernity: A Visual Economy of the Andean Image World*, Deborah Poole addresses this issue and examines the implications of the construct of visual culture through study of prints and photographs from the eighteenth and nineteenth centuries of Peruvian people and places.

Following Crary's thinking, Poole argues that nineteenth-century visual economy is different from Enlightenment or Renaissance predecessors because the domain of vision is organized around continual production and reproduction through lithography and photography and circulation of interchangeable or serialized images and objects (photographs, prints, stereotypes) and visual experiences (exhibition displays). Importantly, within this economy, "the place of the human subject—or observer—is rearticulated to accommodate this highly mobile or fluid field of vision."[39] Her term *visual economy* proposes that the expansion of visual technology and reproductive techniques rearticulated the value of images as no longer limited to their worth as representations of real objects, people, or places but the value these images accrue through the social processes of accumulation, possession, circulation, and exchange.[40] Poole's work results in a lucid understanding of the formation of Peru as a nation and its racial discourses in the context of nineteenth-century Peru's move to modernity.[41] Her insights highlight the productive application of visual culture theory in Latin American contexts, and her study is an important precedent for this study because Mexico as a nation will be created in the context of a similar rearticulation of the value of images.

In sum, nineteenth-century visuality saw the rise of a system of seeing—that is, a scopic regime—that was distinct from Renaissance visual systems and based in substantial changes in the scientific understanding of seeing and, concomitantly, marked by the reconceptualization of the concepts of sight and knowing through vision. New visual technologies expanded the location of meaning from solely the object viewed to proliferating and mobile media—lithographs, stereographs, photos—that could be accumulated, circulated, and exchanged. Lastly, this proliferation of media required display techniques and formats—albums, exhibitions, panoramas, and dioramas—to focus a viewer's seeing or attention. Through his visual concepts of stage scenes and natural vision, García Cubas points us toward exploring his work in this context of nineteenth-century visuality.

Finally, to complete this summary of the theoretical underpinnings of this study, the use of the term *modernity* needs a brief comment, as it is a term that scholars use and define inconsistently. If modernity is only associated with industrialization and technological advancement, then measuring a culture's level of modernity is pretty straightforward. But the presence of railroad service, often cited as an example of modernity, does not make a culture modern; people do. Thus, in the following chapters, I focus not on the indicators of modernity in Mexico but its concomitant processes that are set in motion. As Dilip Parameshwar Gaonkar points out in his introduction to *Alternative Modernities*, because it always unfolds within specific cultural contexts and from different exigencies of history and politics, "modernity is not one, but many" and it is a process through which "people question the present. People 'make' themselves modern, as opposed to being made modern by alien and impersonal forces, where they give themselves identity and destiny."[42] Likewise, modernity in nineteenth-century Mexico is not only identified through increased industrialization or communication systems. Rather, modernity for Mexico was as an ongoing questioning of the configurations of its past as it constructed its present. Such formation of modernity becomes exceptionally apparent in its visual culture.

The theoretical writings on the reorganization and dynamics of visual culture and the processual notion of modernity have significant importance for analyzing and understanding of the diverse mappings of Mexico and the works of Antonio García Cubas. Along with reading maps as actions and mappings as fabricating practices, I suggest that diverse nineteenth-century Mexican mappings are imbricated in nineteenth-century Mexican and European visual culture. Addressed to the attentive observer specifically referenced in his memoirs, García Cubas's work aggregates disparate data, mobile and circulating images, and cumulative observations about Mexico into the hypnotic but illusory wholeness of maps and atlases. While this is a practice that is similar to cartographic bricolage, it is distinct because it also implicates other aspects of nineteenth-century visual culture. These theoretical perspectives lead to the critical awareness that in its complex birth into political independence, Mexico was born into and evolved during a dramatic shift of the dynamics of visuality. Here, modernity is not strictly aligned to chronological time or material indicators. Instead, it is differentiated by its reevaluated understanding and implementing of seeing, which identifies a viewer and citizen who needs to be made an attentive observer and where the value of images accrues through the social processes of accumulation, possession, circulation, and exchange that form and question what Mexico is and could be.

In the following chapters, I examine how the space/s of Mexico are produced from and inserted into this system of exchange and circulation. From his earliest work different kinds of viewing structures inform Antonio García Cubas's mapping projects, especially his atlases. He takes circulating fragments from cartographic, print, lithographic, and photographic images (after 1860) and rematerializes the physical space of Mexico from its dematerialized status promoted by travel literature and illustrations. García Cubas optically reconstructs the nation through map projections as well as associated imagery of physical geography. Thus, Mexico develops as a nation not only through independence or the modernization of political and economic structures, but also through an emerging scopic regime that employs mobile images and mobile viewers who are activated to question and recognize their identity and destiny through intensified methods of visual display.

Consequently, while staying within rough chronological limits, I consciously avoid simple linear narratives and explanation. I examine how diverse and complex textual and visual discourses were functioning concurrently with significant historical events that were taking place in Mexico. Thus, this young nation, in the context of seismic political and social change, adapted to and integrated itself into an evolving international visual culture and economy. The mapping imagination that emerged in nineteenth-century Mexico, while an innovation in scientific thinking, was also a manifestation of new viewing structures and strategies with its shift to physiological optics and construction of the observer. García Cubas's works do not so much accurately map a physical territory as signal the functionality of mapping practices as visual strategies for the explication and exploration of the Mexican nation, Mexican nationalism, and Mexican modernity. Thus, non-Mexican viewers of his atlases become spectators of Mexico, its places, resources, inhabitants, and history. More importantly, however, Mexican viewers—initially elite but by the end of the century middle class—became both spectators and objects in the evolving panoramic exhibit that was the nation of Mexico as they located, observed, and regulated themselves through the visual consumption of ideal and ordered views of themselves.

MAKING THE INVISIBLE VISIBLE

In his preface to the *Atlas geográfico, estadístico é histórico de la República Mexicana* (1858), Antonio García Cubas states that during his preparation of the *Atlas*, he reviewed the works of "los Sres [Señores]. Moral, Humboldt, Garcia Conde, Teran, Rincon, Narvaez, Camargo, Lejarza, Orbegoso, Iberri, Harcort, Mora y Villamil, Robles, Clavijero, Prescot, Alaman, etc. etc."[1] These sources, some going back to the eighteenth century, included cartographers as well as naturalists, political thinkers, and history writers. In addition, indicating his recognition of indigenous mapping traditions, García Cubas included in the *Atlas geográfico*, two supposed indigenous maps showing the migration of ancient peoples in Mexico (figures 49 and 50, chapter 5, p. 157 and p. 158).

While the cartographer and geographer is clear on the immediate sources of his *Atlas geográfico*, the Mexico of Antonio García Cubas's cartographic imagination was deeply embedded in mapping traditions and practices of previous centuries.[2] Such mappings of the New World were initiated in the sixteenth century, continued into the seventeenth century, and, especially, expanded in the eighteenth century.[3] Using diverse information-gathering methods and classifying concepts, these mappings resulted in displays of texts, images, and cartography. Concomitantly, travel was implicit in these mappings. In fact, until the eighteenth century, European readers were most likely to encounter the New World through popular edited collections of voyages or travel-related publications such as Giovanni Battista Ramusio's *Delle navigationi et viaggi* of 1555–65. Europeans mapped the Americas, therefore, through information gathered by processes and materials associated with two distinct and, at times, intertwined types of travel. The first was itinerary travel, associated with a route or path and visualized in cartographic images. Maps created for this kind of travel indicated how to get to and from the Indies, and once there, how to find one's way between points. The second was exploratory or sightseeing travel that was driven not so much by the linearity of a path as by circulation through the space of the Americas to ascertain natural resources and cultural content. Exploratory travel resulted in written texts and collections of flora, fauna, and material culture as well as noncartographic images, such as prints and frontispieces, and would have a great impact on how New Spain / Mexico was understood intellectually.

Through these travel modes, Europeans struggled to understand and discern the Indies as space as well as place. Space may be defined as an unbounded three-dimensional expanse (in which matter exists) that is ascertained through location in a particular rationale, such as a system of measurement providing relative position and direction within the graticule of longitude and latitude of a map.[4] Place, on the other hand, is space that is embedded in the human experience of networks of geosocial and historical relations that produce collective meaning. While a place may be located on a map, its meaning cannot.[5] As a result, the landmass of the Americas was located in physical relationship to other parts of the known world: its boundaries were defined and located within the compass lines of portolan charts and the graticule of longitude and latitude lines on maps and globes. At the same time, the natural and cultural substance of the Americas had to be fabricated, that is, situated in a European discourse that constructed imagined meanings of the Americas, forming it as a foil for European subjectivity.[6] The cumulative effect of this fabrication was the formation of a space named the Americas, with its subset New Spain, into an ostensibly stable and coherent mapped entity.

Excellent cartographic histories of the Americas from the sixteenth century to the eighteenth may be found in numerous scholarly publications.[7] While this is an account familiar to historians of cartography, I provide a general survey of the mapping traditions and visual practices that formed the Americas and New Spain for specialists in other fields. For this overview, I have divided the material into two chapters. This chapter broadly reviews those of non-Spanish-speaking countries of Western Europe from the sixteenth century to the eighteenth century. Although having less access to direct accounts and cartographic data about the Americas, these mappings—in the form of atlases, texts, and images—nevertheless proliferated and circulated broadly. In the following discussion, while highlighting examples of cartographic history, I focus on the development of the discursive imagery associated with maps (usually in frontispieces and marginalia) because these images reappear in mapping practices of the late eighteenth century and nineteenth century. Chapter 2 follows with a summary of the Spanish mapping processes and perspectives; here, access to direct information about the Spanish Americas produced distinctive texts and images.

These non-Spanish and Spanish mappings were neither linear or exclusive in their development nor necessarily summative in their results. It is possible, however, to ascertain broad visual themes resulting from these practices, which demonstrate that America / New Spain gained its meaning as place through iterative processes, erasures, and cumulative effects of multiple mappings. This survey demonstrates that centuries of attempts to represent the physical shape of New Spain, narrate its content, and invent its meanings formed both an enduring epistemological legacy and counterpoint for García Cubas's mapping practices of the nineteenth century.

MAPPING PRACTICES

In 1563, Duke Cosimo I de'Medici (1537–74), head of the Florentine Republic, commissioned the construction of an additional room for his Guardaroba, a storage area for highly valued palace goods such as tapestries, carpets, and sculpture, in his Palazzo Vecchio. Upon entering the Guardaroba Nuova, visitors found themselves surrounded by maps of the known world—fourteen of Europe, eleven of Africa, fourteen of Asia, and fourteen of the West Indies and Americas—painted on cabinet doors. Even more remarkable, the painted doors opened to reveal cupboards containing rare and precious objects from Cosimo's diverse assortment of exotica collected from the areas represented by the maps. These included Chinese porcelains, Turkish weapons, African ivory, and, from New Spain, Aztec feather works, masks, and statuettes. The curious and wondrous objects placed in the cabinets of Cosimo's Guardaroba Nuova were emblematic of distant worlds, at once exhibiting their craftsmanship, strangeness, wonder, and rarity."[8] As Francesca Fiorani explains in *The Marvel of Maps*, having a cabinet of curiosities was not unusual for a European Renaissance prince; however, the association of maps that relied on Ptolemy's geographical order to catalog the artifacts was a unique format.[9] The room exemplified a coming together of two modes—cartography and material culture—through which non-Spanish Western Europeans came to construct knowledge about the New World.

Cosimo's project is also representative of the expansion in mapping practices in the sixteenth century. Spain aggressively thwarted the publication of any unofficial information about the exact location or specific resources of its American kingdoms and limited the travel of foreigners to the Americas. Nevertheless, using available written sources, inexact mapping information, and spurious descriptions of the New World along with random objects (and sometimes indigenous people) sent back from the Americas, cosmographers and cartographers from the Netherlands, England, and France produced extensive maps and images that convincingly fabricated descriptions of the West Indies.

Scholars generally agree that three factors spawned this explosion of mapping practices that occurred in the sixteenth century: the translation of *Geographia* by the second-century AD astronomer and cartographer Claudius Ptolemy; the invention of printing in Europe; and the undertaking of overseas voyages.[10] *Geographia* included instructions for making map projections, proposals for breaking down the world map into larger-scale sectional maps, and the longitude and latitude coordinates for some 8,000 places.[11] Ptolemy's *Geographia* was passed to Roman cartographers, lost in Medieval Europe cultures, and rediscovered in the Renaissance. Over time, new modes of map projection were experimented with and new data added to *Geographia*'s maps and texts. The invention of printing techniques using metal plates provided the potential for increased circulation. In the resulting revised manuscript and printed world maps, America appeared as islands on the fringes of Asia or as a separate land in cartographic works such as those produced in the early part of the sixteenth century by Martin Waldsemüller, Peter Apian, Sebastian Münster, and Battista Agnese.[12] By the last quarter of the sixteenth century, the American continent was confidently placed in coordinate-based spatial relation to Europe, Asia, and Africa.

At the same time, the contents of the Americas also had to be placed in cultural relationship to other lands. As a result, many sixteenth-century maps overflowed with images that depicted flora, fauna, and peoples added within or around the landmasses located by the map. Some of these images were based in pure fantasy; others were derived from explorers' accounts, such as the highly influential letter written by Amerigo Vespucci (1454–1512) in 1508. Under the flag of Portugal, this Florentine cosmographer and cartographer, who became chief pilot in Spain's Casa de la Contratación (House of Trade), which coordinated exploration of the West Indies, traveled during 1501 and 1502 from the African coast to the Brazilian coast. He wrote an account of his voyage in which he described the environment and people he encountered as part of a new continent, not islands on the fringes of Asia. The report was republished extensively in Europe as Vespucci's letter, entitled, *Mundus Novus* (New World).[13] The various editions of his letter were associated with woodcut prints that depicted an environment with diverse flora and fauna and nude people, having an uncivilized culture that included cannibalism.[14] The imagery associated with Vespucci's account furthered the linkage of Americas with natural abundance, diverse flora and fauna, savage people, and primitive cultures.

A growing sixteenth-century demand for maps resulted in the assemblage by Italian publishers of printed maps of different sizes and using different scales. Reconceptualizing the development of these books of maps, in 1570 Abraham Ortelius (1527–98) produced a bound set of maps that were uniform in size and organization of content, and integrated with explicatory texts. The resulting *Theatrum Orbis Terrarum* (Theater of the World) conceived as encyclopedic and the first true atlas became so popular that between 1570 and 1612, thirty-one editions were published in seven different languages. Gerard Mercator's (1512–94) *Atlas sive cosmographicae*, first published posthumously in 1595, used the term *atlas* for the first time in the title of a map collection. Well aware of Ortelius's *Theatrum*, Mercator did not recopy these maps but designed and engraved original maps using a uniform geometric projection based on information he had collected, including Spanish texts.[15] For example, the atlas map of America was based on diverse information he had collected, including some Spanish texts (figure 1).[16]

Mercator envisioned an all-inclusive scope for his *Atlas*. The frontispiece

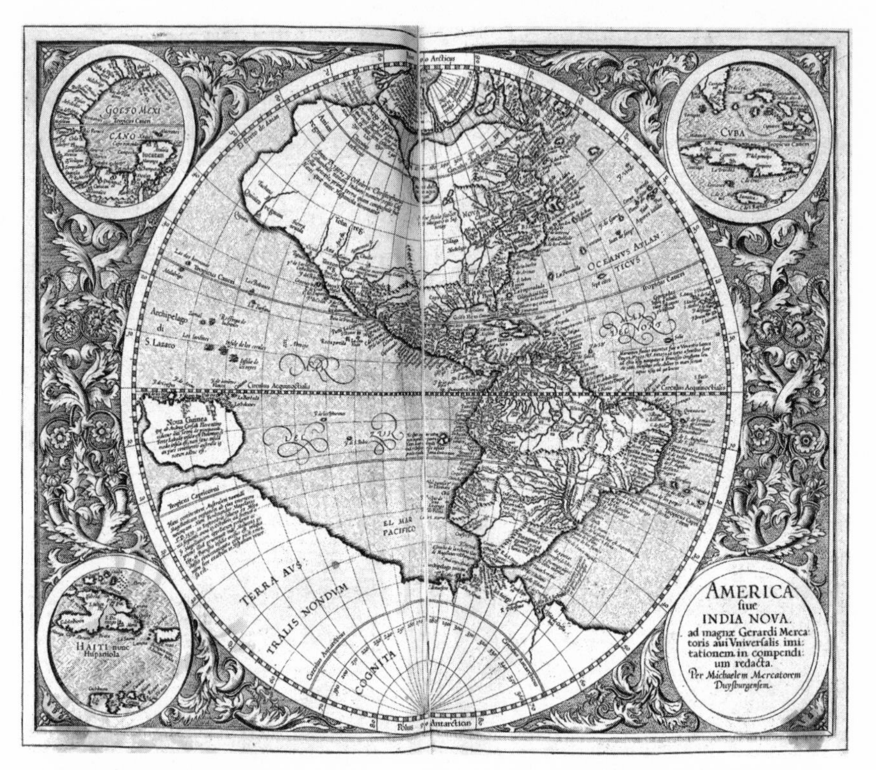

1 Gerard Mercator, *America. Atlas sive cosmographicae* (1607). Geography and Map Reading Room, Library of Congress.

to the 1607 edition is indicative of his comprehensive vision (figure 2).[17] Here, an almost nude male figure, with a globe at his feet and holding a celestial sphere, sits in a central niche within a columned structure that is similar to the Ortelian frontispiece. The upper part of this expanded architectural structure, however, supports the allegorical female figures representing Mexicana and Africa. Four additional niches surround the central figure and hold standing female allegories representing the geographic areas of Europa, Peruana, Asia, and Magalanica, the landmasses located by Magellan. All the figures, mostly nude except Europe and Asia, are associated with objects or animals representing their respective cultures within a highly structured composition. Here, the order and content of the known world are not determined solely by placement and costuming but through a

2 Gerard Mercator, frontispiece, from *Atlas sive cosmographicae* (1607). Geography and Map Reading Room, Library of Congress.

rational order referenced by the architectural format with the niches as well as by the central figure who holds the geometrically measured globe.

While many scholars have noted the scientific contribution of Mercator's work to the history of cartography and geography, José Rabasa appraises the *Atlas sive cosmographicae*'s role in the "invention" of America. He suggests that Mercator "historicized the geographer's eye," identifying geographic description as constituted by writing and history.[18] Rabasa argues that the atlas becomes a cluster of signs, recognizable and comprehensible in discursive configurations and through Mercator's bricolage of diverse information, cartographic and geographic and historical fragments are fabricated into the appearance of wholeness. "We must understand the [Mercator's world] map, and the *Atlas* in general, as simultaneously constituting a stock of information for collective memory and instituting a signaling tool for scrambling previous territorializations. Memory and systematic forgetfulness, fantastic allegories and geometric reason coexist in the *Atlas* without apparent disparity."[19] Explaining geography as a series of erasures and overwritings, Rabasa concludes that through the atlas format, territories could be erased and reconstituted in images and texts to endorse specific points of view, agendas, and so forth. The America found in Mercator's *Atlas* is now the sum total of an allegorical figure, a fabricated map, and a contrived text. The *Atlas sive cosmographicae*, and atlases in general, become mnemonic devices simultaneously instituting a systemic forgetfulness and prescribed remembering through what appears as the naturalness of geometric order.[20]

Gerard Mercator, then, went beyond the Ptolemaic compiling format to structure an illustrated travel narrative that appeared to be a comprehensive revelation of history and its spaces. The *Atlas*'s seemingly irrefutable and overwhelming truth—apparent in its maps, images, and texts—hides its highly subjective and constructed armature, accepted as the objectivity of geometry and historical and cultural facts.

This imagined content and meaning of the Americas continued to proliferate in European maps and atlases of the seventeenth century. For example, in 1634, Willem Janszoon Blaeu (1571–1638) visualized a project to compete with Mercator's production: a monumental folio atlas that was titled *Theatrum Orbis Terrarum* and came to be known as the *Atlas Maior*, consisting of two volumes with 204 maps. After his death in 1638, Willem's sons, Cornelius and Joan, extended the atlas into a six-volume set.[21] In the Blaeus'

3 Cornelius and Joan Blaeu, *America Nuvo Tabula*. From the *Theatrum Orbis sive, Atlas Novus* (1638). Geography and Map Reading Room, Library of Congress.

Theatrum Orbis, the map of the American continent is set in the graticule of longitude and latitude with small ships sailing to and from it and surrounded by phantasmagoric sea monsters still believed to threaten travelers (figure 3). Above the map are cartographic and chorographic images of seaports and cities of America Septentrionalis and America Meridionalis (North America and South America), such as Havana, Santo Domingo, Mexico City, and Cusco. On the sides of the map, ten vignettes illustrate supposed types of indigenous peoples who inhabit the landmass. While perhaps surpassing Mercator's *Atlas sive cosmographicae* in scale and extended content, that of Blaeu repeated the comprehensive ordering of the atlas format as it continues to function as a mnemonic device.

Another example, *America*, a 1638 engraving by Jan van der Straet (1523–1605), illustrates the continued use of allegorical female figures to represent America and New Spain (figure 4). Amerigo Vespucci, recently landed

4 Jan van der Straet, *America* (1638). Courtesy of the Newberry Library, Chicago.

and holding an astrolabe in his left hand and a banner in his right, gazes at America, which is represented metaphorically as an awakening, nude female arising from a hammock. She is surrounded by a landscape filled with the curious peculiarities of the New World: people who eat human flesh, imagined tropical flora, and strange fauna.[22] Here, the presence of the astrolabe indicates the Americas as a space that can be measured and, thus, located. The content of the Americas, however, is imagined as place through the metaphor of a female in an exotic environment.

By the end of the seventeenth century, European atlases firmly established a cultural geography for the Americas that combined recurring tropes such as seminaked, barbarous people; rich lands; and plentiful resources. At the same time, cartographic speculation continued in the form of numerous maps and atlases such as in Nicolas Sanson's (1600–67) *L'Amerique Septentrionale divisée en ses principales parties* (1674). Here, New Spain's exterior boundaries are depicted with California as an island, while the landmass known as the Yucatan peninsula appears as a distorted shape (figure 5).

In the next century, atlases expanded on seventeenth-century traditions.

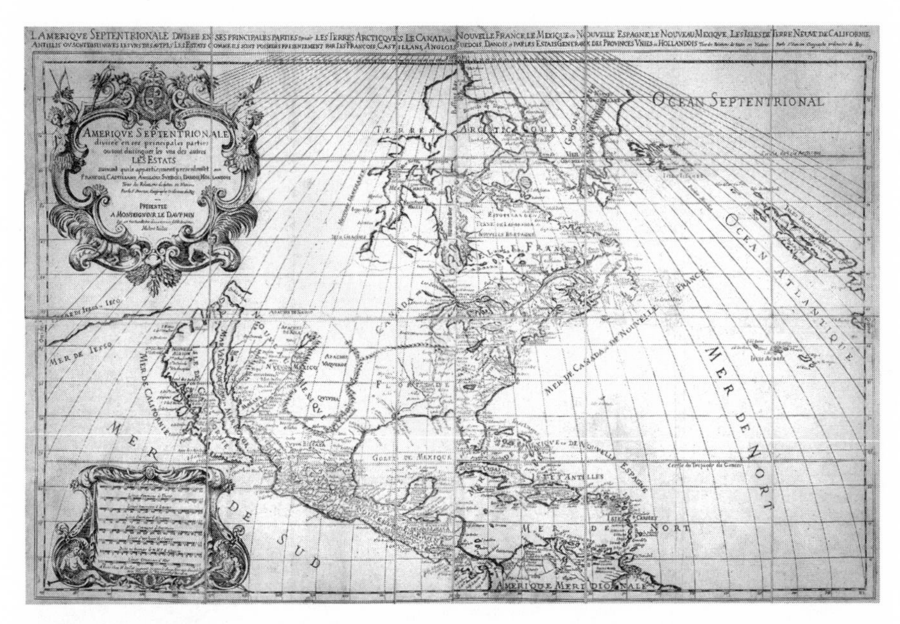

5 Nicolas Sanson, *L'Amerique Septentrionale divisée en ses principales parties* (1674). Courtesy of the Newberry Library, Chicago.

For example, the frontispiece of Robert de Vaugondy's (1688–1766) *Atlas Universel* (1757) synthesized cartographic data and repeats the imagery found in Mercator's frontispieces, although missing an architectural framing (figure 6). Female figures of the continents are associated with objects, flora, and fauna. Europe and Asia are finely dressed, while Africa and America are depicted as scantily clothed. America, a bare-breasted woman in the upper right-hand corner, wears feathered garments and holds a bow.

Another curious visualization phenomenon that marks the eighteenth century is the proliferation of images of America that seem to burst out of atlases onto diverse two- and three-dimensional formats that may be best designated as knickknacks or bric-a-brac. For example, the ceramic piece *Continent of America* (ca. 1760) depicts a female figure who wears a scanty, feathered costume and holds a green parrot and cornucopia while sitting on a caiman (figure 7). Such imagery, clearly similar to the imagery of van der Straet's *America* print and Vaugondy's *Atlas Universel* frontispiece, proliferated in other European domestic objects such as table clocks, tankards, gob-

6 Robert de Vaugondy, frontispiece, from *Atlas Universel* (1757). Courtesy of the Newberry Library, Chicago.

7 *Continent of America*, Germany, Meissen Porcelain Factory (ca. 1760). Wadsworth Atheneum Museum of Art, Hartford. Gift of J. Pierpont Morgan.

lets, teapots, salt cellars, plates, tapestries, and room screens—items associated with elite popular culture. Identified as "Americainerie" imagery and comparable to "Chinoiserie," the interest in Asian exotica, these objects step out of texts and prints into elite domestic settings.[23]

At the same time, the eighteenth century also saw the rise of new kinds of visual mappings of America / New Spain in the form of systematic natural histories. In the eighteenth century, many cabinets of curiosity—with their exotic, precious, and rare objects—were transformed from collections generally organized by oddity, uniqueness, or geography (e.g., Cosimo I's

Guardaroba Nuova) to collections ordered through specific classification.[24] Systematized presentations of objects emerged based on specific underlying principles. Scholars correlate this interest in scientific classification, in part, with the appearance of *Systema naturae* first published in 1735, in which the botanist Carl von Linne (1707–78), also known as Carl Linnaeus, developed simple tables to identify flora, fauna, and minerals within an ordered scheme according to class, order, genus, species, and variety. His resulting handbooks were brief, easily read, and small enough to be carried. Using his tables, it was Linnaeus's contention that plant, animals, and minerals could be identified systematically by the learned as well as the layperson—and even women. Summarily, he envisioned this project as a *Geographia naturae*, a geography of nature, ultimately to be displayed like a map.[25]

Linnaeus's reasoning for his project was in part economic. In 1746 he wrote, "Nature has arranged itself in such a way, that each country produces something especially useful; the task of economics is to collect [plants] from other places and cultivate such things that don't want to grow [at home] but can grow [there]."[26] He believed that European countries could grow important cash crops from collected and studied species. Such a project required specimens—real and illustrated—that in turn required travel for collection and observation. To accomplish this, Linnaeus sent students to travel throughout the world, collecting and annotating specimens; their success was minimal—many lost their specimens and their lives.

In her dissertation study, *Visual Culture in Eighteenth-century Natural History: Botanical Illustrations and Expeditions in the Spanish Atlantic* (2005), Daniela Bleichmar explains that this intensification of interest in natural history increased the importance of travel and collecting as significant ways of knowing nature not just by scientists but also by laypersons. Associated descriptive practices proliferated along with circulated objects or specimens, textual explanations, and printed images, resulting in a dense web of information.[27] Due to this increased circulation, natural history became a highly visual discipline, and the reading of visualized specimens changed such that emblematic readings of nature—as seen in the Vaugondy frontispiece and van der Straet print—gave way to schematized readings through which meaning was located within the orderly, global context of natural histories.[28] Bleichmar explains that "The culture of collecting implied a relationship with natural objects that centered on their materiality as specimens, and this affected their pictorial representation tilting toward truth."[29] In natural history books, the white page with a single specimen on it promised to capture and communicate the true essence of that specimen without ambiguity or room for erroneous interpretation.[30] Such images bridged the gap between the world of objects out there in the field and "here" in collections by offering a hybrid domesticated space, a paper nature that was always perfectly available for virtual exploration: flowers in bloom, or ripe fruit, as seen in this image of a cacao plant (figure 8). Assessing the impact of Linnaeus's work, Bleichmar argues that by mid-century an obsessive interest in images appeared in European natural histories and was associated with the domestication of foreign nature.[31]

From these expanding natural history pursuits, two different but related approaches to knowledge are evident. The first is a wandering impulse to know by direct experience in personal and lived ways. This impulse manifests itself as travel texts and associated images that emphasize the local context. The second is a sedentary impulse to have knowledge come from afar to be examined, compared, and incorporated into the corpus of what is known. This second inclination is manifest in images of specimens set in the presumably neutral and global white space of botanical books. These illustrations act as visual incarnations that define nature as a series of transportable objects whose identity and importance is divorced from the environment. This stationary impulse also rejects the traveler's visualization of the local context as contingent, subjective, and translatable, favoring instead the dislocated global as objective, truthful, and permanent. This European appropriation of non-European nature was predicated upon a selective vision that searched for the globally produced scientific facts through the erasure of the non-European culture.[32] Through the natural history book, one could view New Spain / America never leaving one's own home.

This new way of systematically mapping the natural world had important repercussions for New Spain, soon to be independent Mexico. In his study of the sedentary and the mobile natural history travelers, the historian Jorge Cañizares-Esguerra identifies the appearance of the philosophical traveler. As he explains it, "New 'philosophical' compilations of travel narratives also called into question the credibility of reports and chronicles by conquistadors, missionaries, and bureaucrats who had described the grandeur of the Aztecs and Incas. Editors and travelers began to read all earlier

8 Luigi Castiglioni, *Il Cacao*, from *Storia delle piante forestiere le più importanti nell'uso medico od economico*. Milan: Nella Stamperia di Giuseppe Marelli (1791–94). Courtesy of the John Carter Brown Library at Brown University.

eyewitness accounts of the New World through the lens of contemporary social theory."[33] This skepticism is manifest in the works of such writers as Georges-Louis Leclerc, compte de Buffon (1707–88); Guillaume-Thomas-François Raynal (1713–96), an *abbé*; Corneilus de Pauw (1734–99), a French naturalist; and William Robertson (1721–93), a Scottish historian. By the mid-eighteenth century, these writers were attempting to reassess the previous historical, social, and political evaluations of America and its peoples. To do this, they used methodologies based in specific socioscientific theories to reevaluate sources of information from the sixteenth and seventeenth centuries about the New World and recontextualize American / New Spain's land, resources, and inhabitants within associated theoretical schema.[34]

Georges-Louis Leclerc, compte de Buffon and keeper of the Jardin de Roi (Royal Garden), used natural history notions to argue for the overall degenerate condition of America and its inhabitants. In his *De l'homme* (1747–77), the seeming neutral truth of natural history explains and certifies the deteriorated status of New World flora and fauna. Using another methodology, Guillaume Raynal in *Histoire philosophique et politique, des éstablissemens et du commerce de Européens dans les deux Indes* (1772–74) and Cornelius de Pauw in *Recherches philosophiques sur les américains* (1768–69), certify the unreliability of early accounts due to the nature of the Spanish witnesses, whom they identify as barbarous soldiers, greedy merchants, and missionaries.[35] These authors question the reliability of human perception, mesmerized by marvels and curiosities, and the resulting travel accounts. For Raynal and de Pauw, any truth was to be found by those philosophical travelers who analyzed phenomena through theoretical questions and compared facts within a logical structure. They invalidated histories told through the deeds of great men, such as Antonio Herrera's writings, and authenticated histories that focused on larger questions such as the theory of racial variation or comparative culture through the lens of social theory. Following this thinking, as the historian Jorge Cañizares-Esguerra explains, William Robertson's *The History of America* (1777) endorsed a perspective with the following proposal:

> With the development of modes of production from hunting to herding to agriculture to commerce, individual needs and desires multiplied, and, with them, sociability. As the division of labor increased, so too did mutual dependency, which in turn caused people to refine their social

> skills and put their reason to work in pursuit of their own self interest. In the course of creating commercial societies, violent passions gave way to politeness and prudence.[36]

In this scheme, the world was a "living museum in which different people occupied different levels of a great tableau of emotion and economic development."[37] American Indians represented hunter-gathers and primitive agriculturalist stages of human social and intellectual development.

Whether using the perspectives of natural history or socioeconomic evolution, Buffon, Raynal, de Pauw, Robertson, and other eighteenth-century writers attempted to map New Spain through supposed universal principles. What emerged from these eighteenth-century European discourses was the general view that Spain's explorers and early writers were incapable of understanding what they saw in the Spanish Americas. In addition, America / New Spain was viewed as a basically degenerate place, yet one that offered abundant resources.

Exemplifying this late eighteenth-century thinking is the *Tableaux des Principaux Peuples de l'Amerique* appearing in Paris in 1798 as part of the fascinating work *Tableaux des Principaux Peuples de l'Europe, de l'Asie, de l'Afrique, de l'Amerique e les Decourvertes des Capitaines Cook, La Perouse, etc, etc,* by Jacques Grasset de Saint-Sauveur (1757–1810), a Montreal-born, French vice-consul to Hungary (figure 9). Along with aquatint engravings showing the people of each continent and those described by John Cook in his account of his South Seas travel, the *Tableaux* also includes an explanatory text that describes the mores, customs, religion, and commercial resources of each people. Grasset is adamant about the *Tableaux*'s educational value for the young of both sexes, stating that they will make rapid progress in geographic science: "A nine-year-old child can learn the heart of history of each of these people that make up the Asia tableaux. With forty days of study he can learn a people; with four months, he will know the principal history and geography of the entire globe."[38] While similar to texts and descriptions found in seventeenth-century atlases such as the Blaeus', Grasset assures the reader that his renderings are "exact and true" of the principle peoples of the universe. Echoing Robertson's writings, he states that the imagery is arranged to display progressive social evolution from nomadic Indians to citizens of cities.

In the *Americas* engraving, Grasset depicts the diverse indigenous types

9 Jacques Grasset de Saint-Sauveur, *Tableaux des Principaux Peuples de l'Amerique* (1798). Courtesy of the Newberry Library, Chicago.

and their costumes found in the Americas basing his representations on sixteenth-century and seventeenth-century prints and travel descriptions as well as his imagination. His associated text reiterates Buffon's overall negative assessment of the New World, concluding that in the Americas, the human species was degenerate at the time of the Spanish invasion; the hot temperature causes Americans to have vices and depraved temperament; and only Europeans have sufficient intelligence to read, write, and apply science or art. These attributes, however, do not detract from the commercial importance brought by the discovery of America. For example, his description of "Le Mexique" (Mexico), identifies birds, fertile land, indigo, and cochineal as important natural resources and mentions that Mexico City is spacious with wide streets, a sign of civil order.[39] While Grasset de

Saint-Savueur's *Tableaux* may not be considered a major geographic text of the late eighteenth century, its eccentric and likely highly plagiarized text and images demonstrate the intellectual shift from the emblematic reading of images found in the earlier atlas frontispieces from the beginning of the century to schematized reading of imagery that marks social evolution. While its themes and ideas go back to previous centuries, the *Tableaux* is embedded in a supposed systematic structure, a hallmark of eighteenth-century natural history writing.

Through three centuries, non-Spanish mapping of the Americas were diverse, attempting both cartographic as well as visual constructions. The *Tableaux des Principaux Peuples* offers an important example of Europe's multiple approaches to New Spain / Mexico on the eve of the nineteenth century. Whether through trope or specimen, eighteenth-century Europeans would find in America what they needed to find or invent. Mobile and fluid imagery of objects, prints, and maps fortified mobile and fluid meanings. Knowledge-producing activities formed the Americas for the European imagination.[40] Its content was defined as transportable objects whose meaning and significance were erased from their original environment and incorporated into global schema, identified as reliable and permanent.[41] Further, the writings of Raynal, de Pauw, Buffon, and Robertson illustrate Europe's tremendous power to form and narrate identity. As a result, the Americas would be constructed through information that was always on the way to someplace or something else—Asia, wealth, and empire. As New Spain became Mexico, it would be forced to relocate itself among these images, books, prints, maps, original objects, and Americaineries.

2

LOCATING NEW SPAIN

Spanish Mappings

Spain's interest in mapping the Americas and New Spain was quite different from that of other countries of Europe, in part because it was derived from very distinct intellectual traditions.[1] And, of course, Spain's initial and paramount interest lay in using maps as way-finding devices to mark how to reach, and then return from, the Americas and for administrative purposes. As a result, by 1503, the Casa de la Contratación (House of Trade) in Seville was put into place to oversee the movement of materials, commodities, and people between Spain and its holdings in the Americas; to train ship pilots; and to gather information from voyages to the Americas in order to refine the accuracy of maritime charts and ascertain the resources of newly located lands.[2] As the seat of the royal cosmographer and the chief pilot, the Casa de la Contratación was also responsible for production and maintenance of the *padrón general*, a master chart of the known world.

One of the earliest maps of the New World is the manuscript portolan-style chart of Juan de la Cosa, who likely accompanied Columbus on his first and second voyages (figure 10).[3] Painted on an oblong-shaped animal skin, the landmasses of Europe, Africa, and Asia appear on the right side; while the Americas, a green mass, bends around the semicircular left side. Rhumb lines emanating from compass roses form a web that, along with lines marking the equator and the Tropic of Cancer, locate the Americas

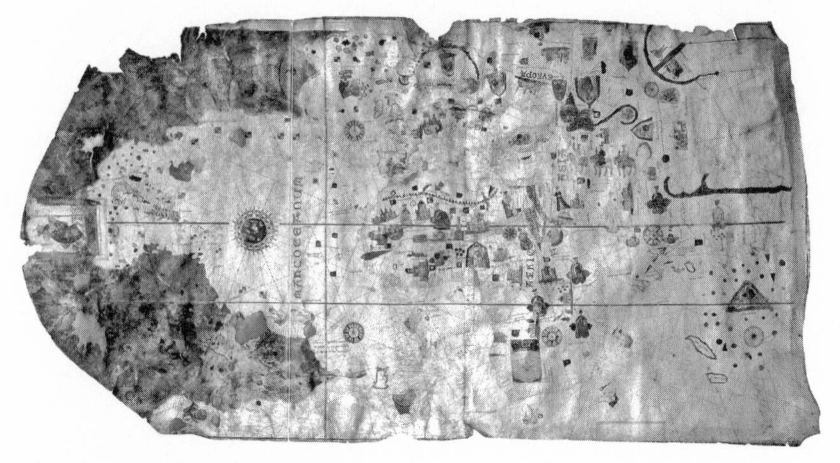

10 Juan de la Cosa, *Carta* (1500). Museo Naval, Madrid.

in correlation to the known world. Along with their spatial location, Cosa's chart depicts Europe, Africa, and Asia as filled with small icons that identify and mix geohistorical data with biblical stories: small buildings locate cities, seated individuals represent kings, and three men on horses represent the biblical Three Kings. In contrast, the American continent materializes as a vacant green space with the image of Saint Christopher, patron saint of travelers, hovering over the area of Central America. Subsequent Spanish maps landscaped this verdant but nebulous territory by adding flora, fauna, monuments, and people.[4]

Spain's charting of the world reached its apex in the sixteenth century under Philip II (reigned 1556–98), who demanded extensive mapping projects of both the Iberian Peninsula and the Americas. This king recognized the usefulness of maps as instruments and symbols of power. A territory that could be *seen* on a map or chart could also be located within the structured order of an empire.[5] Juan López de Velasco (died 1598), chief cosmographer-chronicler of the Consejo de las Indias (Council of the Indies), an oversight body for the New World, completed the *Descripción y demarcación de las Indias Ocidentales* in 1574, a manuscript with 150 folios and 14 maps, including the *Descripción de las Indias Ocidentalis*, showing an outline of the Americas appearing between the Iberian peninsula and the edge of Asia (figure 11). López de Velasco's map divides the Indies into three

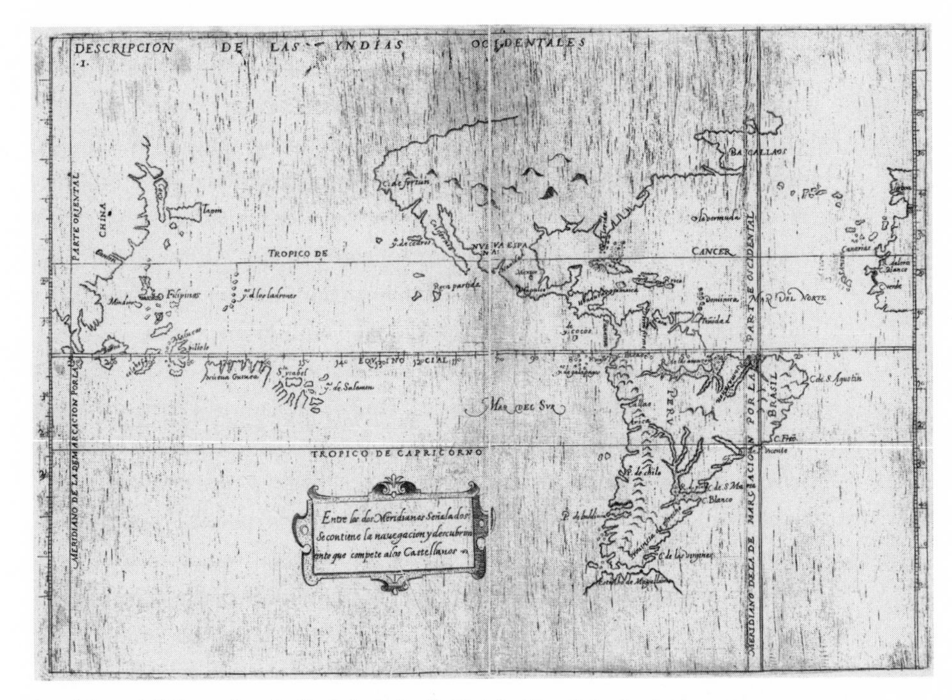

11 Juan López de Velasco, *Descripción de las Indias Ocidentalis*, from Antonio de Herrera y Tordesillas, *Historia general de los hechos de los castellanos en las islas y tierra firme del mar oceano* (1601). Geography and Map Reading Room, Library of Congress.

areas: Indias Septentrionales, the northern section from present-day Florida to Panama; Indias Meridionales, roughly corresponding to South America; and Indias Poniente, which included Spain's Asian possessions.[6] By the end of the century the basic cartographic outlines of the Americas, although rife with discrepancies, were apparent to Spanish authorities.

Also in the sixteenth century, in New Spain, indigenous mappings appeared. For example, *Mapa Sigüenza* illustrates the indigenous approach to mapping that combined time and space.[7] The map records the history of the Aztec-Mexica in the Valley of Mexico (a copy of this is seen in figure 49, chapter 5, p. 157). While likely the response to Spanish inquiries, such mappings were considered to be of little use by Spanish authorities (see discussion below).

Spanish exploratory travel of the sixteenth century was another mode of information gathering resulting in the mapping of the Americas in the form of summaries, chronicles, and histories that were based on primary and secondary sources and that were often illustrated. For example, Gonzalo Fernández de Oviedo y Valdés (1478–1557), historiographer of the Indies, traveled to America six times and published the first book of natural history of the Indies, *Sumario de la natural historia de las Indias*, in 1526. His *Historia general y natural de las Indias* appeared in 1535. Although patriotic in their approach, the publications contain curious information about flora and fauna, which Oviedo stresses he had seen with his own eyes.

Another well-known writer of the first half of the sixteenth century is Francisco López de Gómara (1511?–ca. 1564), chaplain to Hernán Cortés, who never traveled to the Americas but nevertheless published *Hispania victrix* (1553), a general history that retrospectively synthesized early encounters and especially aggrandized Spanish achievements. From the moment of its publication, López de Gómara's work was criticized for its extensive inaccuracies, resulting in Phillip II's prohibition of any reprints and decree in 1553 that all copies were to be collected. Gómara's problematic work, nonetheless, was translated into Italian in 1560 and French in 1578. The publication, however, was influential in other European accounts of the New World in subsequent centuries.[8]

As an extension of this exploratory travel, Spanish kings initiated substantial primary information-gathering practices seeking empirical data on the flora, fauna, geography, demography, natural resources, and climate of their American territories.[9] These efforts took the form of periodic *reales cédulas* (royal edicts) instructing the authorities in the Spanish Americas to gather information in order to answer a specified set of questions. Between 1530 and 1812, about thirty of these questionnaires were issued from Spain to the Indies and were formulated into natural histories and an administrative literature known as *relaciones geográficas*.[10] The questions changed over time, with the earliest inquiries requesting general information about geography, the establishment of religious institutions, the number of indigenous people, the indigenous system of government, and the status of the ongoing process of Christianization.[11]

In the second half of the century, Alonso de Santa Cruz (ca. 1505–67), a royal cosmographer from 1553 to 1567, formulated a plan to use a questionnaire format to bring together itinerary travel with exploratory travel. He sought to resolve issues of the longitudinal and latitudinal placement of the landmass of the West Indies. He also intended to complement the resulting revised map with a textual account that provided brief descriptions of the histories of discovery and settlement.[12] Santa Cruz planned to assemble this chronicle-atlas using information from accounts of conquistadors and travelers, as well as from reports in response to a questionnaire to be completed by officials of the New World. As the art historian Barbara Mundy adroitly summarizes, Santa Cruz's proposal

> was part of a large shift that was tipping the balance of authority away from classical models and toward the eyewitness accounts of humble local officials, whose numerous responses would have a collective authority that other individual accounts lacked. Using scores of responses culled from questionnaires as the basis for his new chronicle-atlas was Santa Cruz's novel way of reviving an old and authoritative form, the encyclopedia combination of texts and maps. The questionnaire replies would fuel him with the fresh, sharp insights of contemporary observation and give him enough data to avoid the pitfalls of the individual account.[13]

Santa Cruz's chronicle-atlas project was only partially completed at the time of his death in 1567 but was continued by Juan López de Velasco, who expanded Santa Cruz's questionnaire and developed a fifty-question inquiry in 1577. Analyzing the revised questionnaire, one finds it apparent that the cosmographer-chronicler sought very specific data. Twenty-seven of his fifty questions request information related to cartography, geography, or navigation, such as

6. What is the attitude, or the altitude of the Pole Star, at each Spanish town, if it has been taken and is known, or if anyone knows how to observe it? Or on which days of the year does the sun not cast a shadow at midday?
16. For all towns, both Spanish and native, describe the sites where they are established. Are they in the mountains, in valleys, or on open flat land? Give the names of the mountains, valleys, and districts, and for each, tell what the name means in the indigenous language.
42. What are the ports and landings along the coast? Make a map showing their shape and layout as can be drawn on a sheet of paper, in which form and proportion can be seen.

Eleven questions seek descriptions of natural history and resources or commerce:

22. Which wild trees and their fruits are commonly found in the district? What are the uses of them and their woods, and to what good are they or could they be put?
33. Through what dealings, trade, and profits do both Spaniards and Indians live and sustain themselves? What are the items involved and with what do they pay their tribute?

The remaining questions solicit data on local history, climate, demography, ethnography, or ecclesiastic information:

2. Who was the discoverer and conqueror of this province? By whose order was it discovered? In what year was it discovered and conquered, as far as is readily known?
5. Are there many or few Indians? Were there more or fewer at other times, and what are the known causes of this? Are they presently settled in planned and permanent towns? Describe the degree and quality of their intelligence, inclinations, and way of life. Are there different languages in the province or a general language that all speak?

In conclusion, question 49 allows the respondent to add any further information: "Describe any other of the notable aspects of nature, and any notable qualities of the soil, air, and sky in any part of the region." The final item, number 50, instructs the respondent: "Having completed the account, the persons who have collaborated on it will sign it. It must be returned without delay, along with these directions, to the person from whom it was received."[14]

Presumably the answers to these questions should have provided extensive data allowing López de Velasco to produce accurate maps showing not only the landmass within the coordinates of latitude and longitude but also the location of cities and villages and detailed descriptions of flora, fauna, and human inhabitants. The responses to the questionnaire were not as expected, because the questions reflected European priorities and practices. Instead, by 1583, López de Velasco had received ninety-eight responses containing what appeared to be rather discordant information. For example,

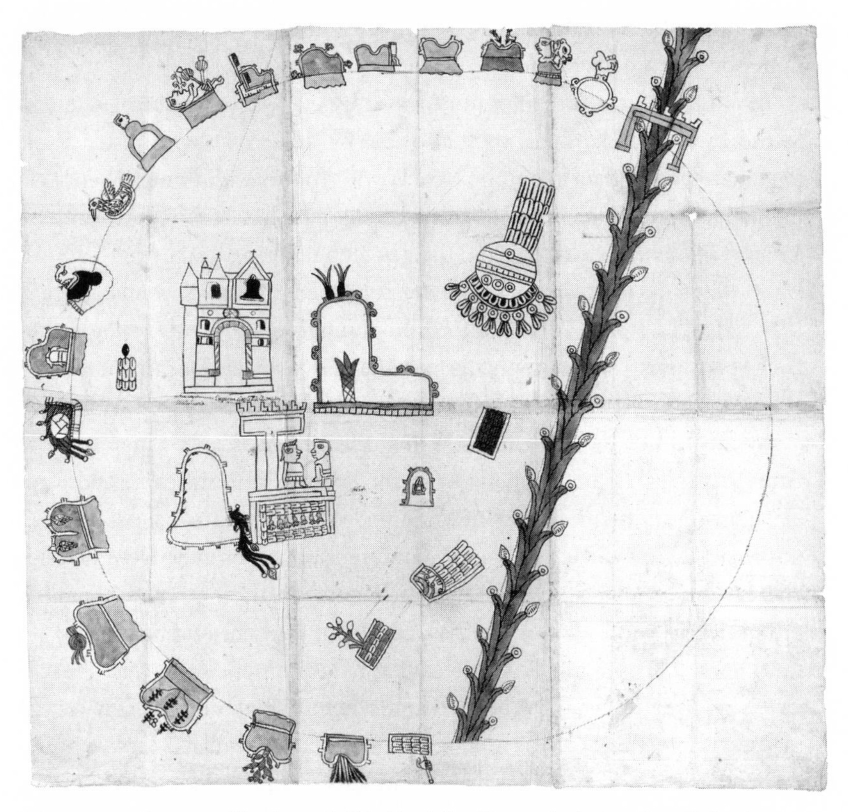

12 *Relación geográfica map of Amoltepec* (1580). Nettie Lee Benson Latin American Collection, University of Texas Libraries, University of Texas, Austin.

in figure 12, the *Relación geográfica map of Amoltepec* (1580), a local artist displays indigenous priorities using visual vocabulary to describe space. A church, denoted by European architectural elements, appears above the ruler's palace, marked with indigenous architectural elements. The structures, located next to a river marked by the indigenous sign for water, are surrounded by a semicircle of toponyms written in logographs that identify the villages that are part of Amoltepec territory.[15] This indigenous map of Amoltepec represents place; that is, the community's network of historical, kinship, and religious relationships to the land. Mappings such as those of Amoltepec could not be collated into a comprehensive map of New Spain using European standards, because indigenous maps used cultural coordinates rather than mathematical coordinates to describe space. Ultimately, López de Velasco abandoned his duties and was dismissed by the king.

In another form of exploratory travel by eyewitnesses during the sixteenth century, individuals were also sent by the Crown with instructions to gather specific data about the New World. To this end, Philip II sent Fernando Hernández, a medical doctor, to New Spain for six months to prepare a report on natural resources and their potential for medical uses. This work was collated into sixteen books but not published until the seventeenth century.[16] He also sent a Portuguese cosmographer, Francisco Domínguez, to New Spain in 1571 with orders to complete survey maps. Domínguez, however, did not produce any known maps of New Spain.[17]

By the end of the sixteenth century, diverse practices for mapping the Americas / New Spain, evolving from the Spanish authorities' varied ways of gathering and synthesizing information, were firmly established. Primary information was garnered from the different points of view from traveler and informant eyewitnesses. These included, on the one hand, Spanish protoscientists and pilots as well as merchants and entrepreneurs; and on the other hand, individuals who lived permanently in the Americas. The collected information was synthesized and synchronized into summary texts, maps, and illustrations. The Spanish Crown institutionalized these empirical practices through the Casa de la Contratación and the Consejo de Indias, linking information-gathering practices to good government of the American kingdoms.[18]

In the seventeenth century, there was a less intense, perhaps almost complacent, interest in the Americas under the late Hapsburg reign.[19] Materials from the preceding century were printed or synthesized into new texts. Thus, Fernando Hernández's report on natural resources for medical uses, commissioned by Philip II in the sixteenth century, was printed in 1648 as the *Rerum medicarum Novae Hispaniae Thesaurus*. At the beginning of the seventeenth century, Antonio de Herrera y Tordesillas (1559–1625), who replaced López de Velasco as royal chronicler, published *Historia general de los hechos de los castellanos en las islas y tierra firme del Mar Oceano* (General History of the Deeds of the Castilians on the Islands and Mainland of the Ocean Sea)(1601–15). Expanding on sixteenth-century works and using the map produced by López de Velasco, Herrera y Tordesillas introduces his work with the idea that the Indies were regions "outside the imagination of the men."[20] Claiming an enlightened overview, Herrera filled four volumes with long texts fabricating a coherent history of the Spanish encounter and domination of the Americas.[21]

Most interesting are the title pages of his volumes, which visually synthesize the texts. For example, in *Decada Terzera* (Volume Three), ten picture blocks frame the center space, which contains the title, dedication, and royal insignia (figure 13). In the upper left-hand corner, a portrait of Hernán Cortés is set in a medallion facing a map of Tenochtitlán. Scenes associated with the conquest follow below this portrait: the retaking of Tenochtitlán and imprisonment of the Mexica king, *Cuauhtémoc*; Cortés's meeting with the king of the Mechoacan; a battle encounter; and the rebuilding of Tenochtitlán. In the upper right corner, an image of Hernando (Ferdinand) Magellan, also set in a medallion, oversees the straits leading from the Atlantic to the Pacific that he discovered. The portrait is followed by scenes from his voyage, Magellan's death at the hands of the Indians, and the return of his ship to Seville. This section of the page concludes with an image depicting the dispute over the partitioning of the world. Continuing the tradition of Oviedo and Gómara, Herrera's illustrative title page demonstrates the seventeenth-century Spanish interest in a synthetic understanding of the Americas / New Spain. Here American space is depicted through historical events that incorporated the place of the Indies and of New Spain into Spanish history through the deeds of great men.

Along with such history writing, data collection through royal questionnaires continued although with fewer requests. Distinctive from the López de Velasco fifty-question inquiry, a 1604 survey contained 355 questions regarding the natural resources, flora, fauna, geography, demography, and climate. The questions, however, were more detailed. For example, fifteen questions sought data about the indigenous population and their tributary status: "53. How many tribute-paying Indians are there in this territorial unit?; 54. How many Indians administrate and have the office of *cacique* [an Indian leader]?; and 55. How many are married, single, old, or youth who do not pay tribute?" This questionnaire introduces ethnographic questions as well: "75. What is the standard food and drink of the Indians of this town?"; and "77. What kind of clothing do they make in the town and livestock do they grow?"

Subsequent questions recognize that, after four generations of Spanish

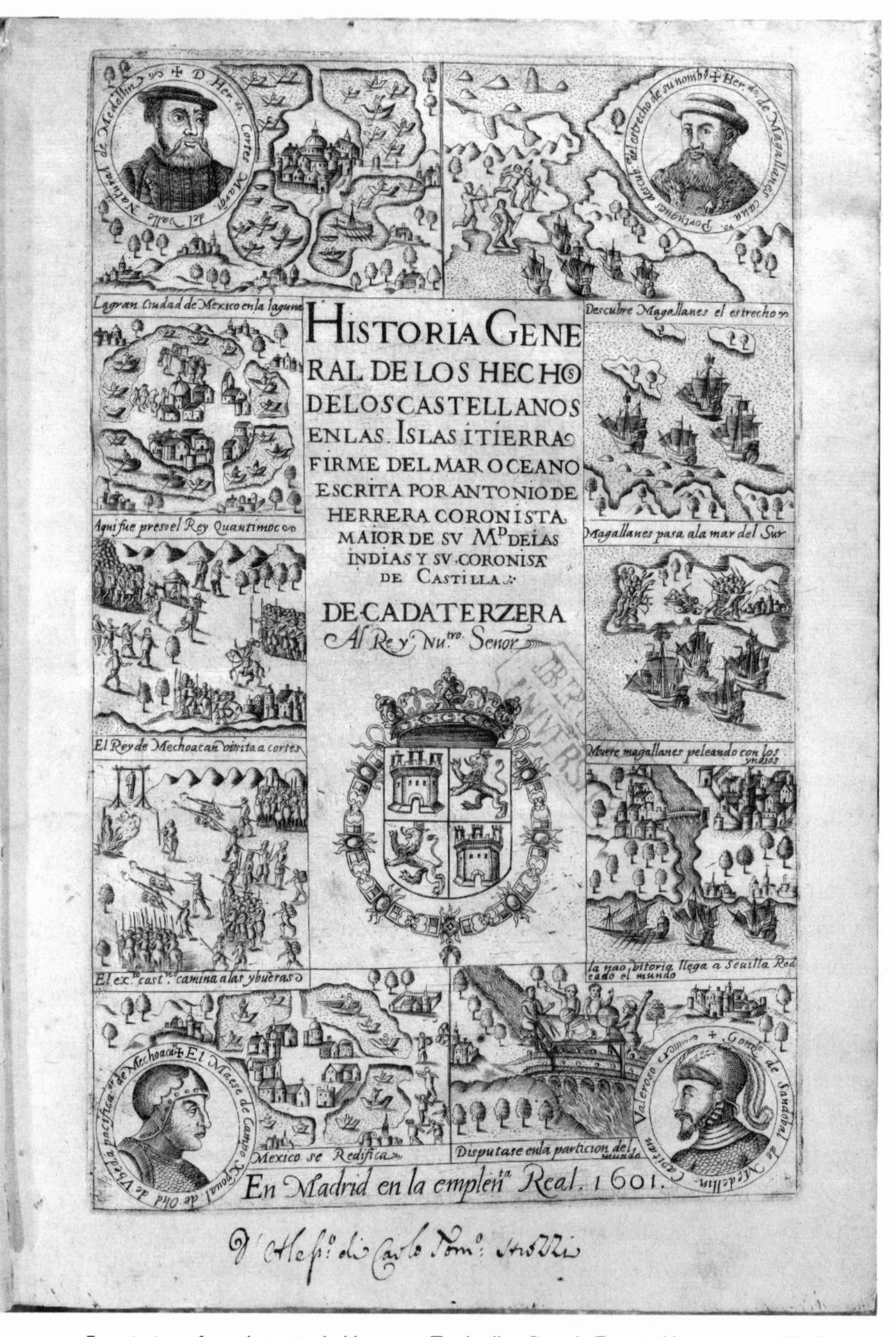

13 Frontispiece, from Antonio de Herrera y Tordesillas, *Decada Terzera: Historia general de los hechos de los castellanos en las islas y tierra firme del mar oceano* (1601). Brown University Library.

contact, new bodies that were neither Spaniard nor Indian were appearing in the population. These include criollos, American-born Spaniards, and *castas*, mixed-blooded individuals. Thus, their demographic information was also sought: "102. How many are Spaniards?; Of the Spaniards, how many men, how many women with the identification of ages and estates and how many are criollos, and how many were born in Spain, and the provinces of Spain?"; "104. How many are mulattos (a Spanish and black blood mix) and *zambaygos* (an Indian and black blood mix), declaring the number of men and the number of women, with identification of age and estates; and of the mulattos, how many are free and how many are slaves?"[22] Overall, the early seventeenth-century surveys emphasized the gathering of quantitative information rather than qualitative narration.

By the end of the century, the Spanish crown began to recognize the economic importance of nonmineral raw materials, establishing a Royal Pharmacy in 1695 and chemistry labs in 1700 to study this potential. Such institutions "demonstrate the way in which scientific pursuits served to reinforce Spanish imperial hegemony by leading to new products for commercial exploitation and to the appropriation of American natural history specimens for European uses and classification."[23] They provided the support and expansion of natural history research in the next century.

In the eighteenth century, the change from the Hapsburg dynastic control of the Spanish throne to the Bourbon dynasty brought Philip V (reigned 1700–46) and reformist perspectives on the administration of American territories. The century was also marked by a series of wars, which resulted in the loss of territory in Europe (War of Succession, 1701–14) and the Americas (Seven Years' War, 1756) and the depletion of Spanish resources and revenue from lost territory. Spain's expansive power of earlier centuries seemed to ebb globally. Despite the setbacks and turmoil, or perhaps because of them, Spanish interest in its non-European territories increased. In the last quarter of the century, Pedro Rodríguez de Campomanes (1723–1802), an economic advisor to the king, published *Reflexiones sobre el comercio español a Indias* (1762), which described Spain—especially its economy—as in a state of decline. To improve this situation, Campomanes sought to refocus Spain's economic interests from precious metals to raw materials, promoting the viceroyalties as sources of raw materials. He recommended increased production of these materials, and cessation of industry that might compete with Spain and levy-free trade with Spain.[24] Campomanes reformatted the abstraction of America's abundance of previous centuries to emphasize the reality of raw resources as the focus of economic development.

Throughout the eighteenth century, and especially after Campomanes's writings, the Bourbon's strong interest in its American territory, and especially its viceroyalties of Peru and New Spain, sought "to renew Spain economically and politically through a radical transformation of its domestic and colonial policies that would allow it to compete with other European nations."[25] As a result of the reforms, New Spain began to gain its geographic and economic shape as Mexico. At the same time, criollos saw these renovated governmental practices, and the resulting revision of the relationship between peninsula and American territories, as threatening to their existing positions of authority in New Spanish society.

Under the Bourbon government, information-gathering practices also intensified. The questionnaires' format expanded—435 questions in a 1730s' inquiry—and information was requested approximately every decade. In 1741, Philip V lamented that the Consejo de Indias lacked critical information about the viceroyalties and mandated that the viceroys produce "noticias más individuales, y distintas del verdadero estado de aquellas provincias" (very detailed and specific reports about the true conditions of those provinces).[26] Such comprehensive reports were used to exploit more effectively the economic and human resources of New Spain as well as identify resources for military advantage against possible aggression by other nations. Also, the Enlightenment interest in natural history and scientific classification, while not an originating impetus, is also reflected in these inquires. Overall, the themes of the questions suggest multiple and probably overlapping concerns and imperatives of the eighteenth-century Bourbon regime's move toward a pragmatic philosophy of governance based on the comprehensive management and administration of people and commercial resources. The Bourbon kings were to be as powerful in practice as they were in theory, and they were to be present in every aspect of New Spanish life.[27]

In 1743, Don Pedro de Cebrián Agustín, conde de Fuenclara, the viceroy of New Spain, responded to Philip V's royal mandate by selecting Juan Francisco Sahagún de Arévalo Ladrón de Guevara and José Antonio de Villaseñor y Sánchez to prepare a comprehensive description of New Spain's regions, resources, and peoples. Sahagún de Arévalo Ladrón de Guevara, a priest, was a chronicler of New Spain and editor of the *Gazeta de México*.[28] José Antonio de Villaseñor y Sánchez, contador general de la Real Azogues (ca. 1700–59), who completed the project, was born in Mexico City and likely entered the viceregal service around 1725. He is considered to belong to the "generación 1730," which represented a growing criollo, New Spanish, sense of cultural identity.[29] The resulting text would be a two-volume compendium of geographic and demographic information, which includes a beautiful frontispiece. In terms of its format and content, Villaseñor's *Theatro Americano, descripcion general de los reynos y provincias de la Nueva España y sus juridicciones* looks backward and forward: in fulfilling the king's demand for comprehensive data about New Spain, it is based in the questionnaire traditions of earlier centuries. At the same time, the compilation also visualizes New Spain as place, that is, a space embedded in networks of social and historical relations. As such, the *Theatro Americano* requires closer consideration because the project envisioned by Villaseñor would be completed by García Cubas just over 100 years later.

While similar to the earlier inquiries of Santa Cruz and López de Velasco, here two inhabitants of New Spain, rather than bureaucrats from the Casa de la Contratación, developed the eight queries for the mandated investigation. These questions requested very specific information about the location and climate of administrative towns and their associated towns and villages; the demographics of local inhabitants; the natural products used for commercial purposes; mineral resources; ecclesiastic missionary information; and finally, the exact distance of the village or town from a major administrative center.[30] The questions formed by Sahagún and Villaseñor elicited significant data about mid-century New Spain. Respondents provided narrative answers to each of the eight questions varying in specificity of detail from thorough to rather general. They also included small maps showing the town's relationship to surrounding topography and other towns.

Examples of the responses to these questions are found in the *Relaciones geográficas de Arzobispado de Mexico. 1743*.[31] The response from the town of Ixmiquilpan is an example of one of the more detailed responses. Following the questions closely, three respondents—Salvador González (mestizo), Lorenzo de Montufar (español), and Lorenzo Rangel (castizo)—certify information about Ixmiquilpan and surrounding villages.[32] Lorenzo Rangel, writing on 7 May 1743, provides the most detailed information. Estimating a total of 12,258 inhabitants, Rangel provides detailed lists of who lived in the town and its pueblos and haciendas. In his statement, he cites: "Casa de don Benito González, español, de 60 años, casado con doña Gertrudis

Moreno, española, de 40 años. Juana, doncella, de 15 años, Ana, doncella, de 13 años, María, de 11 años, sus hijas. Doña Juana Moreno, doncella, de 50 años" (House of don Benito González, Spaniard, 60 years old, married to Doña Gertrudis Moreno, Spaniard [woman], 40 years old. Juana, maiden [unmarried woman], 15 years old, Ana, maiden, 13 years old, María, 11 years old, their daughters. Doña Juana Moreno, maiden, 50 years old).[33] By and large, the maps and descriptions of the *Relaciones geográficas de Arzobispado de Mexico. 1743* demonstrate the varying quantity and quality of the data about local geography and demography provided in response to the questionnaire of Sahagún and Villaseñor.

It was left to Villaseñor to collate these data into the two-volume compendium, *Theatro Americano*. The first volume of this important administrative report, written by a criollo or New Spaniard, was printed in 1746, followed by the second in 1748. Selecting the term *Theatro* for his title, Villaseñor indicates his knowledge of geographic and cartographic practices found in seventeenth-century and eighteenth-century European *theatrum*. In fact, in a preface document endorsing the book, the Marquès de Altamira, *oidor* (judge), explains the theatrum format citing the work of the geographer Abraham Ortelius and his *Theatrum Orbis Terrum*. For this oidor, the *Theatro Americano* "prometan dar à la vista (que esso significa Theatro) Reynos, Provincias, Ciudades, y toda esta Nueva-España" (promises to give an overview [which Theater means] of the Kingdoms, Provinces, Cities and all that is New Spain).[34] Further, in addressing the king in his introduction, Villaseñor states that as in a living theater, it is his intention to represent "á los ojos de V. M. toda la consisté[n]cia de un nuevo Mundo" (to the eyes of Your Majesty, all that constitutes a new World).[35] Villaseñor's *Theatro Americano* continued the perspective of "eyewitness and traveler" of earlier centuries; it was the king who was to be the eyewitness.

The reader is also visually introduced to the content of the *Theatro* through a frontispiece, signed by "Balbas," depicting Philip V elegantly dressed and holding a walking stick, standing atop the earth, in the form of a globe, while two figures below each kneel on one knee (figure 14). Because the globe is tilted on its pedestal with the north–south axis placed on an almost horizontal plane, the king is positioned on Western Europe rather than the North Pole. Across the ocean, the landmass of New Spain is outlined and shaded with an *X* marking the location of Mexico City. Below the image of the earth surmounted by the king, two figures kneel on a precipice

14 Frontispiece, from José Antonio de Villaseñor y Sánchez, *Theatro Americano* (1746–48). Geography and Map Reading Room, Library of Congress.

overlooking the ocean. On the right, a male figure, likely José Antonio de Villaseñor y Sánchez, looks up to the standing sovereign and offers a book—his *Theatro Americano*. Behind this figure, an ocean stretches to the horizon. Opposite Villaseñor kneels a female dressed in Western garb and wearing a crown. A feathered fan rests on the ground in front of her, annotating indigenous reference points. She points to a chest filled with coins and overflowing with (pearl?) necklaces. Behind this figure, a landscape includes a cactus, palm tree, and vaguely outlined mountains. Above the king, a banner displays the Latin words *Digna Orbis*, followed by *Imperio Virtus*; below this scene is the book's title, *Theatro* (here spelled Teatro) *Americano* in an ornate frame.

In the preface, Juan Francisco López, calificador del Santo Oficio de la Inquisición, remarks on the frontispiece: "To the dedication has been added a refined engraving which in my judgment summarizes the overall plan of the work and the intention of the Sovereign, who mandated its production."[36] The print portrays New Spain in three ways. First, with Spain as the functional north of the world, New Spain is a secondary mapped landmass submissive to Spain's orienting structure. Secondly, New Spain is visualized metaphorically as an elegantly dressed female, synonymous with wealth, abundance, and curious natural environments. And, thirdly, New Spain is synthesized into an administrative text, that is the *Theatro Americano* offered by the kneeling figure. In this print, New Spain becomes an ancillary geographic space in the world, *Digna Orbis*, locatable in relation to Spain, *Imperio Virtus*. New Spain becomes a space that may be reduced to a text, readable and, thus, knowable. Here, the king witnesses and oversees the space of New Spain through three mapping practices: cartography—represented by the globe; visual metaphor—represented by the female New Spain; and text—represented by the book. In addition, a map associated with the *Theatro* was praised by the Marquès de Altamira as "en la sola oja de su Mapa, donde se vee reducida sin necessidad de perigrinarla" (in only one page of his Map, which one sees it reduced without the necessity of traveling it).[37]

The compendium's texts consist of two books with chapter subdivisions. In the third chapter of book 1, "Del Reyno en comun, sus distancias y Clima" (loosely, Of the Kingdom in general, its Distances and Climate) initiates the geographic description of New Spain. Villaseñor explains the structure of his text, stating that "as in understanding the parts of a body, it is necessary to have first pondered the whole anatomically in order to distinguish its members" and continues that the parts of the whole must be studied, too "in order to discern its places, and to satisfactory understanding."[38] The subsequent chapters synthesize material provided by respondents, following closely the order of the original questions regarding the distance from Mexico City, climate, and so forth. Each of Villaseñor's descriptions condenses the detailed information provided by the original respondents. For example, in book 1, the thirty-second chapter, "De la Jurisdiccion de Ixmiquilpan," y sus Pueblos, Villaseñor takes the information provided by Salvador González, Lorenzo de Montufar, and Lorenzo Rangel discussed above and compresses the report into just a few paragraphs. Excluding detailed information, Villaseñor summarizes the location, climate, and populations of Ixmiquilpan and its surrounding pueblos, noting the number of "familias de españoles, indios, mulattos, etc."[39]

Villaseñor's *Theatro Americano* was well received by Spanish authorities and used by the Bourbon government to assist in the administrative consolidation of New Spain, including dividing the territory into administrative units called intendancies.[40] As he points out in his introduction, Villaseñor would help to visualize New Spain for the Bourbon crown by recounting the spatial distribution of social and economic data.[41] It was considered a confidential document, and only thirty copies of the book were printed in Mexico City and judiciously distributed. In 1750, its reproduction was banned because the *Theatro* was considered to contain strategic information about Spain's territory.[42] Even as late as 1791, forty-five years after its completion, the viceroy Revillagigedo, cited the Villaseñor map of New Spain as the best available.[43]

The questions used for the reports presented in the *Theatro*, as well as the secretive treatment of the finished work, reflect the awareness that empirical data was a valuable commodity and tool. Villaseñor's references and allusions in the nongeographical sections of the *Theatro*, however, point to an undercurrent within the text that connects data to social and historical networks—specifically, a criollo vision of New Spain as a place. This vision is exemplified in the first and second chapters of book 1, which have received little attention by scholars. The first chapter assembles a history of New Spain from prehistoric times to the sixteenth-century conquest.[44] Villaseñor begins this text in the Old Testament postdiluvian time and states that at that time the land was occupied by Giants, which the indigenous people, the Toltecs, found when they arrived. He goes on to explain the indigenous use of "Mapas, y pinturas con que escribian sus Historias" (maps and pictures with which they wrote their histories). He then reflects that, nevertheless, the barbarous and idolatrous content of this culture required that it be replaced by Christianity, and, thus, the necessity of the Spanish conquest.[45]

In the second chapter, Villaseñor delineates the establishment of the Spanish reign in New Spain, listing the various viceroys and the creation of *audiencias*, and citing the Real Cédula of 1741.[46] Villaseñor initiates the chapter, however, by recounting the story of the Virgin of Guadalupe who appeared to Juan Diego in 1531, emphasizing that "no vista de otra Nacion alguna" (it was not seen in any other nation).[47] Here, Villaseñor refers to the phrase "non facit taliter omni natione" (It was not done thus to all nations) taken from Psalm 147, which appeared in reproductions of the image of the Guadalupe beginning in the last quarter of the seventeenth century.[48]

Thus, in these first chapters, Villaseñor introduces a historical chronicle that contextualizes the *Theatro Americano*'s geographic synthesis. He traces the history of the New World not to the Spanish conquest but back to prebiblical times. In this way, he locates New Spain's origins in an autochthonous history not solely within Spanish historical chronicles as found in López de Gómara's sixteenth-century writings. His reference to the appearance of the Virgin of Guadalupe alludes to criollo belief that her miraculous apparition marked New Spain as a unique place among nations. Thus, Villaseñor's *Theatro Americano* endeavors to produce a catalogue or enumeration of geographic facts to define the boundaries of New Spain's space as formulated in relaciones geográficas of earlier centuries. At the same time, however, in what appears initially to be the inconsequential introductory sections of the *Theatro*, he commences to explain the interrelatedness of the whole to its parts, contextualizing geography with history. Thus, *Theatro Americano* is an important example of the status of New Spanish—not Spanish—geography and cartography of New Spain at mid-eighteenth century. Here, Villaseñor articulates a New Spanish place that is distinct from Spain. From the late 1820s through the 1840s, this articulation was developed further by various authors and in the 1850s visually consolidated by García Cubas.

Related to these eighteenth-century constructions of New Spain as place is a secular visual art known as *cuadros de casta*, casta paintings. Appearing at the beginning of the eighteenth century, this genre of painting depicts a speculative taxonomy based on lineage of the kinds of people who inhabited New Spain. Usually in the format of sixteen to twenty-two separate panels, each casta series begins by illustrating that a Spaniard and an Indian produce a mestizo; the series then proceeds to portray the progeny resulting from further miscegenation of Spanish blood with Indian and black blood. Cuadros de casta have received significant scholarly attention over the past fifteen years resulting in substantial research and promulgating provisional conclusions about their origin, production, and meaning.[49]

The paintings appear at the beginning of the eighteenth century, proliferate after the middle of the century, and conclude production by the first decade of the nineteenth century. Casta paintings produced in the earlier part of the eighteenth century stress the prosperity of New Spain and colonial self-pride. At mid-century, production of cuadros de casta increases and, importantly, they change their format to resituate the casta groups from nondescript backgrounds of the early part of the century to more specific domestic, cityscape, or landscape settings (figure 15). These later works, reflecting Bourbon reforms, focus on social stratification and New Spain's commercial resources.

A cuadro de casta signed by Luis de Mena appeared at about the same time as the *Theatro Americano*.[50] Using the less standard format of a single panel, Mena painted eight vignettes in the center of a single panel portraying the intermixing of peoples that include from left to right:

First row:
From a Spanish woman and Indian, a Mestizo is born
From a Spanish woman and Mestizo, a Castiza is born
From a Castiza and Spanish man, a Spaniard is born
From an African woman and Spanish man, a Mulatto is born
Second Row
From a Spanish woman and Mulatto, a Morisca is born
From a Morisca and Spanish man, an Albino tornatrás is born
From a Mestiza and Indian, a Lobo is born
From an Indian and Lobo, an Indian is born[51]

Depicted below these groups is a tray holding fruit, numbered and identified in the caption at the bottom. The panel also depicts specific landscapes; above and to the left of the vignettes is a scene identified as "Dansa de Mata-

15 Luis de Mena, *Casta panel* (ca. 1750). Oil on canvas. 120 × 104 cm. Museo de América, Madrid.

chines que asen a N. Sra. de Guadalupe" that depicts a traditional masked dance. To the right, a landscape entitled "Paseo Jamayca" illustrates a recreational scene of strolling couples and gliding canoes on a lake. Finally, between the two landscape scenes, floats an image of the Virgin of Guadalupe. This panel depicts specific landscapes, particular kinds of people, and detailed plant products.

The representation of the Virgin of Guadalupe associated with this casta panel emphasizes the distinctiveness of criollo place. During the previous centuries, written sources tied together religion and patriotism associated with the growth of a criollo sense of a New Spain as patria rather than Spain as patria.[52] Like the *Theatro Americano*'s text, the cuadros de casta represent a form of eyewitness accounts that stages a display of the imagined geography of New Spain. The increase of information associated with later casta representations may be seen as concomitant with the increased Spanish information gathering and codification of botanical taxonomy.[53] At the same time, like Villaseñor's compendium, these images illustrate an emerging eighteenth-century discourse on New Spain as a place distinct from Spain.

Likewise, cartographic efforts of the last quarter of this century would echo the notion of New Spain as place.[54] José Antonio de Alzate y Ramírez was a Jesuit priest and well-regarded scientist based in Mexico City. In addition, he was editor of a series of periodicals, which attempted to provide a metropolitan voice for New Spain, highlighting trends of thought and scientific interests.[55] Based on an earlier map attributed to Carlos Sigüenza y Góngora, a highly respected late-seventeenth century polymath, Alzate produced the *Nuevo mapa geográfico de la América septentrional española* in 1767 (figure 16).[56] In December of 1772, he remarked on the poor state of geography and cartography of New Spain.[57] His map was published by the Royal Academy of Science in Paris in 1775 and appeared in New Spain in 1792.[58] It depicts North America, west to east from present-day California to Florida, and north to south from about Wisconsin to Nicaragua. The interior of the map locates the geographic divisions of the landmass and various towns and cities.

The map contains various cartographic errors, reflective of both the extant information available to Alzate and his own mathematical miscalculations. Interestingly, the map is bordered on four sides with small images set in undulating floral frames. Many of the images depict flora and fauna. In addition, ten of the border images depict types of people, including Indians in various stages of nudity, men on horseback, and men and women laboring. The image in the lower right-hand corner depicts the metaphorical image of America as a seminude female found in many frontispieces and cartouches from earlier and contemporary maps. Alzate's combination of map with the vignette images can be seen as a reflection of atlas page models

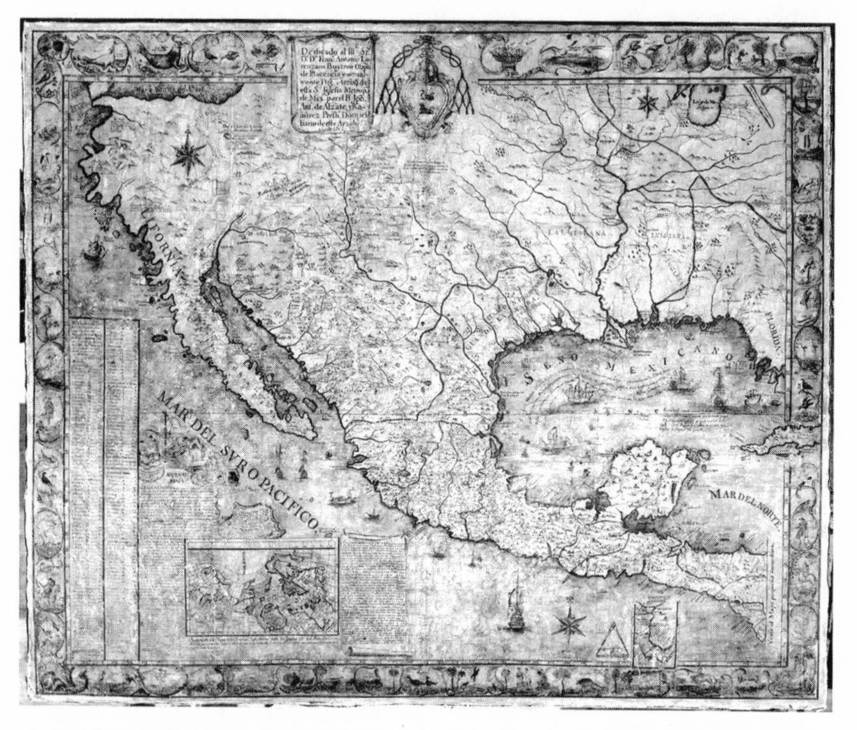

16 José Antonio de Alzate y Ramírez, *Nuevo mapa geográfico de la América septentrional española* (1767). 177 × 210 cm. Museo Naval, Madrid.

of contemporary Europe, such as the Blaeus' *Theatrum Orbis Terrarum* page (figure 3, chapter 1, p. 27); however, it is also similar to the cuadros de casta representations, such as the Mena panel, in its attempt to display a sense of place through types of people. Here, itinerary and exploratory travel combine as both the space and content of New Spain are laid out in a single image.[59]

Finally, in the late eighteenth century, the antiquities of New Spain became of interest to Spanish and New Spanish authorities. Carlos III sponsored a 1786 expedition lead by Antonio del Río, an artillery captain, who authored a report to the king describing the ruins of Palenque that was later translated and published as *Description of the ruins of an ancient city discovered near Palenque*. In his narrative, del Río marvels at "these truly interesting and valuable remnants of the remotest antiquity" and suggests to the king that "the glory of the Spanish arms would be exalted, and . . . credit would accrue to the national refinement," if sculptures and other remains of Palenque were brought to Madrid.[60]

Interest in antiquities was further stimulated in 1790, when renovations near Mexico City's Plaza Mayor (Central Plaza), resulted in the excavation of two large monuments of the Indian past. One sculpture was a flat disc-shaped piece with a human face surrounded by concentric bands of relief elements carved on one side. Known as the "Sun Stone," or "Calendar Stone," the image was identified as representing a male god and was placed on display for public viewing in the Plaza Mayor on an exterior wall of the Mexico City Cathedral. The second monolith was semianthropomorphic in form; observing the pendulous breasts of this freestanding statue, it was identified as the "Teoyaomiqui," representing an Aztec-Mexica goddess of death. Subsequently, it was also known as the "Coatlicue," after an Aztec mother goddess. Fearful that this sculpture might resurrect old idolatries, it was removed from the plaza at midnight, transported to the Royal and Pontifical University in Mexico City by order of the Spanish Viceroy, and buried under a corridor pavement of the university.[61]

Nevertheless, the Calendar Stone and the Teoyaomiqui spawned extended discussions locating an originating past for New Spain. Antonio de León y Gama, a well-regarded and erudite mathematician and astronomer with a scholarly interest in preconquest antiquity, articulated this perspective. Immediately after their excavation, he began to measure, draw, and analyze the sculptures. His finished work, *Descripcion histórica y cronológica de las dos piedras* [stones], was published in June 1792.[62] In it León y Gama, like Villaseñor, argues for an original and authentic Indian, and Indian culture, and thus, the existence of an original and authentic preconquest (that is, pre-Spaniard) history. In fact, in a side discussion, León y Gama states that the authenticity of a true Indian is not found in their present appearances (that is, living Indians), but in the original condition emanating from historical context.[63] In this way, he postulates an authentic and unified Indian identity based in history. León y Gama anticipated the nineteenth-century liberal ideas of national identity as formulated through indigenous history, and not in estates, corporations, or Spanish history.

Finally, between 1805 and 1807, Carlos IV sponsored Guillermo Dupaix, a military captain, to undertake three archaeological expeditions, with the artist José Luciano Castañeda. The group surveyed the areas of Chiapas, Oaxaca, and Yucatán and explored ancient sites such as Xochicalco, Monte Albán, Mitla, and Palenque, which resulted in the illustrated two-volume publication, *Viages sobre las antiguedades mejicanas*, published in Dupaix's *Antiquités mexicaines*. By the end of the eighteenth century, antiquities collections became another way to map the content of New Spain.

In sum, through the extended collection of data, Spanish mappings, in contrast to non-Spanish ones, focused on administrative purposes. Bourbon reforms focused on Atlantic trade, the strengthening of the domestic industry, restructuring of the Spanish economy, and increased competitiveness with European nations.[64] There was also increased attention to constructing history through the excavation and collection of antiquities. By the beginning of the nineteenth century, New Spain—soon to be independent Mexico—would come to know itself through this Spanish economic lens and cultural themes. Its basic cartographic outlines, while still imprecise, were established, but maps were not widely circulated due to Spanish authorities' concern with revealing critical information to other nations. Nevertheless, New Spain's content was filled with a geography that identified its resources through the perspective of commercial development. Further, works like those of Villaseñor, Alzate, and León y Gama exemplified an emergent structure for fabricating a New Spanish identity that was separate from Spain through distinct history, objects, and people. By the end of the eighteenth century, this New Spanish cultural identity could be displayed through texts, print images, and antiquities—rather than maps. Subsequently, such display fed a nascent nationalism, touting a unique geography, culture, and history—real or imagined—upon which to build an independent Mexico.

3

TOURING MEXICO

A Journey to the Land of the Aztecs

Salt cellars, figurines, historical narratives, atlas frontispieces and cartouches, cabinets of curious objects, natural history specimens, archaeological objects, and, of course, maps: in the course of three centuries of such mappings, America / New Spain became visible for Europeans. As a result, dissonant, "and/or," readings of the space called New Spain coexisted by the end of the eighteenth century. Depending on the reader's or viewer's inclination, selective vision, or desired outcome, America / New Spain could be imagined as an inkblot-shaped mass embedded in the graticule of longitude and latitude *and/or* the trope of seductive female, seminude and accessible; a plant or animal could be at once a curiosity *and/or* a taxonomically classified specimen *and/or* potential commercial material; an indigenous-crafted object could be a wonder *and/or* an indicator of absent industry; and New Spain's inhabitants and their societies could be viewed as degenerate cultures *and/or* representatives of a specified level of social evolution *and/or* productive subjects of the king. Through texts, images, and maps, America / New Spain was appropriated; indigenous memories were erased and new ones fabricated. Truth and meaning were mobile and fluid, that is, randomly aggregated and disaggregated from different fields according to need or desire.

The cumulative effect of these mappings was to make visible to Europeans a territory that was conceived of as invisible, and to provide a desultory assortment of conjectural narratives explaining the context and significance of the people who inhabited the continent. Whether through trope or specimen, eighteenth-century Europeans invented in America / New Spain what they needed to find. In the early part of the next century, travelers from continental Europe to Mexico embedded these dissonant understandings, sometimes latently and sometimes overtly, into their travel narratives.

The first half of the nineteenth century would also see a trickle of visitors and explorers become an influx of travelers as Mexico's borders opened up.[1] This was due to dramatic political changes resulting from Napoleon Bonaparte's occupation of Spain and the forced abdication of King Carlos IV; the renunciation and captivity of his successor, Ferdinand VII; and the appointment of Joseph Napoleon to the Spanish throne in 1807. These events led to greater autonomy for Mexico and, ultimately, complete independence from Spain in 1821. The interests of these travelers ranged from curious adventure, to demographic overview, to economic inspection, to scientific description.

The results of these travels, however, were quite distinct from those of previous centuries. This chapter examines samples of travel writing and travel images that appeared in the first half of the nineteenth century. The disaggregated ideas of salt cellars, natural history specimens, atlas frontispieces and cartouches, figurines, and so on, permeated travel-writing perspectives and, at the same time, began to cohere into intensely visual fabrications of Mexico for readers as well as viewers, that is, stationary travelers. The themes, ideas, and perspectives advanced by these early nineteenth-century travel narratives also provided a means through which postindependence, Mexican intellectuals—including Antonio García Cubas—traveled from New Spain to Mexico.

TRAVEL PERSPECTIVES

The literary, historical, and anthropological analyses of travel and travel literatures of the eighteenth and nineteenth centuries constitute a large scholarly corpus. Generally, these works look at Europeans' travel in Europe, for example, the Grand Tour, or outside of Europe to Asia or Africa. There are few critical analyses of travel literature's impacts on Mexico as it forms itself as a nation, despite the fact that there are numerous examples of travel literature of New Spain / Mexico. Thus, it is necessary to begin the discussion of early travelers to New Spain / Mexico within an overview of certain conceptual underpinnings of travel writing and illustration that guide my subsequent analysis and assist in charting common perspectives among disparate texts.

A significant issue faced by travelers of the eighteenth century and early nineteenth, or for that matter, any traveler, was the intellectual comprehension of the foreign and unfamiliar in relative terms. Travelers, always extracultural in their identity, find what is different from the familiar and elucidate it in relevant written and visual terms and through schema that erase or overwrite the nonfamiliar original. James Duncan and Derek Gregory, geographers, focus specifically on this process of making the unfamiliar familiar. They stress that there is a need to be aware of the "multiple sites at which travel writing takes place and hence the spatiality of representation."[2] This is to say that while a reader may assume that a travel narrative reflects writing or an image made at the time of the journey, these narratives are in fact assembled translations of field notes, sketches, and sometimes daguerreotypes or photographs compiled in quiet studies or other passive spaces when the traveler is actually quite sedentary. As a result, travel writing is inherently domesticating, an act of translation that is both a recognition and recuperation of difference as travel writing translates one place with another and creates a "place in-between."[3] Ultimately, these translations start to form an archive of texts and images that subsequent travelers will add to as well as use.

Chloe Chard, a scholar of travel writing, further argues that the first-person narrative of travel allowed "the subject of the commentary to move easily between one domain of objects and another, to shift back and forth at will between specific objects of commentary, and to pause in the account of a particular place, in order to reflect at length on some idea that springs to mind."[4] As discussed in the previous chapters, seventeenth-century travel descriptions emphasize curiosity to arouse responses of wonder and enthrall the reader or spectator. In contrast, in the next century, travelers claimed unmediated eyewitness experience as a new methodology to verify their own authority to describe and comment. These travelers, who saw earlier travel as purposeless movement, sought to inform readers and extract benefit from the topography.[5]

This benefit assumes finding otherness and grasping its difference through the use of rhetorical strategies of "opposition" and "intensification."[6] Chard explains that the strategy of opposition employs asymmetrical descriptions, such as the contrast of Mexican idleness with European industry. As opposition strategies in travel writing attempt to link the unfamiliar to the familiar, the strategy of intensification distinguishes and narrates the unfamiliar through particular themes that accentuate dramatic difference. One such thematic is "incomparability" through which distinctive foreign objects, landscapes, or peoples are found to be without a rival or equivalent—travelers found no volcanoes comparable to the twin volcanoes of Popocatépetl and Iztaccíhuatl. Another common thread of this strategy is "profusion," which highlights the immeasurable abundance of the foreign resources—the immense natural wealth of Mexico. Intensification also narrates the unfamiliar through the theme of "excess," which focuses on unrestraint in the consumption of resources by local populations—Mexican misuse of its natural resources, for example.[7]

The strategies of opposition and intensification were consistently employed in early nineteenth-century travel writing about Mexico as part of the eyewitness methodology. These rhetorical strategies structured not only the texts but also informed the emerging inventory of travel illustrations that came to represent Mexico. As a result, by the middle of the nineteenth century, Mexico was seen and known, that is, translated, by Western European readers and observers through text *and* images—maps, charts, and illustrations of kinds of people, ancient ruins, astounding landscape, and archaeological objects. The imagery of this archive circulated widely to a consuming audience due to the easier and cheaper reproductive processes of lithography. Images were aggregated into new display formats—highly circulated travel books, as well as exhibitions, dioramas, and panoramas. Through the intensified visuality of travel literature, as the various travelers discussed below demonstrated and the famous historian William Prescott stated emphatically, there was no need to travel to Mexico for Mexico came to the reader or viewer.

ALEXANDER VON HUMBOLDT

Beyond a doubt, the most prominent of early nineteenth-century travelers to New Spain was Alexander von Humboldt (1769–1859), a Prussian mining official who resigned his position to pursue what he called "terrestrial physics" (physique du monde).[8] Humboldt originally wished to work in Egypt as part of Napoleon's expedition of conquest (1798–1801), which included military personnel as well as numerous cartographers, scientists, and artists. While Napoleon lost the campaign, it did result in comprehensive, detailed information and images about Egyptian archaeology, history, astronomy, geodesy, urban mapping, zoology, mineralogy, botany, demography, climatology, anthropology, and so forth.[9] This expedition material, published as the twenty-one folio volume, *Description de l'Egypte* (1809–28), generated intense public interest—"Egyptomania." Unable to participate but still eager for foreign exploration, Humboldt found his way to Madrid, where he received permission to journey to the Spanish Americas from King Carlos IV. He traveled from 1799 to 1804 with Aimé Bonpland (1773–1858), a French naturalist. Returning to Paris with his data, specimens, sketches, and notes, Humboldt spent the next twenty-two years writing and publishing *Voyage aux régions équinoxiales [*equinoctial or equatorial*] du Nouveau Continent*, consisting of thirty volumes, heavily illustrated, with over 1,400 maps, images, graphs, and charts. They were published separately, depleting Humboldt's personal financial resources as well as those of three publishers by the time of their completion in 1834.[10]

Throughout the nineteenth century, Humboldt's scientific and popular recognition and influence was far reaching on the European and American continents.[11] Nineteenth-century travelers to Mexico often acknowledge having read Humboldt's works as well. García Cubas, who held Humboldt and his work in the highest of esteem, incorporated the Prussian traveler's comprehensive approach into his own geographic study and included, often by copying directly, Humboldt's maps, graphs, and charts into both the *Atlas geográfico* and the *Atlas pintoresco*.[12]

An important objective of Humboldt's overall study was correction of the ideas of philosophical travelers such as de Pauw, Raynal, and Robertson, whom he believed were misguided in their writings about the Spanish Americas because preconceived notions and schema predisposed their descriptions and conclusions. In contrast, and reflecting criticism of the unmediated eyewitness approach, Humboldt asserts, "Biased by no system, I shall point out those analogies that naturally present themselves."[13] He further elaborates that

> Until now, little has been achieved by travelling naturalists for terrestrial physics [*Physik der Erde*] (*physique du monde*), because they are almost always concerned exclusively with the descriptive [*naturbeschreibenden*]

> sciences and collecting, and have neglected *to track the great and constant laws of nature manifested in the rapid flux of phenomena, and to trace the reciprocal interaction*, the struggle, as it were, of the divided physical forces.[14] (emphasis mine)

He firmly believed that the patterns found through the analysis of collected data should lead one to a conclusion and broader concepts. Along with description, then, Humboldt also sought to integrate reflective practices, that is, the identification of the multiple dimensions of data. Thus, in this search of such natural laws, Humboldt insisted, for example, that plants needed to be studied in their natural environment as part of geographically interrelated and systematic wholes rather than as isolated botanical specimens uprooted from their natural context.[15]

To accomplish this holistic approach as well as produce a comprehensive study similar to the *Description de l'Egypte*, Humboldt utilized a significant assortment of instruments to collect data, including chronometers to indicate the passage of time; telescopes, quadrants, and sextants to mark angular distance of stars; thermometers to specify temperature; hygrometers to show humidity; barometers to indicate atmospheric pressure; electrometers to identify electrical tension; and eudiometers to ascertain the chemical composition of the air. His use of these and other measuring devices provided overlapping information that could be evaluated and compared to determine errors and to identify patterns.[16]

Like Villaseñor's *Theatro Americano*, Humboldt's amassed information resulted in maps, long explanatory texts, and data tables as seen in figure 17, which lists the value of the gold and silver mined and minted in New Spain between 1690 and 1803. While valuable, such data was overwhelming and difficult to understand and appreciate unless summarized in some synthetic way. Thus, Humboldt insisted on associating his maps, tables, and texts with emerging mapping practices—graphs and charts—which he understood as "animated pictures of distant regions."[17] He explains that

> It would be ridiculous to try to express by curved lines moral ideas, the prosperity of peoples, or the decadence of their literature. But anything that has to do with extent or quantity can be represented geometrically. Statistical projections [graphs and charts], *which speak to the senses without tiring the intellect, have the advantage of bringing attention to a large number of important facts*."[18] (emphasis mine)

ESTADO N°. I.

Oro y plata sacados de las minas de Mégico y acuñados en Mégico desde 1690, *hasta* 1803.

Años.	Valor en pesos.	Años.	Valor en pesos.	Años.	Valor en pesos.	Años.	Valor en pesos.
1690	5,285,580	1720	7,874,323	1750	13,209,000	1780	17,514,263
1691	6,213,709	1721	9,460,734	1751	12,631,000	1781	20,335,842
1692	5,252,729	1722	8,824,432	1752	13,627,500	1782	17,581,490
1693	2,802,378	1723	8,107,348	1753	11,594,000	1783	23,716,657
1694	5,840,529	1724	7,872,822	1754	11,594,000	1784	21,037,374
1695	4,001,293	1725	7,370,815	1755	12,486,500	1785	18,575,308
1696	3,190,618	1726	8,466,146	1756	12,299,500	1786	17,257,104
1697	4,459,947	1727	8,133,088	1757	12,529,000	1787	16,110,340
1698	3,319,765	1728	9,228,545	1758	12,757,594	1788	20,146,365
1699	3,504,787	1729	8,814,970	1759	13,022,000	1789	21,229,911
1700	3,379,122	1730	9,745,870	1760	11,968,000	1790	18,063,688
1701	4,019,093	1731	8,439,891	1761	11,731,000	1791	21,121,713
1702	5,022,550	1732	8,726,465	1762	10,114,492	1792	24,195,041
1703	6,079,254	1733	10,009,795	1763	11,775,041	1793	24,312,942
1704	5,627,027	1734	8,506,553	1764	9,792,575	1794	22,011,031
1705	4,747,175	1735	7,922,001	1765	11,604,845	1795	24,593,481
1706	6,172,037	1736	11,016,000	1766	11,210,050	1796	25,644,566
1707	5,735,032	1737	8,122,140	1767	10,415,116	1797	25,080,038
1708	5,735,601	1738	9,490,250	1768	12,278,957	1798	24,004,589
1709	5,214,143	1739	8,350,785	1769	11,938,784	1799	22,053,125
1710	6,710,587	1740	9,554,040	1770	13,926,329	1800	18,685,674
1711	5,666,085	1741	8,663,000	1771	13,803,196	1801	16,568,000
1712	6,613,425	1742	16,677,000	1772	16,971,857	1802	18,798,600
1713	6,487,872	1743	9,384,000	1773	18,932,766	1803	23,166,906
1714	6,220,822	1744	10,285,500	1774	12,892,074		
1715	6,368,918	1745	10,327,000	1775	14,245,286		
1716	6,496,288	1746	11,509,000	1776	16,463,282		
1717	6,750,734	1747	12,002,000	1777	21,600,020		
1718	7,173,590	1748	11,628,000	1778	16,911,462		
1719	7,258,706	1749	11,823,500	1779	19,435,457		

Total desde 1760 hasta 1803 en oro y plata, 1,353,452,020 pesos.

17 Alexander von Humboldt, *Oro y plata sacados de las minas de Mégico y acuñados en México desde 1690, hasta 1803*, from *Ensayo politico sobre el reino de la Nueva-España* (1822), tomo 3:205. Author's collection.

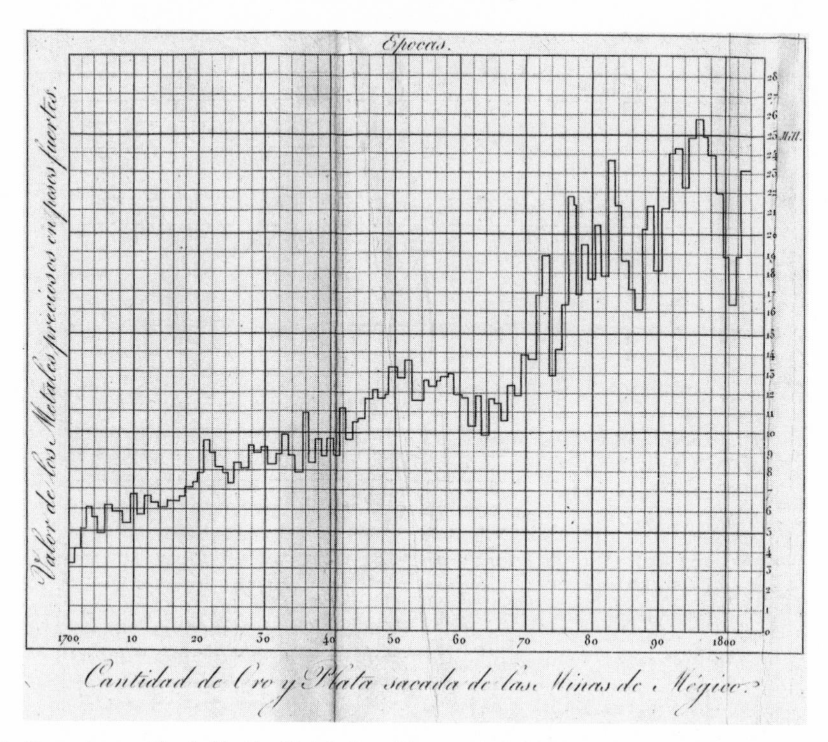

18 Alexander von Humboldt, *Cantidad de Oro y Plata sacada de las Minas de Mégico*. From *Ensayo politico sobre el reino de la Nueva-España* (1822), tomo 3:439. Author's collection.

Thus, the data table of mined gold and silver was translated into *Cantidad de Oro y Plata* (Quantity of Gold and Silver), a graph which converts the table's blur of numbers into a quick visualization depicting the increase of mining production over time (figure 18). Through analysis of change, distribution, interactivity, and interior functioning of natural and social phenomena, Humboldt sought to reveal—that is, track and trace patterns—not visible to the naked eye.[19]

Humboldt applied these innovative mapping techniques to the data he collected on the natural environment as well. In particular, he used charts and graphs to expose the variation of particular phenomena across space, time, or both. One of the most effective examples of this form of Humboldt's visual analysis is his summary of New World plants (figure 19). Based on his research in Peru, the chart, *Géographie des plantes Équinoxiales*, uses a

19 Alexander von Humboldt, *Géographie des plantes Équinoxiales*, from *Voyage de Humboldt et Bonpland* (1814). Brown University Library.

cutaway section of Chimborazo, a mountain located in northern Peru, to visualize tables of data that classify the types of vegetation located at specific altitudes. Rather than using a taxonomic catalogue of species and genera in the tabular (Linnaean) layout, this format also visually illustrates the total environmental distribution of individual plants. Here, Humboldt provides an intensified comparative visuality asking his readers to ponder further the "character of vegetation over the entire surface of the globe, and the impression produced on the minds of the beholder by the grouping of contrasted forms in different zones of latitude or elevation."[20]

Humboldt spent only twelve months in New Spain and did not travel extensively through its intendancies as he had in other parts of the Spanish Americas. In fact, much of the data he used for his publication on New Spain were gathered from existing sources provided by the Spanish authorities and New Spanish intellectuals. Humboldt produced a draft document in Spanish of what he called a statistical essay and circulated it to local intellectuals and the viceroy.[21] He published the final text in French as the third section of the *Voyage*, entitled *Essai politique sur le royaume [kingdom] de la Nouvelle-*

Espagne, from which he garnered great international renown and respect.[22] The *Essai politique*, consisting of the folio volume *Atlas géographique et physique du Royaume de la Nouvelle-Espagne* (1811), provides geographical data and images, and two quarto volumes, divided into six sections:

I. General consideration of the physical extension and physical aspect of New Spain;
II. Overview of the general population of New Spain;
III. Statistical information about the intendancies;
IV. General overview of agricultural and mineral resources;
V. State of manufacturing and commerce; and
VI. Investigation of income and military defense of the country.

Vues de Cordillères et monumens des peoples indigènes de l'Amérique (1810–13) was published separately. This folio included maps along with images of flora, fauna, and archaeological remains, including the Calendar Stone and the Teoyaomiqui sculpture, as well as incomparable physical features such as the mountain Orizaba, and the volcanoes Popocatépetl and Iztaccíhuatl. In his image *Rochers basaltiques et Cascade de Regla*, cliffs of magnificent stone prisms that surround a waterfall dwarf the seated figures of Humboldt and Bonpland (likely) (figure 20). This inclusion of a figure—usually the traveler—recurs in these early travel landscape illustrations and may be seen to certify the eyewitness authority of the writer, as well as to intensify the scale of the surrounding environment. In this image, the two figures appear to be in an animated discussion. One figure points up to the cliffs and waterfall as the other figure turns his head toward the first figure and gesticulates down toward the plants and flowing water; behind the two figures, an Indian gazes at them. Demonstrating the visualization of the strategy of opposition, the indigenous figure may be read as merely an observer; it is the scientist and illustrator who identify with the landscape's significance.

Another major objective of Humboldt's research was the production of accurate maps of New Spain. He was acutely aware that current cartographic examples were based on older, mostly incorrect, maps. Humboldt collected these older maps and compared their cartographic data. He explains that

> A labour [map] of this nature was in fact very necessary, both for the administration of the country, and for those who wish to know its national industry. Instead of merely inserting in my map the names of three hun-

20 Alexander von Humboldt, *Rochers basaltiques et Cascade de Regla*, from *Vues de Cordillères et monumens des peoples indigènes de l'Amérique* (1810). Special Collections Research Center, Earl Gregg Swem Library, The College of William and Mary.

dred places known for considerable mining undertaking, I proposed to unite together all the material which I could procure, and to discuss the difference of position which these heterogeneous materials at every instant presented. . . . I ought to have foreseen, that, notwithstanding the most assiduous labour during three or four months, I could only give a very imperfect map of Mexico, compared with the maps of the most civilized countries of Europe. [This is] preferable to what has yet been offered to the public on the geography of New Spain.[23]

Here, he confirms the poor state of cartography of America and New Spain. The maps Humboldt produced for his texts reconciled latitude and longitude data he collected while traveling with about thirty different cartographic descriptions of New Spain, such as the *Mapa del Arzobispado de Mexico, por Don Jose Antonio de Alzate* of 1768, which he evaluates as "très-mauvaise" (very bad); a map of the environs of Mexico City by Carlos Sigüenza reprinted in 1786; and an undated *Carte manuscrite d'une partie de la Nouvelle-Espagne*, made for Viceroy Revillagigedo, which, while also inaccurate, nevertheless showed the division of New Spain into intendancies.[24] He also mentions that when possible, he retained the Aztec place names and did not "disfigure" the names as did earlier writers such as Robertson, Raynal, and de Pauw.[25] The end products were to be up-to-date maps copied and recopied throughout the nineteenth century, such as the *Carte Du Mexique* (figure 21). Along with this map, Humboldt drew other maps, including a *Plan of the Port of Acapulco*; *Map of the Valley of Mexico*; *Plan of the Port of Veracruz*; and a thematic map showing the trade patterns of precious metals among continents.

In addition, Humboldt associated his maps with images that visualized data of certain physical features. For example, *Plano fisico de la Nueva-España. Perfil del Camino de Acapulco á Mégico, y de Mégico á Veracruz*—Humboldt's profile of the road from Veracruz to Acapulco—demonstrates the variation in elevation between Mexico's east and west coasts (figure 22). These images added another visual dimension to the cartographic information laid out in Humboldt's maps by visualizing the space. Humboldt's writings also contained extended data about geography, demography, and natural resources, which he also displayed in table format in his texts. Again, trying not to "tire the mind," he took such data and converted it into visual images. For example, *Tableau comparatif de l'étendue [extent] territoriale de*

21 Alexander von Humboldt, *Carte Du Mexique* (1804), from *Essai politique sur le royaume de la Nouvelle-Espagne* (1811), vol. 1. General Collection, Library of Congress.

Intendances de Nouvelle-Espagne, a table enumerating the area of certain territories, was converted into image using proportional squares to compare visually the area of each of the intendancies and that of Spain.

Distinct from the comparison of single sets of geographic data such as elevation or area, Humboldt also used charts and graphs to track the interaction of two or more sets of data and trace the variation of certain phenomena. For example, *Cantidad de Oro y Plata* (Quantity of Gold and Silver), discussed above (figure 18), dramatically illustrates the increasing extraction of mineral resources from New Spain over time.[26] While in the graph entitled *Étendue territoriale et Population des Métropoles et des Colonies en 1804*, Humboldt produced a comparative analysis of New Spanish demographics, not as family groups as found in the *cuadro de casta* paintings but as population divisions using different-sized blocks to compare the population in various areas of the Spanish Americas to Spain's population; the inhabitants of the colonies of other European countries are also graphed (figure 23).[27] Here, census information collected in the late eighteenth century was given graphic form, allowing the viewer to have a comprehensive

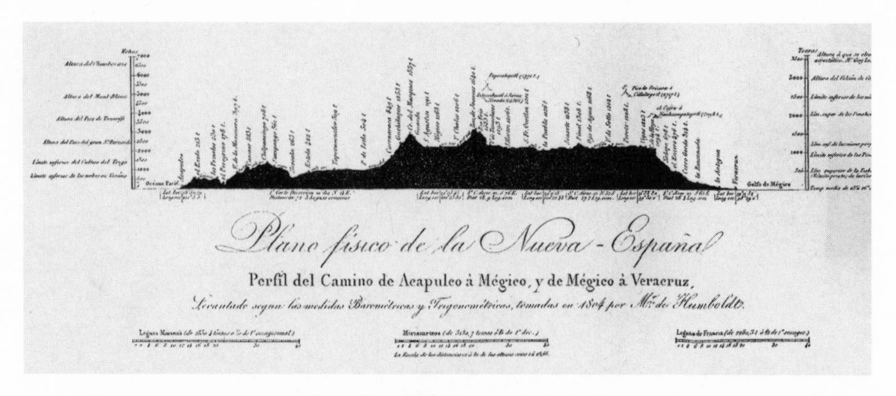

22 Alexander von Humboldt, *Plano fisico de la Nueva-España. Perfil del Camino de Acapulco á Mégico, y de Mégico á Veracruz*, from *Ensayo politico sobre el reino de la Nueva-España* (1822), vol. 2. General Collection, Library of Congress.

overview of the inhabitants of New Spain. Utilizing the relatively new construct of "population," Humboldt directed the viewer away from the overwhelming, detailed demographic data or the specificity of individuals presented through texts or tables and toward the grouped patterns of people. The viewer could see New Spain's population without seeing a New Spanish person. In contrast to the familial groups of cuadro de casta paintings that represented members of the social units according to bloodline, Humboldt visualized a New Spanish—soon to be Mexican—population, people grouped by geographic area and social characteristics.[28]

When compared to the seventeenth- and eighteenth-century imagery of New Spanish people by Blaeu or Grasset, for example, it is clear that like Villaseñor, Humboldt sought to visualize how parts fit into the whole, rather than how a part represents a whole. It is through this comparative visual analysis of sets of data that Humboldt truly initiated a way to "stimulate the reader's attention" to understand the interaction of data—quantity of silver mined over time or kinds of people across space—that would be difficult to reconcile and understand quickly as written text or in a table format. These graphs formed a new synthetic and intensified visuality for New Spain / Mexico.

The quarto size and cost of the original *Voyage* and its subsections, such as the *Essai politique* and *Vues de Cordillères*, made it highly inaccessible to

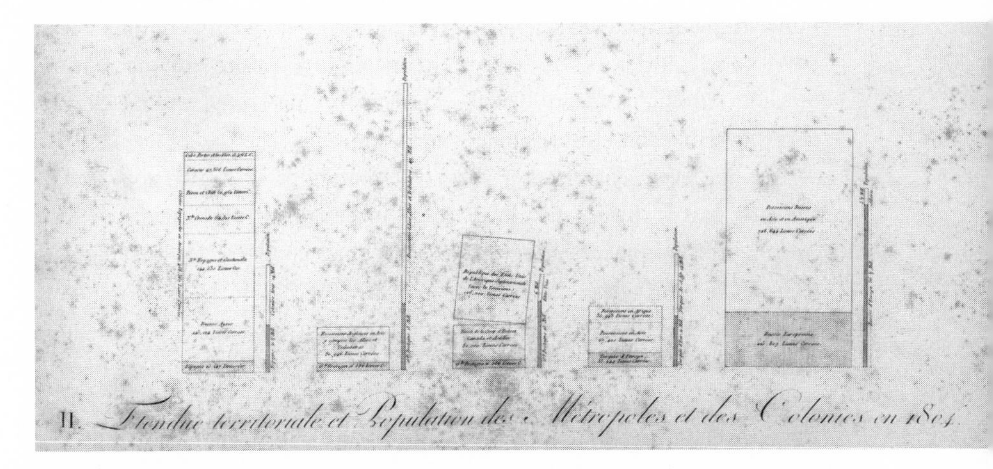

23 Alexander von Humboldt, *Etendue territoriale et Population des Métropoles et des Colonies en 1804*, from *Atlas géographique et physique du Royaume de la Nouvelle-Espagne* (1811). General Collection, Library of Congress.

most readers. As a result, almost immediately after Humboldt published a volume, condensed octavo versions appeared in Spanish and English. For example, an English version appeared in 1811 published as the *Political essay on the Kingdom of New Spain*, translated by John Black and published as four volumes in London and two volumes in New York. A Spanish version was published in Madrid in 1818 as *Minerva. Ensayo político sobre el reyno de Nueva España*, . . . translated by Pedro María de Olive in two volumes. Another Spanish version, in four volumes, was published in Paris in 1822 under the title *Ensayo político sobre el reino de Nueva-España* and translated by Vicente Gonzalez Arnao.[29]

In the introductions to the foreign language editions of the *Essai politique*, the translators complain of Humboldt's loquaciousness. In the 1811 English version, John Black explains his reasons for abridging the original material: "Yet it is to be regretted that the author could not throw occasionally more rapidity into his descriptions, and give somewhat more condensation to his materials.[30] In the 1818 Spanish version published in Madrid, Gonzalez Arnao also claims that Humboldt's account is overly detailed for the general reader and that he uses only the most substantial, curious, and important information, presenting the general results and omitting the less interesting and detailed scientific explanation.[31] Along with editing of text, editors and translators selected certain images to associate with the texts. As a result, Humboldt's work became known through edited and condensed versions rather than the original oversize volumes.

Abridged versions of other parts of Humboldt's work also appeared in the early part of the nineteenth century. The expansion of the less expensive printing technology of lithography, rather than copperplates or steel plate techniques, allowed his publications to overflow with illustrations, thematic maps, graphs, and charts, and to be reproduced inexpensively. These include *Researches concerning the institutions & monuments of the ancient inhabitants of America: with descriptions and views of some of the most striking scenes in the Cordilleras!* translated into English by Helen Maria Williams and published in London in 1814; the *Vues des Cordillêres e Monumens des Peuples indigènes de l'Amérique* published in Paris in 1816; the *Vistas de las Cordilleras y Monumentos de los Pueblos indígenas de América* published in Mexico in 1822; and *The Travels and Researches of Alexander Von Humboldt being a condensed narrative of his journeys in the equinoctial regions of America, . . . (with map and engravings)*, by W. Macgillivray, part of a publication entitled *Harper's Family Library*, no. 54, published in New York in 1833. These shortened versions—and there were others—which read as adventure novels, included sixteen to twenty small plates instead of the original sixty-nine plates of the *Vues de Cordillères* and were priced for a more general audience. The plates emphasize a cataloguing of images, including the mountain Orizaba and the volcanoes Popocatépetl and Iztaccíhuatl. Also, just as Humboldt's texts were edited and recontextualized, so were the associated images. The *Rochers basaltiques et Cascade de Regla* reappears in Macgillivray's edition as almost a direct copy of Humboldt's original. However, closer inspection confirms that the scale of the figures is slightly smaller in relation to the stone cliff and vegetation, and the standing indigenous figure is deleted (figure 24). Further, there is an intensification of the density of the tropical vegetation on top of the cliff to make it appear to be a more generic South American jungle rather than the more arid central Mexico plateau.[32]

Like his texts, graphs, and images, references to Humboldt's maps reappear in atlases such as H. C. Carey's *A Complete historical, chronological, and geographic American atlas being a guide to the history of north and South*

TRAVELS AND RESEARCHES

OF

BARON HUMBOLDT.

NEW-YORK:

J. & J. HARPER, 82 CLIFF-STREET.

1833.

24 Title page, from *The travels and researches of Alexander von Humboldt: being a condensed narrative of his journeys in the equinoctial regions of America . . . by W. Macgillivray* (1833).

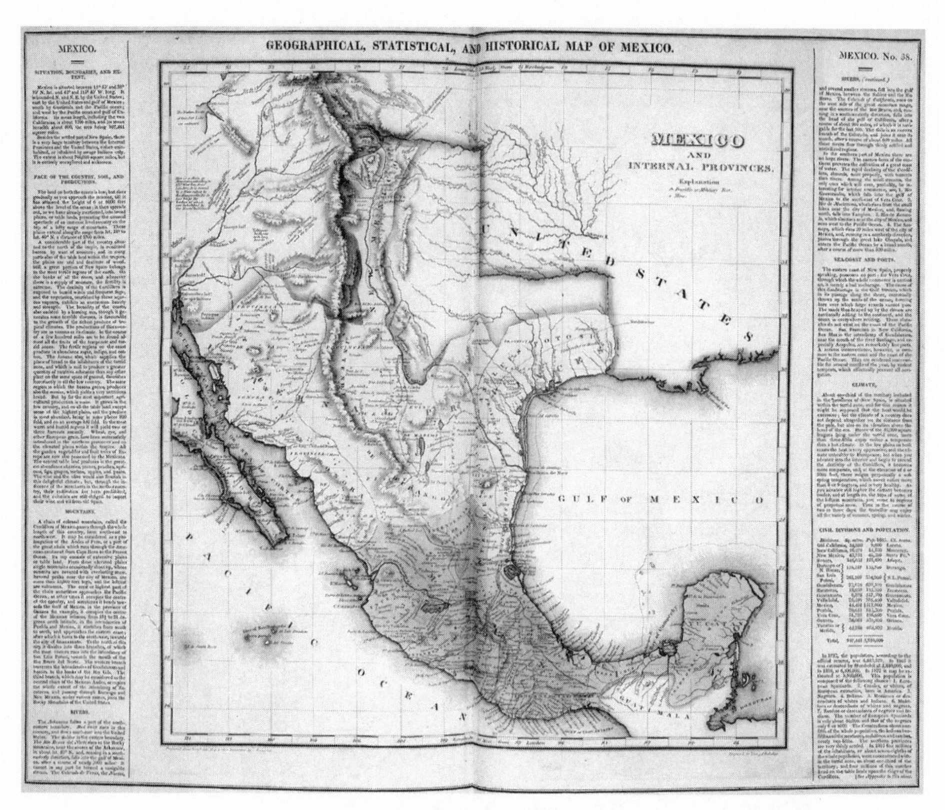

25 H. C. Carey, *Geographical, Statistical and Historical Map of Mexico*, from *A Complete historical, chronological, and geographic American atlas being a guide to the history of north and South America, and the West Indies* . . . (1822). Courtesy of the Newberry Library, Chicago.

America, and the West Indies . . . , published in 1822 (figure 25). Instead of separating the map from the data or associating it with surrounding images, Carey's atlas page shows the territory of Mexico surrounded by text that provides information about its "Situation" [location], "Boundaries and Extent," "Face of the Country," "Soil and Productions," "Mountains, Rivers, Seacoast and Ports," "Climate," and "Civil, Divisions, and Populations."

In sum, Humboldt translated data into summative works that shaped a reimagined American continent, as well as New Spain, for both learned and popular international audiences.[33] Instead of disparate and disjointed glimpses visualized by figurines, atlas cartouches, and so on, of earlier centuries, he fabricated a holistic space of New Spain. Here, data about resources, people, and geographic features were conceptually connected not through metaphorical images, such as in frontispieces with allegorical figures, but more importantly through mesmerizing graphs and charts that aggregated data and seemed to narrate only incontrovertible truth to the attentive, although stationary, readers or viewers. For Humboldt, these visual images were not merely illustrations but visual ways to summarize data and direct the attention of the viewer to see and understand interconnected relationships.[34]

In his writings, Humboldt tended to inverted rhetorical strategies of travel to manifest a familiar New Spain out of unfamiliar data. Instead of overwhelming the reader with opposition and intensification strategies, he reversed their impact to illustrate continuity across context—terrestrial physics. He did not read dramatic differences in New World plants as signs of a degenerate environment but as part of a continuum that constituted the natural environment. That is, Humboldt fabricated familiarity through demonstrating the transferability of data and analysis across fields and contextualized difference into a multidimensional holistic discourse on commonality across environments.[35]

Finally, as is to be expected, there is a cautionary tale in Humboldt's works. This is evident in the frontispiece of Humboldt's 1814 *Atlas géographique et physique du nouveau continent*, which depicts Mercury, the god of commerce, assisting a highly costumed, Indian noble to his feet while Minerva, the goddess of letters, offers an olive branch; Chimborazo rises prominently in the background (figure 26).[36] Here, as a counter to the disparaging critique of New Spain during previous centuries, the image visualizes the study of the physical environment as a new awakening for the Americas. At the same time, the frontispiece, although using a renovated visual vocabulary, is based on the same premise found in van der Straet's 1638 print, *America*, which illustrates Vespucci awakening America (figure 4, chapter 1, p. 28). Set in geographic context like the *America* print, this image substitutes Minerva and Mercury for Vespucci; the wondrous flora and fauna and savage behaviors (cannibalism) are replaced by Chimborazo as a metaphor for terrestrial physics. Most importantly, the noble male Indian replaces the reclining nude female. The premise of this image remains the same: America needs European culture to provide it with information about itself.

By including this frontispiece in an atlas containing geographic and topographic maps, Humboldt metaphorically envisions that his research

26 François Gérard (engraving by Barthèlemy Roger). *Humanitas, Literae, Fruges*, frontispiece engraving, from Alexander von Humboldt, *Atlas géographique et physique du nouveau continent* (1814). Geography and Map Reading Room, Library of Congress.

would allow America to participate in European commerce and learning. Thus, while not denying Humboldt's achievements, it is important to be aware that aspects of Humboldt's writings are problematic as well. Mary Louise Pratt vigorously argues that while Humboldt gave Europeans and criollos new ways to imagine America, it was through "imperial" eyes.[37] Jorge Cañizares-Esguerra's research demonstrates that Humboldt borrowed extensively from and expanded ongoing research by Spanish American researchers.[38] While Nicolaas Rupke reminds us that Humboldt rose to prominence in his role as a surveyor of New Spain through *Essai politique*; his image as a polymath and great intellect of the European Enlightenment tradition was constructed later in the century.[39] Aware of ongoing criticism, nevertheless, Humboldt's notion of "looking" remains critical because the intensified visuality proposed by his ideas and images reverberated throughout the subsequent nationalist texts and images of Mexican writers.

TRAVELING THROUGH MEXICO

During the decades following Humboldt's and Bonpland's travels, numerous European travelers journeyed to Mexico to sketch and describe Mexican landscapes, people, and resources in their notebooks and journals. By the early 1840s, lithography allowed quick and relatively inexpensive production and the emerging technology of the daguerreotype was also used. Following the model of Humboldt, their travel materials were sources for subsequent publications and exhibitions, putting numerous lithographic images into circulation and further stimulating growing popular European interest in exotic cultures and lands. Generally, most of these travel accounts were considerably less intellectually comprehensive and methodically systematic than Humboldt's; in fact, they often lapsed back to the ideas of Raynal, de Pauw, and Robertson and even the seventeenth-century themes of curiosity and wonder.

A comprehensive analysis of European travel accounts of Mexico between 1800 and the 1850s is beyond the scope of this study.[40] Therefore, in the following sections, I examine various kinds of materials and themes developed by several travelers to Mexico during this period. The samples illustrate the diverse types of early to mid-nineteenth-century travel writing about and illustration of Mexico. I also review in greater detail the very influential writings of the historian William Prescott. These examples are not meant to cohere into a linear account; rather their disparateness demonstrates the diverse intentions and outcomes of travel writing. These writings, however, share commonality in their use of the rhetorical strategies of opposition and intensification, their methods of display and circulation, and their subsequent influence on contemporary and subsequent Mexican authors, including García Cubas.

WILLIAM BULLOCK: PANORAMIC MEXICO IN LONDON

William Bullock (ca. 1771–1849), an English entrepreneur, prefaced his travel narrative *Six months' residence and travels in Mexico; containing remarks on the present state New Spain, its natural products, state of society, manufacturers, trade, agriculture, and antiquities, &c.: with plates and maps* (1824) with these words:

> But the jealousy of the Government of Old Spain had so fully succeeded in shutting out Europeans from the knowledge of Mexico, that, since the period of Carlos I, I am acquainted with no book of travels by an Englishman in that country. There is, consequently, much of novelty to attract even the most indifferent visiter [*sic*]; and I have simply endeavored to relate what fell under my personal observation in the plainest manner. The rising interest attached to this portion of the world, and the growing importance of Mexico to the commercial enterprise of Britain, will, I trust, give that degree of value to my statements which they may want on the score of authorship. I have looked most at those objects which are most intimately connected with the new relations springing up, and daily strengthening, between the countries; and, relying solely on the patriotism of my intentions, I humbly submit my best endeavours to that public, through whose kindness and patronage I have been enabled to perform this voyage—thus adding another to the many efforts I have successfully made to obtain its countenance and favour. W. B.[41]

While the title of Bullock's work states the economic purpose of his writing, one phrase in particular signals the unmediated eyewitness approach embedded in European narratives about Mexico: "I have simply endeavored to relate what fell under my personal observation in the plainest manner."[42] His travel narrative lapsed into descriptive and collecting activities that emphasized the exotic and the curious.

In a six-month period, Bullock dashed through Mexico garnering information about mining resources in Mexico, as well as sketching sites and people. In his description, he published illustrations of Indians going to market. Furthermore, whenever possible, he zealously collected objects, often unscrupulously. In one passage, Bullock laments that after visiting the home of one family, "I expected to have found something worth bringing home [to England]: but all my research done this way only produced for me two small pictures."[43] He also gathered casts of major archaeological monuments such as the Calendar Stone, Teoyaomiqui, and the Stone of Tizoc.

Upon his return to England, Bullock gathered the materials into a large exhibition in London entitled *Ancient and Modern Mexico* and described in his catalogue, *A description of the unique exhibition called Ancient Mexico*. Bullock contextualizes his contribution by recognizing Humboldt's achievement and censuring Robertson's history writing. He begins the catalogue texts with a harangue that derides Spain for its destruction of indigenous culture and monuments. Bullock proclaims, "From the wreck thus made, the following Catalogue will show that much has been saved; and owing to the happy opportunity which new revolutions [of Mexican independence] afford, that a very interesting collection has been formed for the observation of the British public."[44]

Bullock, first and foremost a showman, used the latest techniques of display to present Mexico for popular consumption. His dioramas presented natural habitats, complete with stuffed animals and reproductions of tropical plants (figure 27). Visitors also found themselves in a replica of an Indian hut and peering out at a panorama of a Mexican landscape. Patented in 1787, by the 1820s panoramas were wildly popular in London, Paris, Berlin, and the eastern part of the United States, with thousands of people experiencing this format.[45] Surrounded by the 360 degrees of the panoramic vision, the viewer almost becomes a participant in the scene. Bullock's panorama may have evoked the landscape of the Veracruz to Mexico itinerary, first seen by Cortés, charted by Humboldt, and traveled by Bullock.

Further, casts of major Mexican antiquities were placed in the great Egyptian Hall along with a scale model of a pyramid from an archaeological site from Teotihuacan, a fifth-century city north of Mexico City. He also included Aztec sculptures as well as maps showing the peregrinations of the Aztecs. In another section of the exhibition, the visitor viewed a 360-degree panorama of Mexico City, also with botanical and zoological exemplars and an Indian hut complete with a "real" Mexican Indian, José Cayetano Ponce de León.[46] Bullock's exhibition of monuments, panoramas, and dioramas catapulted Mexico from text and images, including graphs and charts, into a public display.[47]

Re-creating his travel journey in the spaces of museum display, Bullock created an illusion that was Mexico. *The Literary Gazette*, a London periodical, lauded this vicarious travel journey and claimed that the exhibition pro-

27 Agostino Aglio, *Exhibition of Modern Mexico at the Egyptian Hall, Piccadilly*, from William Bullock, *A description of the unique exhibition called Ancient Mexico* (1824). Courtesy of the John Carter Brown Library at Brown University.

vided an experience better than an actual visit to Mexico.[48] Nigel Leask, who analyzes Bullock's work quite thoroughly, criticizes the showman, claiming that he converted Humboldt's erudite analyses into a Mexico that was at once "a travesty and inevitable translation into the language of commodity capitalism."[49] Bullock, in fact, may better be understood as reverting to the mobile and fluid meanings of earlier centuries. He randomly aggregated and disaggregated images and objects, which began the formation of an inventory of images associated with Mexico.

CLAUDIO LINATI DE PRÉVOST: INTENSIFICATION OF SENSUALITY

Claudio Linati de Prévost (1790–1832), an Italian painter, visited Mexico between February and August 1826. Linati, who brought the technology of lithography to Mexico in 1826, produced forty editions of *El Iris*, a literary periodical that included lithographic images of fashions and antiquities.[50]

Upon his return to Europe, he published *Costumes civils, militaires et réligieux du Mexique; dessinés d'après nature* (Civilian, Military and Religious Costumes of Mexico: Drawn from Nature) containing forty-eight colored lithographic images, each showing individuals and details of their costumes.[51] The figures in this compendium include historical personages such as Moctezoma—the Aztec king at the time of Spanish contact—and Miguel Hidalgo—a leader of the fight for Mexican autonomy from Napoleonic Spain—as well as examples of military personnel of various ranks. The majority of the prints, however, depict types of people of early nineteenth-century Mexico and their various occupations and associated costumes. Conspicuously, Linati's figures float on white backgrounds, much like the specimens of popular botanical prints.

Linati's short essays associated with each depiction annotate the costume and comments on the positive, but more often, negative attributes of each character. For example, Linati's *Porteur d'eau* (a water carrier) illustrates a raggedly dressed man carrying a large container of water on his back. Linati remarks, "All countries offer some customs one does not understand either because of their inconveniences or because of their oddity. The water carrier of Mexico is one of those things that most strikes foreigners' eyes: one has difficulty conceiving how to carry 50 pounds of water."[52] His *Tortilleras* (*tortilla* makers) depicts two women who sit in what appears to be the interior of a lean-to shelter (figure 28).[53] One woman, kneeling on the floor, grinds corn into meal; she is depicted on her knees bending forward over the grinding stone with her shirt slipping down to reveal the shape of her breasts. A second women shapes dough into the circular cakes and cooks them on a large circular griddle that sits on top of a fire. Linati's commentary explains, "Wheat was not known to the ancient Mexicans. Regions situated under the Tropics do not favor its cultivation; lack of frosts, excessive heat, periodic rains, and other causes, make it grow too richly, and hinder the development and maturation of the grain."[54] In another example, Linati is particularly disparaging of the *Jeune ouvrière* (young female worker) who is costumed in a pink patterned dress with three layers of ruffles at the bottom with a blue-striped *reboza* (shawl) over her head and shoulders, which Linati notes is made in the city of Puebla (figure 29):

> Charming sex, the nicer half of humankind, under all the climates of the land, despite ignorance and barbarity, no matter under which colors and

28 Claudio Linati, *Tortilleras*, plate 5 from *Costumes civils, militaires et réligieux du Mexique; dessinés d'après nature* (1828). Brown University Library.

in which costume, the empire of your grace extends its beneficial influence, and makes men better by imposing a truce on the hate-filled passions that agitate them. In spite of her pale olive complexion, the young Mexican worker does not renounce the privilege of pleasing and knows how, with her natural vivacity, her quick and graceful movement, to make [one or men] forget the gentle Parisian *grisette*.[55]

29 Claudio Linati, *Jeune ouvrière*, plate 1 from *Costumes civils, militaires et réligieux du Mexique; dessinés d'après nature* (1828). Brown University Library.

He intensifies the image of the women of Mexico by bringing a sensual dimension into his descriptions. Emphasizing himself as a direct observer, the rhetorical strategies of Linati's images and commentaries on Mexican types not only linked people with kinds of work and costume but also identified textual and visual narratives that use categories of people to define the national character of Mexico.

KARL NEBEL: DÉLASSEMENT (DIVERSION)

Subsequently, Karl Nebel, a German designer and artist, traveled to Mexico between 1829 and 1834 and in addition to recording inhabitants, their customs and manners, he recorded ancient ruins, Mexican landscapes, and architecture. Nebel's 1836 publication, *Voyage pittoresque et archéologique*, a travel account, included an introduction by Alexander von Humboldt and fifty hand-colored lithographs.[56] In it, Linati's characters reappear in the Nebel illustrations. For example, in *Tortilleras*, Nebel depicts two women who sit in the interior of a dwelling (figure 30). Repeating the erotic element of Linati's rendering, one woman, her shirt falling forward to reveal the outlines of her right breast, kneels on the floor, grinding corn into meal. A second woman shapes dough into the circular tortilla shapes and cooks them on the circular griddle. Two men, one standing and drinking and the other seated, are located behind and to the side of the women. Forming a comparison, the German traveler comments:

> Here we see women, of which one is Indian and the other is Creole, occupied with making food; the Creole woman grinds maize on a stone; this results in a dough from which the other [woman] forms into a kind of crepe or omelette, which she throws on the stoneware griddle to cook it. This food which uses neither salt or butter, serves as bread for people throughout all the republic. One can see a stew made with a bad piece of sun-dried meat. Add to that some green peppers, that is called Chili, and a drink known as pulque, taken from the juice of the aloe, and you have the standard repast of the lower class people[57]

Nebel also included *poblanas* (women from the state of Puebla) who, with their colorful patterned skirts and rebozas, are versions of Linati's *Jeune ouvrière* (figure 31). In Nebel's image, we see two women standing in a doorway while a third woman, her back to the viewer, wears a scarf around her neck. All three smoke cigarettes and are in some kind of an exchange with a male dressed in a white shirt, pants, a folded serape (blanket), and spurs. Nebel augments this image with a commentary on poblana character.

> These are women of the class worker, from whom the ladies of first class often adopt this attire at home. The essential part of the costume is the *ruboso*, or light shawl, that they wear on the head and that they almost

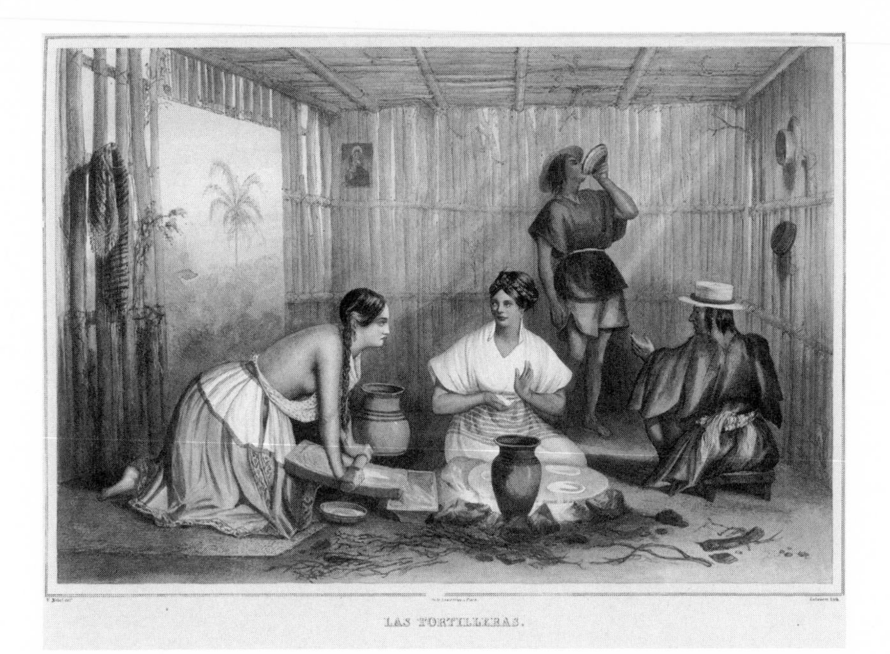

30 Carl Nebel, *Las Tortilleras*, from *Voyage pittoresque et archéologique, dans la partie la plus intéressante du Mexique: 50 planches lithographiées avec texte explicatif* (1836). Brown University Library.

> never take off, even to cook, without bothering them at all. They do not wear any corset, and even the great ladies do not use it during the hours of receiving . . . all wear accessories of gold or silver on the clothing. . . . . It is in the national character to spend the money as it comes as if to think only to one's pleasures. Happy the country where the climate and the ease of access allows such lack of concern![58]

Here Nebel's description of poblana women leads to an explicit reference to national character.

Along with images of national characters, Nebel's plates also include images of archaeological sites, such as *La piràmide de Papantla*, the pyramid at Tajin, Veracruz (figure 32); antiquities such as *Compte du Soleil* (Sun Stone); and various views such as *Vista de Jalapa*, *Bajía de Acapulco* (port of Acapulco), and the *Vista de la Catedral y de la Plaza Mayor de México*.

In the preface to the *Voyage*, Nebel confirms unequivocally that his work

31 Carl Nebel, *Poblanas*, from *Voyage pittoresque et archéologique due mexique dans la partie la plus intéressante du Mexique: 50 planches lithographiées avec texte explicatif* (1836). Brown University Library.

is addressed to the curious and serves the public for *délassement*, diversion, and not scientific study nor does it have the pretension of teaching.[59] Ultimately, he succeeded in developing a popular study that familiarized the unfamiliar people and places of Mexico for the interested and the curious.[60] His images reappear in subsequent texts published in Mexico.

JEAN-FRÉDÉRIC WALDECK: EROTICA

Jean-Frédéric Waldeck (1766–1875), an engineer and artist, arrived in Mexico in 1825 to work in an English mining enterprise for three years and subsequently as a portraitist and engraver in Mexico City for six years. During the final years of his residence in Mexico, he indulged his passion for archaeological ruins by traveling through the Yucatán and recording the ruins of such Mayan sites as Palenque and Uxmal. In 1838 he published *Voyage*

32 Carl Nebel, *La piràmide de Papantla*, from *Voyage pittoresque et archéologique, dans la partie la plus intéressante du Mexique: 50 planches lithographiées avec texte explicatif* (1836). Brown University Library.

pittoresque et archéologique dans la province d'Yucatan pendant les années [during the years] 1834–1836.[61] It includes images of Palenque copied from the expedition report of del Rio and, like Humboldt, Waldeck attempted to make Mexican antiquities familiar by connecting them stylistically to Old World (Egyptian) cultures. Repeating the imagery of Linati and Nebel, Waldeck also depicts the tortilleras lasciviously as they prepare and shape the dough. His sensual reference turns highly erotic in Waldeck's display of a nude female Indian, bathing among the ruins at the archaeological site of Palenque. His work was highly criticized by Humboldt; William Prescott called him a charlatan. Nevertheless, his images of ruins were republished extensively in various texts.

PEDRO GUALDI: CITY SPACES

Prints by Pedro Gualdi (1808–57) were also widely copied. The Italian artist arrived in Mexico City in late 1835 or the beginning of 1836 with an opera company as a painter of stage scenery. While he worked at the Teatro Principal, he proposed a publication that would illustrate the spaces of Mexico City based on a series of his paintings. The resulting *Monumentos de Méjico* was published by J. M. Fernández de Lara in 1841.[62] Revealing his set designer background, his images focused on major monuments and spaces of Mexico City, such as the exterior of Mexico City Cathedral and Colegio de Minería; the Alameda, a city park, and a boulevard; as well as the interior of buildings, such as the Cathedral and Cámara de Diputados (House of Representatives). The meticulous detail of his images as seen in figure 33, *Interior de la Universidad de Méjico*, gives evidence of his considerable skill with perspective drawing. Unlike other travelers, Gualdi remained in Mexico for about twenty years, establishing himself in Mexico City and marrying a Mexican woman. His images, initially directed to a local audience, emphasize the urban intensity of Mexico's capital city as a backdrop to daily activity.

JOHN LLOYD STEPHENS AND FREDERICK CATHERWOOD: APPROPRIATION

John Lloyd Stephens (1805–52) and Frederick Catherwood (1799–1854) traveled through Mexico and Central America between 1839 and 1843. Stephens, a travel writer from New York City, had previously published two travel accounts of his journeys: *Incidents of Travel in Egypt, Arabia Petraea, and the Holy Land* (1838), *Incidents of Travel in Greece, Turkey, Russia and Poland* (1838). Catherwood, an English artist with impressive credentials from the Royal Academy in London and significant experience in architectural renderings, specialized in travel illustration of archaeological ruins and had previously traveled to many of the same countries as Stephens.[63] He applied his skills in 1835, when he drew the images for a 10,000-square-foot panorama of Jerusalem at the Leicester Square Panorama, and established the Catherwood Panorama in New York City.[64] The travel writer and travel illustrator formed a partnership to further their entrepreneurial interests that resulted in *Incidents of Travel in Central America, Chiapas and Yucatan* (1841) and *Incidents of Travel in Yucatan* (1843). In 1844, Frederick Catherwood also published more images in his *Views of ancient monuments in Central America Chiapas, and Yucatan.*

Catherwood's steel-engraved images of Mesoamerican monuments and

33 Pedro Gualdi, *Interior de la Universidad de Méjico*, from *Monumentos de Méjico/tomados del natural y litografiados por Pedro Gualdi* (1841). Courtesy of the Yale University Library.

ruins filled the pages of these travel books. His high standards for exact rendering were assisted in the field by use of a camera lucida, a box or chamber that projected an image onto a surface allowing it to be traced onto a piece of paper. These images were not the general and imprecise sketches of archaeological remains like those of Humboldt or Nebel but highly accurate reproductions showing exact details of the archaeological remains.[65] For example, from *Views of ancient monuments*, Catherwood's image of Uxmal's House of the Nuns, *Portion of the La Casa de las Monjas*, superbly illustrates the complex details of the stone mosaics used in this Puuc architecture (figure 34). From *Incidents of Travel in Central America*, Catherwood's images of Palenque show precise details of the architecture such as the corbel arches (figure 35).

In his excellent analysis of Stephens's and Catherwood's work, the historian R. Tripp Evans points out that the narrative of Stephens and Catherwood broke with previous analyses of Mexican antiquities by Humboldt, Waldeck, and others by proposing indigenous authorship and more recent

34 William Catherwood, *Portion of the La Casa de las Monjas*, from *Views of ancient monuments in Central America, Chiapas and Yucatan* (1844). Brown University Library.

35 William Catherwood, *Palenque*, from John Lloyd Stephens, *Incidents of Travel in Central America, Chiapas and Yucatan*, opposite page 213 (1841). Brown University Library.

archaeological dating rather than associations with the Old World or with Egypt.[66] Their images and text, nevertheless, emphasize the unruly condition and unappreciated context of the ruins. The ruins of Uxmal and Palenque loom out of a background of tropical forest wildness that seems to swallow up the archaeological remains. Further, the indigenous people are posed within the ruins in ways that depict them as disinterested in these vestiges of past civilizations. Thus, Stephens's and Catherwood's visual and textual narrations of the ancient ruins tell two stories: they simultaneously intensify the cultural achievements represented by the magnificent remains and rationalize the exploitation and appropriation of the archaeological materials due to neglect by Mexicans. Within this narrative, the ruins belonged not to Mexico but to the North American continent.[67] In this way, Stephens justified buying archaeological sites and stealing monuments to be shipped back to the United States in hopes of setting up a museum of American antiquities.

WILLIAM HICKLING PRESCOTT: STATIONARY TRAVEL

Perhaps the most influential nineteenth-century writing on Mexico was *The Conquest of Mexico, with a preliminary view of the ancient Mexican civilization, and the life of the conqueror, Hernando Cortés*, by William Hickling Prescott (1796–1859), a three-volume, 1,200-page work published in 1843. Using numerous primary and secondary sources, this highly respected scholar from the United States conjures a history of Mexico as a divinely determined story of the Spanish conquest. The original publication includes six lithographic images: Volume 1 includes a three-quarter-length portrait of Cortés and a map showing the area of the conquistador's journey from Veracruz to Tenochtitlán / Mexico City; volume 2 includes a three-quarter-length portrait of Moctezoma and a map of the Valley of Mexico (figure 36); and volume 3 includes a one-half-length portrait of Cortés and a facsimile of his signature. In the front matter of volume 1, Prescott notes that the maps have been adopted from Humboldt's, which "are the only maps of New Spain which can lay claim to the credit even of tolerable accuracy."[68]

Prescott's personal notes and letters reveal that he began to formulate the ideas for the work in 1838.[69] In preparation for writing, he read various works, including those by Robertson, Gómara, Dupaix, Waldeck, Bullock, Giovanni Francesco Gemelli Careri, León y Gama, and Bernal Díaz, as well as geographical works. He also read travel writing, commenting that these gave "life & truth to the picture."[70] Admiring Humboldt's essays and atlases, Prescott comments that the naturalist combined, "to a curious degree, the characteristics of science treatise—& fashionable itinerary—& as such is now at the distance of so many years resorted to for information equally by the scholar and the tourist."[71] Also impressed with *Incidents of Travel*, Prescott also wrote directly to Stephens,

> I cannot well express to you the great satisfaction and delight I have received from your volumes. . . . Your researches in Palenque have made some important additions to the collections of Dupaix. . . . Too much praise cannot be given to Mr. Catherwood's drawings in this connection. They carry with them a perfect assurance of his fidelity, in this how different from his predecessors, who have never failed by some over-finish or by their touches of effect to throw an air of improbability, or best uncertainty over the whole.[72]

36 William Prescott, *Map of the Valley of Mexico*, from *The Conquest of Mexico, with a preliminary view of the ancient Mexican civilization, and the life of the conqueror, Hernando Cortés* (1843). Courtesy of the John Carter Brown Library at Brown University.

Overall, Prescott's notes and letters demonstrate that he believed that he was working with the best available primary and secondary documents on Mexico and his belief in the visual veracity of Catherwood's images.

Prescott had outlined his history by July of 1839, deciding that the narrative was to be a beautiful epic with Hernando Cortés, the Spanish conqueror, cast as its hero. "It has all the interest which daring, chivalrous enterprise, stupendous achievements worthy of an age of knight-errantry, a magical country, the splendors of a rich barbaric court and extraordinary personal qualities in the hero can give."[73] In addition, he decided to give "life to the picture" through detailed descriptions of "the features of those parts of the country which are the scenes of operation, as existing at that time, and in the present. . . . transporting the reader to the country, and to the age."[74] He concludes,

> In short, the true way of conceiving the subject, is not as a philosophical theme, but as an *epic in prose*, a romance of chivalry; . . . while it combines all the picturesque features of the romantic school . . . , the restless march of *destiny* is more discernible than in the sad fortunes of the dynasty of Montezuma. It is, without doubt, the most poetic subject ever offered to the pen of the historian.[75]

Prescott's interest in visual accuracy is certified when he disparages the accuracy of Bernal Díaz's history but admires his writing style because "Bernal Diaz, the untutored child of nature is a most true and literal copyist of nature. He transfers the scenes of real life by a sort of daguerreotype process, if I may say so, to his pages."[76] Through his work, Prescott envisioned Mexico as a larger-than-life panorama for display of his fabricated historical drama.

Prescott's poor health and declining eyesight, however, prohibited him from undertaking extensive travel for the research on this project. Nevertheless, he was able to garner extensive primary and secondary materials through an extended network of friends and acquaintances who gathered or transcribed documents from foreign archives as well as located images and other materials he requested. For example, through his friendship with Ángel Calderón de la Barca, the Spanish ambassador to Mexico, and his wife, Fanny Calderón de la Barca, he was able to obtain numerous documents, images, and descriptions from Mexico. In one request, he solicited two Aztec skulls (a male and a female), the signature and a portrait of Cortés, and detailed geographic descriptions of the journey from Veracruz to Mexico City, that is, the itinerary of Cortés's march to the Aztec capital. In a letter dated 14 October 1840, Madame Calderón updates Prescott:

> Nevertheless, he [Ambassador Calderón] is constantly remembering your affairs, and speaking and writing to [Lucas] Alamán [a Mexican historian and diplomat] on the subject; indeed, I have at this moment two skulls beside me, not of Aztecs however but of Ottimiti Indians,

> who little thought that you were destined to disturb their repose. These Calderón desires me to say he sends to-morrow by the diligence. . . . It appears impossible to find an authentic Aztec skull—nevertheless we continue to try. Calderón sends you the autograph of Cortés. The portrait is unfortunately begun, it is to be full length, and will cost $120. . . . As for the appearance of the country the *tierra caliente*, you may boldly dip your pen in the most glowing colors—we passed through it, from Vera Cruz to Jalapa in *December*. The road lay through a succession of woody country—the trees were covered with every variety of flower and blossom, loaded with delicious fruit, and bending down under the weight of the beautiful creepers that twined around them. . . . But I am inclined to imagine that the landscape is *less* beautiful than in the time of Cortés. . . . The first sight of the valley of Mexico is magnificent, and I wish I could transport you to the point where it first bursts upon the view.[77]

Madame Calderón concludes her letter by thanking Prescott for the daguerreotype equipment and promising that he will receive "its first fruits."

In a December letter of response, Prescott thanks Madame Calderón for the update and excellent description of the tierra caliente. He states that he received the skulls and they were suitable, explaining "In the cause of science however, as you know I intended them for an anatomical collection here, in which the Aztec link is alone wanting." Prescott also acknowledges the receipt of the daguerreotype and concludes by mentioning, "I received a message from Humboldt, last week, impressing on me the necessity of a journey to the land of the Aztecs. But with such good friends as I have in Mexico, although it would add immeasurably to the pleasure, I have the less occasion for it, for Mexico comes to me."[78]

The Conquest of Mexico, the result of Prescott's vision and synthesis of collected resources, resonated with readers. For example, describing Cortés and his company's journey to Mexico City, he writes:

> They had not advanced far, when, turning an angle of the sierra, they suddenly came on a view which more than compensated the toils of the preceding day. It was of the Valley of Mexico or Tenochtitlan, as more commonly called by the natives; which with its picturesque assemblage of water, woodland, and cultivated plains, its shining cities and shadowy hills, spread out like some gay and gorgeous panorama before them. In

> the highly rarefied atmosphere of these upper regions, even remote objects have a brilliance of coloring and a distinctiveness of outline which seem to annihilate distance. Stretching far away at their feet, were seen noble forests of oak, sycamore, and cedar, and beyond yellow fields of maize and the towering maguey, intermingled with orchards and blooming gardens. . . . .
>
> Such was the beautiful vision which broke on the eyes of the Conquerors. And even now, when so sad a change has come over the scene; when the stately forests have been laid low, and the soil, unsheltered from the fierce radiance of tropical sun, is in many places abandoned to sterility; . . . . .even now that desolation brood over the landscape, so indestructible are the lines of beauty which Nature has traced on its features, that no traveler, however cold, can gaze on them with any other emotion than those of astonishment and rapture.[79]

Here, Prescott writes this description as a stationary traveler through Mexico, constantly deploying the rhetorical strategies of opposition and intensification. Reading this completed text, it is clear that Prescott weaves together the scene described by Bernal Díaz and the description of Madame Calderón. In footnotes, he also cites information from Humboldt's *Essai politique*.[80] His reference to a "gay and gorgeous panorama" recalls Bullock's vision for his Piccadilly display.

The precision and cadence of the descriptions of Cortés's journey fulfilled Prescott's desire to give "*life and coloring to the picture*" and transport "*the reader to the country, and to the age*." Prescott's history is fabricated not only as a series of actions and facts; it is also an intensely visual, almost cinematographic, display.[81] The reader journeys through a diorama-like display in which historical characters act out their roles in the epic tale. Like Bullock and Humboldt, he materializes Mexico through visual imagery. Thus, Prescott's work is a narrated travel album where time and space collapse as the reader tours through a highly Eurocentric account of the conquest of Mexico. His account was positively received in Mexico as a comprehensive history. Like Humboldt's writings, *The Conquest of Mexico* was republished extensively in various languages. In the United States, it added momentum to a growing expansionist interest.[82] Further, two translations in Spanish were published in Mexico in 1844. A version published by Vicente García Torres was translated by José María González de la Vega and entitled *His-*

toria de la conquista de México, con bosquejo preliminar de la civilización de los antiguos mejicanos y la vida de su conquistador Hernán Cortés, escrita en inglés por William H. Prescott, autor de la Historia de Fernando e Isabel (History of the Conquest of Mexico, with a Preliminary Outline of the Civilization of the Ancient Mexicans [Indians] and the Life of Its conquistador Hernán Cortés, written in English by William H. Prescott, Author of the History of Fernando and Isabel). Appearing almost simultaneously was *Historia de la conquista de México, con una ojeada sobre la Antigua civilizacion de los Mexicanos, y con la vida de su conquistador, Fernando Cortes* (History of the Conquest of Mexico, with an Overview of the Ancient Civilization of the Mexican [Indians], and with the Life of Its Conquistador, Fernando Cortes) translated by Joaquín Navarro and published by Ignacio Cumplido. These extended titles indicate that the translations refocused the conquest theme of Prescott's original writings to include the history of ancient Mexicans.

CORYDON DONNAVAN AND JOHN DISTURNELL: MANIFEST DESTINY

In 1847, Corydon Donnavan (no dates), a journalist and United States soldier captured in Mexico, published his experiences in Mexico in *Adventures in Mexico; experienced during a captivity of seven months*, which went through twelve editions within less than two years. In 1848, United States audiences in the Midwest and in the eastern states were presented with an active version of the journalist's memoir through a 21,000-square-foot panorama mounted by Donnavan ostensibly for educational purposes. This presentation was unique in that instead of the viewer being surrounded by the images, this panorama consisted of the scenes painted on the extended canvas that was wound onto two spools. The viewer began his or her journey in Corpus Christi, Texas, moving south, across the Colorado River, to witness the Battle of Palo Alto, with "one of the most grand and imposing views that can possibly be transferred to canvass."[83] Proceeding across the Rio Grande and through the cities of Monterey and Saltillo, the viewer could see the Battle of Buena Vista, where General Taylor "met and defeated about 20,000, the flower of the Mexican Army, under Gen. Santa Anna."[84] The journey continued through the city of Zacatecas; a Mexican ranch; and San Luis Potosí, Querétaro, and Valladolid. In the second part, the journey followed, of course, the path of Cortés, from the City of Tampico, through Veracruz and Jalapa, stopping for views of Orizaba and the Cofre de Perote. It proceeded to the city of Puebla, to the archaeological site of Cholula, to the Castle of Chapultepec, and concluded with the arrival in Mexico City, which Donnavan describes as incomparable.

> Perhaps no city in the world offers a more striking and beautiful panoramic view, than Mexico. . . . the warm mellowness of a tropical atmosphere, throws over the scene of fairy-like delusion; with the calm, glassy lakes, encircled by the high ridge of mountains which surrounds the valley, groves of the orange and cypress, aqueducts, statuary, and forests of church spires, impart a general gorgeousness, unsurpassed in the wide world.[85]

Here, in moving-picture fashion, the landscapes, cityscapes, and war scenes, were then reeled past the immobile spectators. In the appendix to the *Adventures in Mexico*, Donnavan, aware of Prescott's *History of the Conquest*, attempts to rewrite the conquest story as an Anglo-Saxon conquest. In the text describing the Veracruz–Mexico City route, he proclaims,

> Almost the same route trod by the Spanish cavalier three centuries since, as a superior being, sent from a better sphere, has been retraced by the Anglo Saxon; and whether it ultimately proves for the interest of mankind or not, manifest destiny seems suddenly to have brought us in direct collision and more familiar communication with a people to whom we have been comparative strangers.[86]

Again, very much like Prescott's history, stationary travelers could see this manifest destiny literally and figuratively unroll before their eyes.

Additionally, it should be noted that this interest in the United States in the Mexican-American War motivated John Disturnell (1801–77), a businessman who published directories, guidebooks, and maps, to publish the *Mapa de los Estados Unidos de Méjico* in 1847 (figure 37). Like Donnavan's moving panorama, Disturnell's map was an approximation of Mexican space. The map, based on an 1822 map produced by Henry S. Tanner, reproduced the full extent of Mexican territory, including some erroneous information.[87] In the lower-left corner, tables indicate the distances between Mexican towns and cities; an inset map depicts the now legendary territory between Veracruz and Mexico City; and two images repeat Humboldt's

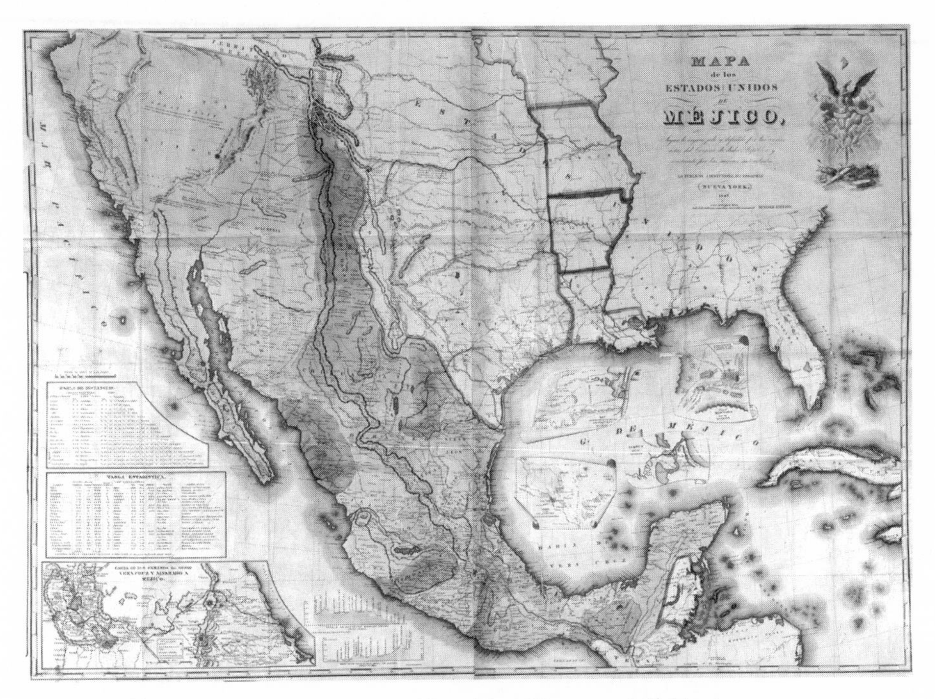

37 John Disturnell, *Mapa de los Estados Unidos de Méjico* (1847). Courtesy of the Newberry Library, Chicago.

image of the elevation of the road from Veracruz to Mexico City and Acapulco to Mexico City. Unlike Carey's 1822 atlas map, Disturnell's map visually emphasized the physical contiguity of Mexico and the United States. This map, considered to be the most up-to-date cartography at the time, was used to negotiate the Treaty of Peace, Friendship, Limits and Settlement, or the Treaty of Guadalupe Hidalgo, at the conclusion of the Mexican-American War. For García Cubas, the Disturnell map came to mark a low point in Mexican cartography because it marked the inability of Mexico to map itself at this critical juncture of Mexican history.

KARL SARTORIUS: VOYEURISM

Carl (or Carlos or Karl) Sartorius (1796–1872) was a successful German mining entrepreneur who settled in Mexico. His *Mexico: Landscapes and popular sketches* was published in 1858, the same year as García Cubas's *Atlas geo-*

gráfico. In fact, García Cubas later identifies both Humboldt and Sartorius as "profound observers" of Mexico.[88] In its preface, Sartorius recommends reading Alexander von Humboldt's work to become acquainted with the country because, he explains,

> In the following pages, the indulgent reader must expect neither a book of travels, conscientiously detailing every event from day to day, with the customary adjunct of the bill of fare, nor geographical-ethnological-statistical treaties, nor even a systematical enumeration of the natural history of Mexico: but views of the country, sometimes a mere outline taken at a distance, sometimes a more complete picture, drawn in the immediate vicinity, adorned with foliage and creeping lianas—sketches taken from the life—in the palace, or in the cottage, on the far-extending savannah, or in the mine. . . . With great interest I have devoted myself to the study of the history and monuments, the habits and mode of life of the Indians, and am enabled to present much wanting, neither in freshness, nor in close and careful observation.[89]

Sartorius's publication presents different landscapes: evergreen forests, savannahs and coastal regions, as well as geographic sights such as volcanoes and waterfalls "in order that the reader might picture to himself in the surrounding landscape."[90]

He also pictures and describes the people of Mexico. Visiting a Creole house he writes, "The ladies [female readers] would perhaps like to take a peep behind the curtains; but I scarcely know if I may indulge thus far without being deemed indiscreet."[91] Overcoming his squeamishness, Sartorius continues, "A half-opened door affords us a sight of the interior of a sleeping apartment; . . . [where he sees] A somewhat stout señora sits on the bed on a fringed tiger-skin in the Turkish fashion. . . . Her morning-gown too is not plaited, but hangs about her much like a sack."[92] Sartorius's voyeurism sets up a comparison and opposition.

> [Mexican] Domestic life is very different from that of the Germanic races. The life led by the ladies in their boudoirs savours somewhat of the Oriental; they work beautifully with the needle, weave and embroider, play and sing; the intellectual element, however, is wanting, the understanding and the heart are uncultivated, and sensuality, therefore easily obtains the upper hand.[93]

Like Bullock, the narrative includes both public and private spaces. He also employs Raynal's and de Pauw's ideas of degeneration, which he weaves into a summative overview of Mexico. He explains:

> In the course of three centuries, this [Spanish custom] has been developed in a manner peculiar to the climate and soil, and the character of the Creole is no longer that of the Spaniard. A fertile country, producing abundance almost unasked, a clear sky, a mild climate, where hardships of winter are unknown, have spoiled the Creole and render him more indolent and thoughtless than his transmarine relations; but he has retained the liveliness, the excitability, and the romantic sentiments of the latter. . . . the Mexican earns with facility, but just as easily lets it slip through his fingers; he seeks to enjoy the fleeting moment and leaves Providence to care for the future.[94]

Like Nebel and Linati, Sartorius emphasizes an intensity of difference highlighting both profusion and waste of natural resources that mark an assumed inherent character and traits of nation.

CONCLUSION

Surveying these early nineteenth-century travel accounts, a question arises: Where is the real New Spain / Mexico? Is it in Humboldt's graphs and charts; or Bullock's diorama and panorama of Mexican landscape; or Linati's and Nebel's Mexican types, tortilleras, water carriers, and poblanas; or Stephen's and Catherwood's descriptions and illustrations of ancient ruins; or Prescott's historical epic; or Donnavan's moving panorama; or Sartorius's domestic views and landscapes? The fact of the matter is that early nineteenth-century travel writings continued the tradition of previous centuries of aggregating the multiple and mobile mappings of New Spain /Mexico. Unlike in previous centuries, however, the emphasis on eyewitness methodology structured these writings through rhetorical strategies and resulted in visual intensification. Maps, charts, and illustrations of kinds of people, ancient ruins, the Veracruz to Mexico City landscape, and archaeological objects formed an archive that both circulated widely to a consuming audience due to expanding reproduction processes and display formats—highly circulated books, as well as exhibitions, dioramas, and panoramas.

Where is the real New Spain / Mexico to be found? In fact, there was no real Mexico, only a fabricated, in-between place situated in this circulating inventory of images consumed by historically located observers. This is to say, in the nineteenth century, seeing Mexico was not a neutral phenomenon. Rather, Mexico was located historically, ideologically, and culturally through travel texts and associated images. By the mid-century, the distinctive Mexicos of Humboldt, Linati, Nebel, Sartorius, Waldeck, Stephens, Catherwood, Prescott, Donnavan, and Disturnell were manifestations of a fragmented and still emerging international nineteenth-century culture that emphasized visuality and claimed unequivocal veracity.

Consequently, these travelers, and others I have not reviewed, visually mapped certain boundaries of the emerging Mexican nation—cartographically and epistemologically—for an observing non-Mexican audience. Mexico's exact physical boundaries became visible through updated cartography, as did the shape of its demography, the extent of its resources, the content of its national character, and the epic of its history, as well as its unlimited potential for appropriation. By mid-century, this visual inventory was not constituted through desultory objects of wonder or curiosity or allegorical depictions of early centuries. Instead, visual intensification became a strategy to bring together diverse eyewitness data into a visual intradiscursiveness that fabricated a liminal Mexico. Mexico was not so much a place as a journey that required no travel.

4

IMAGINING THE NATION AND FORGING THE STATE

Mexican Nationalist Imagery—1810–1860

In 1836, José Maria Luis Mora, a lawyer and liberal writer, wrote, "In the current state of things, it is still difficult to form an exact idea of the Mexican character, which is still developing and is impossible to define. . . . Mexican society, still embryonic, does not present anything but a confused mixture of habits, practices and customs derived from the metropolis [Spain], France and England."[1] His words confirm that in the early nineteenth century, Mexico struggled to represent its evolving national identity and content as it shifted from being the kingdom of New Spain, a part of the Spanish empire, to Mexico, a nation. With independence, Mexico had to be formed within the contradictory contexts of the fabricated understandings of the Americas in general and New Spain in particular that were inherited from previous centuries. These struggles of identity formation were compounded by the ongoing proliferation of written and visual travel narratives by foreigners and, of course, Mexico's continuing political travails.

Mexican independence required both new and reconfigured allegorical imagery to destabilize and replace those images related to Spain and Spanish rule. In the creation of Mexico, such allegorical images promoted *affective* relationships for Mexican citizens. At the same time, while such revised imagery gave form to the idea of the nation, it could not manifest the Mexican state, the entity that administers the nation. As a result, images depicting scenes and objects of geography and history, initially formed by foreign travelers, were recontextualized and circulated through print culture by state institutions. Such geohistorical imagery helped citizens remember what could not be remembered, giving legibility and functionality to the state and structuring *effective* relationships, that is, manifesting the ability of the state to realize new social and political conditions.

In this chapter, I elucidate the comprehensiveness and interconnectedness of visuality and visual production. An ever-growing inventory of images accrued value through the processes of accumulation, possession, circulation, and exchange. Further, recognizing the power of visual display, state institutions expanded their use of this inventory. Antonio García Cubas matured into adulthood during an epoch of national formation, and this increased production and circulation of images. This chapter establishes the complex, multifaceted contexts from which García Cubas came to conceptualize his early work, as Mexico was imagined as a nation and forged as a state.

LOCATING A NEW BODY

At the end of a three-day celebration in Mexico City in December 1796 that marked the birthday of Maria Louisa de Borbón, the queen of Spain, an equestrian statue of Carlos IV was unveiled in the city's central plaza.[2] In announcing this sculpture project, the viceroy asserted that the magnificent metropolis of Mexico City had enjoyed many benefits of royal patronage and now possessed what it had previously lacked—a great image of the king placed in the center of this capital, which illustrated his kingly virtues.[3] Subsequently, in December of 1803, the project was completed when a massive bronze equestrian statue of Carlos IV was installed, replacing the temporary wooden version that had been used for the 1796 celebration (figure 38). Again, the ostensible purpose of the permanent statue was to honor the king and bring his physical presence to New Spain and into the hearts and minds of New Spanish subjects. His overwhelming physical presence reminded royal subjects that they belonged to the corporate body of monarchy—the ancien régime—and that their New Spanish identity was derived from the king. Further, for the subject or viewer, the equestrian sculpture emphasized the king's dual power to bestow beneficence as well as inflict punishment.

Curiously, the monumental statue was initially dedicated about three

38 Manuel Tolsá, *Equestrian Statue of Carlos IV* (1803). Photograph by author.

years after the king of France, Louis XIV, also a Bourbon, was beheaded and the French monarchical system—figuratively and literally—was dismembered as a result of the French Revolution. Within this political context, the equestrian statue may be understood as a reification of the paramount trope of the ancien régime. This is to say, that the body of the Spanish king was *assembled* in New Spain as the sovereign body of the French king was *disassembled*. Embedded in a broader corporate and corporeal discourse of the ancien régime, the production of the equestrian statue of Carlos IV may be seen as an attempt to propose absolutist stability and continuity in New Spain at a time when the conceptual tenet of the king's body as a sign of sovereignty was highly unstable in Europe and its continuity in jeopardy.[4]

In the first decades of the nineteenth century, New Spain saw dramatic political changes as the French Revolution's purging of the ancien régime trope spread to Spain in 1807, when Napoleon Bonaparte invaded Spain and forced the abdication of King Carlos IV; the renunciation and captivity of his successor, Ferdinand VII; and the appointment of his brother Joseph Napoleon to the Spanish throne. New Spain protested the Spanish king's forced confinement and moved initially to be autonomous from Spain and, eventually, independent. During this time, New Spain had to confront radical questions about the definition and imagery of sovereignty: If sovereignty was no longer located in the body of the king, where was it to be located and how could it be visualized through a new body?

Alegoría de las autoridades españoles e indígenas by Patricio Suárez de Peredo, an 1809 oil painting, reflects these early nineteenth-century historical events and exemplifies how corporeal imagery was reformed and incorporated into the visual language of Mexican nation building (figure 39). In this panel, three oval-shaped medallions and two groups of figures surround the central image of Ferdinand VII, a captive of the French. An image of the Virgin of Guadalupe floats in a white cloud above the king; the medallion to the left of the king bears the Spanish crest of Castille y León, while the one to the king's right illustrates the insignia of Mexico City. The abraded legend in the lower section of the panel begins with the phrase "Long Live the King" and identifies the standing figures.[5] While the painting emphasizes loyalty to the captive king, its imagery does not depict Ferdinand VII as a singular, powerful figure like the equestrian statue of Carlos IV. Instead, he is protected by a bulwark of emblems that, with the exception of the Castille y León standard, emphasize New Spanish identity. Further, sets of bodies are also put into this constellation with that of the king; these include Spanish and Indian subjects. Particularly striking, the Virgin of Guadalupe, larger in scale than that of the king and hovering over the king, becomes an image associated with issues of sovereignty facing New Spain at this time.

"Long live religion! Long live our most holy Mother of Guadalupe! Long live Ferdinand VII! Long live America and death to bad government!" These insurgent words, pronounced on 15 September 1810 by Miguel Hidalgo y Costilla, a parish priest of the town Dolores, elucidates the context of the Suárez painting. Spoken seven years after the inauguration of the bronze equestrian statue, the words called for a revolution from the Napoleonic tyranny that had beset Spain. Hidalgo reinforced his "Grito [shout?] de Dolores," as it is known, by adopting a standard emblazoned with the image of the Virgin of Guadalupe.[6] Thus, although the Grito demanded Ferdi-

39 Patricio Suárez de Peredo, *Alegoría de las autoridades españoles e indígenas* (1809). Oil. 170 × 90 cm approx. Museo Nacional del Virreinato, Tepotzotlán. Reproduction authorized by the Instituto Nacional de Antropología e Historia.

nand's freedom from captivity, Hidalgo did not call on the imagery of the absolutist body of the king to visualize his declaration; rather, he called on Guadalupe—an image unique to New Spain. Hidalgo's use of the image of the Virgin of Guadalupe signals that the relationship of New Spanish subjects to Spain was being radically restructured. At the same time, it marks the continued importance of the use of the corporeal imagery of females to mark a shifting of the signifying site of sovereignty from the body of the king to a new body.

Napoleon's actions in Spain marked the transition in Spain's political culture away from the ancien régime, where legitimacy of the state was based in corporate institutions and their pact with the king to republicanism and secular society.[7] In Mexico, this transition was certified in September 1813, when José María Morelos—a priest, student of Hidalgo, and prominent military leader of the Mexican independence movement—convoked a Constitutional Congress at Chilpancingo, Guerrero, and declared Mexico's independence from Spain. His declaration emphatically stated that sovereignty springs from the people and established that the "twelfth of December be celebrated in all villages in honor of the patroness of our liberty, the Most Holy Mary of Guadalupe."[8] Thus, as New Spain became the "Empire of Mexico," the imagery of the Virgin of Guadalupe in early nationalist verbal and visual discourse marked this passage between the old regime and the emerging Mexican nation. While introducing new imagery for the incipient stages of Mexican independence, however, the Virgin of Guadalupe imagery could not become the central signifier of the Mexican nation in subsequent years. In the following decades, a new set of females reflecting the changing political landscape appeared across nationalist painting.

At the beginning of the third decade of the nineteenth century, Mexico gained its independence from Spain through the Treaty of Córdoba (1821) and constituted itself as an independent state through the Plan of Iguala, known as the Plan Trigarante (1821). The Plan of Iguala attempted to bring together diverse groups—liberal and conservative, rebels and royalists, criollos and Spaniards—into an uneasy union and declared all inhabitants of Mexico to be citizens.[9] Establishing the founding principles of Religion, Independence, and Union, the Plan called for a constitutional monarchy; and Agustín Iturbide, a hero of the war of independence, was proclaimed Emperor Agustín I of Mexico in May of 1822. Iturbide attempted "to preserve the best features of the social administrative structure of the Bourbon monarchy, believing that national unity required centralism."[10] These ideas were paralleled by the relocation of the equestrian statue of Carlos IV, also in 1822, to the patio of the Royal and Pontifical University, indicating its demotion as an emblem of monarchy and its promotion as an object or site of history, as specified in Gualdi's depiction of the statue in the interior of the university (figure 33, chapter 3, p. 95). Iturbide was forced to abdicate in March of 1823, however, because he had tried to impose a centralized system on an empire that was in fact an aggregate of provinces that sought significant autonomy and separation from centralized government. In the end, the Congress claimed that Iturbide did not possess the right to exercise sovereignty because such power belonged to the Congress as representative of the regional states.[11]

By 1823, the Mexican Congress had adopted a constitution founding the United Mexican States under Guadalupe Victoria, Mexico's first elected president.[12] Known as the First Republic, this fragile government suffered from continuing political upheavals. These struggles in the early 1830s brought into office Antonio Lopéz de Santa Anna, who subsequently abrogated the federal constitution in an attempt to centralize government. Santa Anna's government faced various insurgencies, including the cessation of part of Mexico's northern territory to become the Republic of Texas in 1836.[13]

Various scholars have followed a historical narrative of the early years of independence promoted by historians of the late nineteenth century, which justified Porfirio Díaz's government agenda of "order and progress" by pronouncing the first decades of Mexican independence as chaotic and anarchic. The historian Timothy Anna questions this formulation and argues that the years following independence may be better understood as a fragmented and discontinuous *process* of transition, in which regional entities that made up Mexico had to form themselves into, if not a cohesive entity, at least a confederated one. The political disarray that marked the period was a means to social, economic, and political progress. Will Fowler, also a historian, expands on this idea, stating that "it was a period of change, uncertainty and experimentation, which of necessity meant that no faction remained static in its demands, and everybody's political stance evolved in response to the different states of hope, disenchantment and profound disillusion and despair."[14] This approach to early postindependence history does not advocate discounting the turmoil of the period. It focuses, instead, on a comprehensive understanding of incipient and emerging processes that were galvanized as a result of the United States' invasion of Mexico in 1846.

In this political landscape, two distinct points of view framed arguments about the best kind of government for Mexico: centralist and traditionalist ideals represented by men such as Lucas Alamán, and more liberal values promoted by individuals such as Carlos María de Bustamante.[15] Both men understood the importance of history in the construction of nation. Alamán, a correspondent with William Prescott, saw the viceregal period as the true origin of the nation of Mexico; he and his followers envisioned a Hispanic nation with a Catholic foundation. In contrast, Bustamante and others argued that Mexico's beginnings were to be located in its pre-Columbian past and the colonial era was a period of degradation; here, the nation was a resurrection of the strong and independent Anahuac, the ancient civilization of the Aztecs.[16] Thus, between 1821 and 1847, Mexico formed national governments, testing out traditionalist and liberal ideological stances and their potential to promote effective government and form the Mexican state while precariously balancing national with regional interests and demands. This is to say that political independence announced the possibility of the Mexican nation. It did not assure a Mexican state, that mechanism that gives functionality to the imagined nation. Mexico had to be not only imagined but also forged from preexisting parts, governmental experiments, tests, successes, and failures.[17]

As Mexico moved from the immediate struggle for independence to the creation of a nation later in the century, revised female imagery reflected this process and the changing relationships of the citizen to a formative nation. Distinct from metaphorical female figures seen in atlas frontispieces such as Vaugondy's (figure 6, chapter 1, p. 30), Villaseñor's (figure 14, chapter 2, p. 53), or from the Virgin of Guadalupe imagery, a young, frail-looking, seated female seen in the painting *Alegoría de la Independencia* (1834) is identified as Mexico through her crown of red, white, and green feathers marking Mexico's national colors (figure 40). No longer seminude, she is dressed in a long gown, carries a quiver—a reference to indigenous culture—and holds a Phrygian cap on a stick, which signifies liberty through its reference to the hat worn by French insurgents and symbol of the French

40 Anon. *Alegoría de la Independencia* (1834). Oil. 169 × 196 cm. Museo Casa de Hidalgo, Centro INAH-Guanajuato. Reproduction authorized by INAH.

Revolution. This allegorical Mexico is also placed between two male historical figures: Agustín Iturbide, who clutches broken chains representing freedom from Spain, and Miguel Hidalgo, who crowns her with a laurel wreath representing victory.

Such metaphorical images of Mexico as an elaborately bedecked female were repeated in other paintings and prints throughout the early part of the century as independence from Spain became a reality.[18] These figures are also often associated with a cornucopia overflowing with fruits and vegetables, denoting the natural fecundity of the land and a bow and arrow with tricolored feathered costume elements, such as skirts and headpieces, referencing indigenous cultures. An eagle, a flag, drums, and a *macuahuitl*, an indigenous weapon, which refer to different aspects of an emerging story of the independence of Mexico, sometimes surround the figure as well. No longer the body of the Virgin of Guadalupe, instead, these allegorical figures certify that a new phase of corporeal imagery of Mexico had initiated.[19]

Particularly distinctive about these new female bodies is their erotic dimension. The seductive attraction of this metaphorical image of Mexico, the need for protection of *Alegoría*, or Mexico, are essential qualities needed to lure citizens to love, protect, and even give up their lives for the nation. The Virgin of Guadalupe may mark the process of independence but she did not have the requisite seductive qualities to represent the nation. Thus, while the Virgin of Guadalupe marked a transitional boundary, as New Spain became Mexico, the allegorical female figures of Mexico marked the transformation of Mexico into a nation. These allegorical images with their patriotic reference formed a mapping practice for the idea of the nation. They could not, however, manifest the Mexican state, the entity that administered the nation.

"¿QUIENES SOMOS? ¿DE DONDE VENIMOS? ¿PARA DONDE CAMINAMOS?"

Various early nineteenth-century texts and discourses written by Mexican authors would chart how the nation could become a state. One of the most interesting of these is *Mañanas de la Alameda de México. Publícalas para facilitar a las señoritas el estudio de la historia de su pais* (Mornings at the Alameda. Published to Assist Young Women in the Study of the History of their Country), written by Carlos María de Bustamante and published in 1835 and 1836. In the fictional text, during an early morning stroll through the Alameda, a large public park in Mexico City, the narrator and Doña Margarita—an intelligent, educated, and knowledgeable woman—happen to observe Mr. Jorge and Milady, English travelers, making disapproving comments about the *Diosa Libertad* (Goddess of Liberty), a statue depicting liberty located in the Alameda. Overhearing these comments, Doña Margarita approaches the couple and initiates a series of daily morning conversations with them in the public setting of the Alameda to provide the two foreigners with a better background to understand the sculpture and Mexico. Using Doña Margarita's voice, Bustamante uses these occasions to expound in detail on the history of Mexico beginning with the origins of indigenous culture and concluding, in tomo 2, with the appearance of the Spaniards on the coast of Veracruz. The narration is made authoritative through the heavy use of footnoted sources from previous centuries, including the writings of Bernal Díaz; the letters of Cortés; the texts of Antonio de Herrera, Francisco Clavijero, and López de Gómara; Villaseñor's *Theatro Americano*; and the excavation descriptions of León y Gama.[20]

Using these sources, the conversations, enumerated as twenty-three chapters, elucidate the events of the precontact cultures of Mexico, which implicitly argue that a Mexican state can be formed through the construction of a history. Occasionally, there are detours in this historical itinerary; for example, in the third conversation, or chapter, Mr. Jorge asks the doña if she has seen the lithographs of the map showing the peregrinations of the Aztecs in London, perhaps referring to remnants of Bullock's exhibition. Having explained the function and validity of these kinds of maps earlier, Doña Margarita answers positively and then launches into a discourse on the importance of antiquities as Mexican patrimony, decrying their recent looting by European nations.[21] Overall, however, Doña Margarita, representing Bustamante, verbally exhibits Mexico through an aggregation of information from indigenous maps and texts as well as primary and secondary sources. For Bustamante, independence marked the reversal of the conquest and viceregal New Spain was a 300-year period of Mexican degradation.[22]

In his preface to *Mañanas de la Alameda*, Bustamante poses three critical questions for Mexicans to answer: ¿Quienes somos? ¿de donde venimos? ¿para donde caminamos? (Who are we? Where do we come from? To where are we going?).[23] His queries identify issues arising at the end of the first decade of Mexican independence, which were addressed from both the traditionalist and liberal points of view. In its emergence as the Mexican nation, gendered corporeality in the form of allegorical Mexico could manifest the yearned-for collective body of the nation. Bustamante understands this when he initiates his essays around an imagined conversation about the misinterpretation of the statue *Diosa Libertad*. While such allegorical imagery was successful in its affective results, it could be misread and, more importantly, could not visualize the content of the nation. Bustamante affirms the need to "consultarse las memorias que nos dejaron nuestros mayores" (consult the memories left by our elders). That is, knowledge of history will answer critical questions about the content of national identity and, thus, create the state through remembering and make the nation legible.

The formation of a functional state requires such ongoing projects of legibility: the state needs a means of making its presence not just recognizable but actively decipherable in diverse public contexts.[24] Government authority needed to be everywhere in citizens' lives in order to reiterate and implement, overtly and covertly, responses to the questions posed by Bustamante. The Mexican state tangibly materialized in the first half of the century through an array of new institutions whose main functions were to collect, order, display, and preserve the nation. Unlike gendered allegorical figures, imagery associated with state institutions helped citizens remember what cannot be remembered: "Who are we?"—that is, identity through the nation because before Mexican citizens there were only Spanish subjects; "Where do we come from?"—that is, national history because before 1810, there was only Spanish history; and, "To where are we going?"—that is, the itinerary to understanding national territory, because before traveling through Mexico, there were only the itineraries of New Spain.

Institutions that promoted such remembering through visuality included the Museo Nacional and the Sociedad Mexicana de Geografía y Estadística (SMGE). In addition, in the early 1840s, with the impetus of improved public education, the Academy of San Carlos, founded in 1781 as part of Bourbon reforms but closed during the wars of independence, reopened and received increased funding and an influx of European artists who brought new direction to the academy and changes to the curriculum of the Academy of San Carlos in the mid-1850s.[25] These institutions had two functions in common: to aggregate objects and information about Mexico and to contextualize them to create historical and cultural memories for Mexico where there had been none. In this way, the institutions identified for Mexican citizens key ways of belonging to the nation. In each case, visual recollection was a critical process for the efficacy of these institutions.[26] This emerging remembering became a context for fabricating history. Finally, in placing Doña Margarita's conversations in the public space of the Alameda, Bustamante's *Mañanas de la Alameda* also points out that the location of remembering would occur in public venues. This public culture was manifest in the displays by institutions such as the Museo Nacional and the Academy of San Carlos. But it was particularly evident in the tremendous expansion of print culture due to greater press freedom and commercial viability and supported by new technologies of reproduction—lithography and photography.

MUSEO NACIONAL

Mexico's national museum developed out of the intense interest in antiquities evident at the end of the eighteenth century. As we recall, with León y Gama's 1792 writings, the Teoyaomiqui and Sun Stone were seen respectively as confirmation of the pagan beliefs and the high level of Aztec mathematical achievement. Guillermo Dupaix's explorations of Xochicalco, Monte Albán, Mitla, and Palenque between 1805 and 1813, authorized by Carlos IV, further raised awareness of antiquities and their potential for shedding light on the ancient past.

In 1808, the viceroy created a Junta de Antigüedades, which was followed by Conservatorio de Antigüedades and Gabinete de Historia Natural. The Junta de Antigüedades was charged with collecting manuscripts and antiquities from all parts of the Republic. In 1825, the Museo Nacional de México was created to ensure continued collection, investigation, and exhibition of antiquities in order to provide knowledge of ancient cultures, their origins, and achievements.[27] Ignacio Cubas began to gather objects for the Museo's collections, which were located in the university; subsequently, two priests, Isidro Ignacio de Icaza and Isidro Rafael Gondra, were put in charge of the collections. A formal announcement of the museum, included in the *museo*'s inaugural publication, *Colección de las Antigüedades Mexicanas que existen en el Museo Nacional*, recognized the recent European interest in Mexican antiquities and the importance of forming a public museum to be visited by "toda clase de personas."[28] Antiquities laws were put in place in 1828 and in 1834, and under the direction of the newly formed Dirección General de Instrucción Pública, the Conservatorio de Antigüedades and the Gabinete of Historia Natural were joined into a new entity, the Museo Mexicano.[29]

Colección de las Antigüedades, published by Icaza and Gondra in 1827 with twelve lithographs produced by Waldeck, consists of print images of stone sculptures, hieroglyphic examples, and historical images with brief associated descriptive texts.[30] One of the earliest examples of lithography in Mexico, the *Colección* manifests the intention of using the museum's collection to educate people of all classes. Reading the associated text, it is clear that the objects were framed as remnants of past cultures. Through this state-sponsored museum, citizens could be brought together through the commonality of a potential history, which the *Colección* alludes to, but does not quite articulate, because unified historical narratives of Mexicans would be fabricated later in the century.[31]

INSTITUTO NACIONAL DE GEOGRAFÍA Y ESTADÍSTICA

In order to promote the process of remembering, the government also needed to have a clear conceptualization of the physical boundaries and contexts of this emerging past. A university specialization of *ingeniería geográfica* (geographical engineering) was proposed at this time, signifying a belief in the possibility of a scientifically described, homogeneous national territory.[32] To this end, President Guadalupe Victoria ordered preparation of geographic materials in the form of maps, and in 1825 Miguel Bueno produced the *Mapa de la República Mexicana*. Based on older maps, the map identified the early government's recognition of the importance of cartographic activity in marking the political form of the nation.[33]

This interest was institutionalized when Spanish America's first geographic society, the Instituto Nacional de Geografía y Estadística, was established in 1833; in 1839, it became the Comisión de Estadística Militar.[34] Along with cartography, the *instituto* was also charged with the improvement of gathering statistical information. This information, along with scientific articles, was published in the *Boletín del Instituto Nacional de Geografía y Estadística de la República Megicana* in 1839.[35] The excellent research of Héctor Mendoza Vargas, a geographer, provides important insights into Mexico's changing emphasis in gathering geographic data and map production. He notes that through the late 1840s, considerable amounts of these institutions' time and resources were focused on gathering geographic and statistical information on topography, water resources, climate, vegetation, minerals, fauna, population, agriculture, industry, commerce, public administration, and historical data.[36] Under supervision of the instituto, information was accumulated through recopying of information of somewhat unreliable quality from sources dating to viceregal times. The methods and outcomes of these efforts varied between 1833 and 1850.

Subsequently, the successor organization, the Comisión de Estadística Militar, made up of high-ranking military men, assumed the role of producing a state-of-the-art *carta general*, as well as regional maps of Mexico.

Paralleling the aggregating method found in Icaza's and Gondra's *Colección de las Antigüedades*, the *comisión* gathered regional data and produced a geographic and statistical summary of the Mexican Republic. An 1835 *Carta geográfica de México* was updated by Pedro García Conde, an engineer and director of the Colegio Militar between 1838 and 1841, and published in England. The scale of this map made it useless when Mexico faced the Mexican-American War ten years later.[37]

Distinct from the gender allegorical expression of the nation, the history of and examples from these national institutions demonstrate how Bustamante's queries—"¿Quienes somos? ¿de donde venimos? ¿para donde caminamos?"—would be addressed by the Mexican state through ongoing projects of legibility. The state showed its presence through initiating and supporting institutions that made Mexico actively readable through an emerging history and geography. In fact, across these early nineteenth-century institutions, we see the outlines of a system that coordinated Mexican history and geography in a gridlike infrastructure to form an emerging narration of the nation.[38] In this way, people, objects, and maps could be exhibited as narrative elements within these historical-geographical coordinates.

IMAGING TECHNOLOGIES AND THEIR THEMES

The activities of these institutions also point to the fact that Mexico was forming a visual culture that emphasized ordered seeing and display. This visual culture grew exponentially, exhibiting Mexico consistently and repeatedly through an ever-increasing inventory of images. The emergence of this visual culture occurred in conjunction with the appearance of new technologies for image reproduction—specifically, lithography and photographic technologies.[39] The introduction and spread of these technologies in Mexico are less recognized by scholars but important because they simultaneously supported and further stimulated the growth of the historical-geographical system of cultural exhibition in the first decades of Mexican independence.

Prior to the invention of lithography in 1798 by Johann Alois Senefelder, prints were produced through the time-consuming and expensive processes of engraving, woodcut, and so on. These methods could not produce significant numbers of prints, because the metal or wood-printing surface breaks down after repeated use. The process of lithography requires that the designs or images are drawn or painted with greasy ink or crayons on the smooth surface of a specially prepared limestone. Prints are made by pressing paper against the stone's inked surface.[40] There are no raised parts of the printing surface to break down, so numerous prints were made by reapplying ink onto the stone surface. Within a few years of its invention, chromolithography appeared, which created multicolor printed images by using a separate stone for each color.

Claudio Linati brought the technology of lithography to Mexico in 1826, building a lithographic workshop and using the technique in the production of his short-lived journal, *El Iris*.[41] He abandoned his equipment when he left Mexico in 1827. Subsequently, Waldeck used the press to produce *Colección de las Antigüedades Mexicanas* for the Museo Nacional. The press was transferred to the Academy of San Carlos by the beginning of 1830.[42] Lithography, however, was viewed initially by the academy as a low art form and its use was limited and not incorporated into the fine arts curriculum at this time; eventually, the press was transferred to the Colegio Militar, where it was used to print tactical military plans.[43]

Although the academy's artists did not take up lithography immediately, other artists and artisans with commercial interests did so because political independence opened new possibilities for a public commercial market; in viceregal times, the Holy Office of the Inquisition had controlled publications.[44] For example, José Severo Rocha and Carlos Fournier, two French citizens, set up a workshop in 1836 and began to publish books with lithographic illustrations.

Print culture—circulating periodicals, journals, and the like—was well established in the eighteenth century. Attempting to distinguish a metropolitan voice for New Spain, publications such as the *Gazeta de Literatura de México*, edited by José Antonio de Alzate y Ramírez, highlighted period events and provided links to European periodicals.[45] In the 1840s, publications expanded dramatically as numerous publishing houses were established during the next two decades of independence, including those of Ignacio Cumplido, Juan Antonio Dacaen, Ignacio Díaz Triujeque, Vicente García Torres, Mariano Galván Rivera, José Mariano Fernández de Lara, Manuel Murguía, and Alejandro Valdés.[46] These publishers employed highly skilled lithographers, the foremost of whom were Casimiro Castro, Luis Garcés, Joaquín Heredia, Hesiquio Iriarte, Luis García Pimental, Hipólito Salazar, and the artists in the workshop of Agustín Massé and Julio Michaud.[47] Together, these editors and artists envisioned their work in part as the development of national culture, which reinforced national sentiment and educated the public through texts and images. Instead of imitating European models, they favored writers with a Mexican consciousness. Also, nationalist imagery circulated by print culture increased among the literate middle class.[48] To provide some sense of scale of this growth of the press and its circulation, the data indicate that there were eleven periodicals or about one for every 18,181 inhabitants in Mexico City in 1839.[49]

Mexico City publishers initiated a tremendous production of pamphlets, books, reviews, and image albums produced for Mexican consumption, which used lithographic illustrations and extended the emerging historical-geographical system. These publications included serials such as the newspapers *El Siglo Diez y Nueve*, *El Monitor Republicano*, and *El Omnibus*.[50] Also popularized by these publishers were calendars, which contained astronomical information, dates of eclipses, epigrams by various Mexicans, short articles on various topics, horoscopes, geographic notes, biographic and government guides, and images of important contemporary personages. In 1831 Mariano Galván Rivera published *El cocinero mexicano* (The Mexican Chef), the first known printed Mexican cookbook, indicating foodways as another mode of expanding and explicating the content of the nation.[51]

Publications were also often thematic, and some were directed to specific audiences for specific purposes.[52] For example, Bustamante's mid-1830s *Mañanas de la Alameda de México* was dedicated to women's education in the history of their country. By the early 1840s, regular publications for and about women began to appear. They often focused on instructing and improving women, a theme initiated as part of Bourbon economic reforms of the late eighteenth century and reinforced by Mexico's independence movement.[53] *El Semanario de las Señoritas Mejicanas. Educación científica, moral y literaria del bello sexo* (1841–42) provides an example. Published by García Torres and edited by Isidro Gondra with illustrations by Hipólito Salazar, *El Semanario* emphasizes the intellectual growth of women, providing lessons in history, sciences, moral philosophy, math, and discussions of literary heroines, with much of the material translated from English and French. Women were also educated on practical topics such as domestic economy and hygiene, while trivial topics—fashion, embroidery patterns, and puzzles—were avoided.[54] As one *El Semanario* writer argues, "Now is the time to leave the present ignorant and limited conditions that do not allow us to be considered intelligent companions for men . . . ; the world and our own country is transforming around us and all protest for another organization in favor of social improvements. . . . I would be happy if only one young lady upon reading these lines was convinced of the necessity she had to cultivate her education and to perfect her intelligence in order to became a true woman of the house [*muger de casa*]."[55] Through the government's recognition of the importance of education for the advancement of the nation, women's education was further advanced in hopes of making them "good daughters, excellent mothers, and the best and most solid support for the goals of Society."[56]

In her excellent overview of the historiography of women and gender in Latin America, the historian Sueann Caulfield explains that discourses regarding sexuality and female agency constructed the boundaries of normal and pathological female identities. These restrictions, in turn, shaped state regulation of public space, women's work, and private morality. Caulfield concludes that "the family and gender were to be 'rationalized' in the interest of nation-building and development."[57] Consequently, publications such as *El Semanario de las Señoritas Mejicanas* may be seen as another emerging mapping practice through which Mexican women were charted within the intertwining narratives of educational reform and female domesticity.[58] Across these publications, domestic space was defined as national space.

CONSTRUCTING HISTORIES

By the 1840s, Mexico-based publishers also began to produce a genre of illustrated publications about Mexico that focused on the people and land for internal and external consumption.[59] While the format and many images derived from the texts of foreign travelers such as Humboldt, Linati, Nebel, and Gualdi, these publications aspired to locate the diverse cultures and places of Mexico and put them on display.[60] *El Mosaico mexicano, ó Coleccion de amenidades curiosas é instructivas* was one of the first magazines of this genre published regularly by Cumplido in 1836–37 and again in 1840–42 and distributed to diverse parts of Mexico. *El Museo mexicano: ó Misceláea pintoresca de amenidades curiosas é instructivas*, more of a literature review, was subsequently published by Cumplido in 1843–44 and emphasizes national culture.[61] Cumplido's use of the term *museo* (museum) in the title relates to his view that the examples in this collection would someday take their rightful place among other cultures and civilizations.[62] These publications, while clearly national in their intention, nevertheless, borrowed from earlier travel imagery—including *tipos sociales* (social types)—and continued to indulge the European taste for exotica.

Importantly, and paralleling the museums, editorial houses also promoted a consciousness of the history of Mexico. Thus, just as pre-Columbian objects were displayed in the Museo Nacional, primary texts from earlier centuries were published. Some of these included the 1844 *Historia antigua de México*, a history written in 1780 by Francisco Xavier Clavijero and facsimiles of *relaciones geográficas*. Also, Antonio León y Gama's 1792 *Descripción histórica y cronológica de las dos piedras* was republished by Alejandro Valdés in 1832, with notes by Carlos María de Bustamante. These reprinted and facsimile documents created an archive upon which to further develop historical narratives. For example, aware of Prescott's ongoing research for the *History of the Conquest*, Lucas Alamán published his own three-volume history, *Disertaciones sobre la historia de la República Megicana*.

The proliferation of circulating images through lithographic prints was augmented by images made through the mechanical imaging technology of the daguerreotype. Daguerreotype, a photographic technology invented in Europe, arrived in Mexico late in 1839 and its popularity spread quickly.[63] The daguerreotype is a direct-positive process, which creates an image on a sheet of copper plated with a thin coat of silver without the use of a negative.[64] It was used for producing images of landscapes and ruins as we have seen in Prescott's references as well as Stephens's and Catherwood's 1841 works. Early daguerreotype images, dated to 1839 and 1840, depict the architecture of Mexico City, such as the Cathedral and the School of Engineering and Mining, as well as the Sun Stone located on the side of the Cathedral and the equestrian statue of Carlos IV in the University courtyard.[65] Daguerreotype themes parallel those produced by Gualdi in his *Monumentos de Méjico*'s lithographs. Baron Emanuel von Fridrichsthal, of the Austrian delegation to Mexico, established a studio in Merida in 1841 and produced daguerreotype images of ruins found in the Yucatán peninsula.[66] Foreign-born itinerant daguerreotypists traveled through other parts of Mexico, taking and selling images.[67] The visual impressiveness of the daguerreotype is hailed by one 1843 writer, who states that the daguerreotype "came to surprise a man by its originality, and to provide him a medium, the simplest one and the most relevant, to make a portrait of nature with all its properties with an exactitude that is extraordinary."[68] The truthfulness of the images is not to be doubted by this author.

By 1844, Joaquín Díaz González, the first Mexican daguerreotypist, operated a portrait studio in Mexico City.[69] Daguerreotype equipment seems to have been publicly available in Mexico as a brief notice in the *Diario de Gobierno* dated Sunday 1 August 1847 announced a daguerreotype, recently arrived from France, for sale and stating it made portraits in all sizes.[70] By mid-century, seven daguerreotype workshops were operating in Mexico City.[71] And following the popular French format of *carte de visite* (calling cards), *tarjetas de visita*, with photographs of individuals, were very popular in Mexico, too.[72] Men, women, and children of all classes appear in portraits made in the new medium along with images of well-known personages.[73]

Through this lithographic and emerging photographic technology, it is clear that by the mid-1840s there was intensification in the production and circulation of images. The lithographic images found in newspapers, calendars, popular genre books, and reprinted and facsimile publications, as well as the production of daguerreotypes, complemented and extended the move to exhibit the sites, places, and people of cultural Mexico.[74] An account of the publication of the Spanish translation of Prescott's *History of the Conquest* in Mexico provides a compelling example of the growing power of images to display Mexico by the middle of the 1840s.

Recall that Vicente García Torres published a version translated by José María González de la Vega, *Historia de la conquista de México*. García Torres's edition contains notes by the conservative Lucas Alamán and forty-six lithographic images which—he explains in his introduction—were necessary for the understanding of history. Appearing almost simultaneously was *Historia de la conquista de México* translated by Joaquín Navarro and published by Ignacio Cumplido. Cumplido's edition contains seventy-one lithographic prints of antiquities and historical monuments in a separate title, *Esplica-*

ción de las laminas pertenecientes a la "Historia antigua de México y la de su conquista" que se han agregado a la traducción mexicana de la de William H. Prescott (Explication of the Plates [illustrations] pertaining to the *History of Ancient Mexico and its conquest* that Have Been Added to the Mexican Translation of [the work] by William H. Prescott).[75] It includes an appendix and notes by the liberal-oriented José Fernando Ramírez, soon to be the second director of the Museo Nacional. The volume contains seventy-one images with an essay and bibliography of primary documents by Isidro Gondra, at this time the director of the Museo Nacional. The lithographic prints, besides depicting pre-Columbian objects such as the Calendar Stone and reliefs from archaeological sites, also illustrates supposed scenes from the history of Mexico such as *Cortés manda prender a Moctezuma*, in which Cortés shackles the noble Aztec king.

These translations of the *History of the Conquest* with their addition of images demonstrate the construction of a Mexican history that corroborated certain political views.[76] With these editions, Prescott was not just translated and republished in Mexico but became contextualized in this growing interest in creating and displaying a history of Mexico. The inventory of objects developed by foreign travelers and Mexican artists, as well as the illustrations of specific supposed historical events, were used to picture and, thus, substantiate historical viewpoints. We might ask: Does this image of a shackled Moctezoma illustrate the text or display history? This is to say that we have no indisputable facts to confirm if or how Moctezoma was shackled; yet the image identifies a decisive moment that changed the course of Mexican history according to the emerging liberal version of this history. The important point of the illustration of this supposed event, however, is that it casts the Spanish as shackling indigenous culture. These are the same restraints Mexico must release herself from politically as evident in the *Alegoría de la Independencia* (1834), which depicts allegorical Mexico shackled (figure 40). The productions' written history with the use of lithographic illustrations were also moves to certify this liberation.

THE MEXICAN-AMERICAN WAR

The expansion and circulation of images were curtailed in the mid-1840s by one of the most politically and socially traumatic events of Mexico's history: the 1846–48 Mexican-American War. An 1846 essay written anonymously and entitled *Conquest of Mexico! An Appeal to the citizens of the United States, on the justice and expediency of the conquest of Mexico; with historical and descriptive information respecting the country*, published in Boston, reminds us that even as Mexico went about the work of constructing a history and geography of itself, the United States subverted information about Mexico to justify the political imperatives of the expansion of its own territory. As its title underscores, the essay was written as a justification for the invasion and subjugation of Mexico. Its rationale explicitly states that the role of the United States is necessary because of "Her [Mexico's] utter demoralization, her incapacity to govern herself, and discharge her duties to other nations."[77] In the argument, the anonymous writer cites the commercial potential of Mexico's excellent climate, inestimable agriculture resources and abounding mineral sources, certified by quoting directly from Humboldt's *Essai politique*: "There is scarcely a plant in the rest of the world, which is not susceptible of cultivation in one or other part of Mexico; nor would it be an easy matter for the botanist to obtain even a tolerable acquaintance with the multitudes of plants scattered over the mountains, or crowded together in the vast forests at the foot of the Cordilleras"[78] Subverting Humboldt, the writer contends that all this potential, however, is wasted: "And yet what is the actual condition of this finest section of our continent? Abject in morals, degraded in intelligence, ruined in business, and anarchical in government, it is second only to the savages of our wilderness."[79] With such rhetoric and incontrovertible facts, the writer ineluctably concludes with the question: "Should not civilized nations, especially those which by their trade or proximity have suffered from it commercial injury and national insult, reduce it to order?[80] Thus, the North American or United States' war on Mexico was not just justified, but required!

The Mexican-American War, resulting in the subsequent loss of over half of Mexico's sovereign territory to the United States, physically and emotionally devastated the nation and forced self-reflection. A print entitled *Progressos de la República Mexicana* and published in the *Calendario de Galván para 1848*, illustrates the disillusionment of the period. The image depicts a well-dressed and confident allegorical Mexico seated near a precipice, the date of independence, 1821, next to her.[81] In the lower section of the print, associated with the date 1847 and referring to the invasion by the United States, Mexico is shown plunging over the precipice and plummeting into an abyss; her clothes are shredded and she is stripped of her attributes.[82] The war and its resulting territorial loss literally and figuratively stripped Mexico of its prewar identity.

While the February 1848 Treaty of Guadalupe Hidalgo concluded the war, response to its repercussions continued for the next decades. Writers such as Lucas Alamán responded by attempting to structure a wholeness for the recently dismembered nation through his five-volume work *Historia de Méjico desde* (from) *los primeros movimientos que prepararon su independencia en el año de 1808, hasta* (until) *la época presente* in 1849–52. State institutions also responded to disruption and disjunction caused by United States imperialism. In 1852, José Fernando Ramírez became director of the Museo Nacional and in 1857 published the catalogue *Descripción de algunos objetos del Museo Nacional*, which included forty-two lithographs by the artist Casimiro Castro.[83] An associated text, written by Ramírez, attempts to give order to the archaeological remains by describing each numbered object, its materials, and possible meaning. While a more detailed analysis, nevertheless—such as Icaza's and Gondra's earlier *Colección de las Antigüedades Mexicanas*—the objects hint at a past but are not yet fully integrated into a narrative history. Ramírez situates the Mexican museum collection, however, within larger European discussions about antiquities. Specifically, he mentions that the Castro image was inspired by an illustration in volume 27, page 176 of *L'Illustration*, a weekly French journal, which incorrectly interpreted certain Mexican remains. At the same time, Ramírez argues for their similarity to Egyptian objects in the Turin Museum in Italy, the British Museum in London, and the Louvre in Paris.[84]

In 1849, the Sociedad Mexicana de Geografía y Estadística succeeded the Comisión de Estadística Militar. Its renewed geographic and cartographic efforts included reviewing and evaluating three hundred maps and an equal number of geographic coordinates from throughout the Republic. Its membership expanded to include civilian scientists and intellectuals from Mexico City, who formed a scientific community that shared a common public social life of Mexico City.[85] In 1850, the SMGE produced the *Carta General de la República Mexicana formada por la sección geografía de la Sociedad Mexicana de Geografía y Estadística. . . .* Here, for the first time, Mexico visualized the full extent of its territorial loss to the United States.[86]

In January 1851, the SMGE announced the *Atlas y Portulano* [referring to a portolan, a navigational map] *de los Estados Unidos Mexicanos*, an atlas with forty-six updated regional and coastal maps. Mendoza Vargas points out that the atlas represents a notable achievement; however, due to the penury of the Mexican economy, the work was limited to originals and was not published for mass distribution.[87] The *Atlas y Portulano* was a summative effort similar to Ramírez's 1857 catalogue. Finally, the *Boletín de la Sociedad Mexicana de Geografía y Estadística*, the institution's publication, was also established at this time. Thus, after the Mexican-American War, the SMGE functioned to aggregate, synthesize, and display geographic and statistical data, intensifying the process of remembering Mexico.

THE ACADEMY OF SAN CARLOS

The postwar period also brought changes to the established institution of the Academy of San Carlos, whose media specialties included painting, sculpture, and engraving. An 1848 article in the popular newspaper, *El Siglo XIX*, emphasizes the importance of the academy's education of the public and states that the fine arts truly promote the glories of the nation without the cost of tears and blood.[88] Visitors attending the academy's inaugural exhibition in December of 1848 saw these glories in sculpture, painting, engravings, and drawings along with architectural drawings and ornaments which emphasized classical topics and excluded art from the viceregal period.[89]

In the early 1840s, an influx of European artists had brought new direction and rigor to the academy. One of the initiatives of this time brought Eugenio Landesio (1810–76), an accomplished Italian landscape artist, to the academy in 1855 as an instructor. Prior to Landesio's arrival, landscape had been a background element in Mexican and European art, as seen, for example, in Bullock's panorama as well as the work of Waldeck and Catherwood.[90] It also appears in Mexican-American War images made by Carl Nebel.[91] Under Landesio, Mexico's diverse landscapes would be represented in full splendid spectacle. To achieve this, he put into place a rigorous and thorough curriculum that required knowledge and application of geometry, linear perspective, light theory, and aerial perspective.[92] The addition of landscape as a genre specialty in the academy, while following European trends, also marks the recognition of Mexican land as a valid subject of academic art; more importantly, it once again put Mexico on display through a government intervention.[93]

Visitors attending the academy's exhibitions in the mid-1850s saw expanded image content that included landscapes.[94] This postwar commitment to public display was particularly manifest when, in 1852, the bronze equestrian statue of Carlos IV was relocated from the University to the Paseo Bucareli, a road on the western outskirts of the city, as an example of excellent art of the past.[95] Here, a statue that was once the sign of corporatist monarchy, then demoted to a sign of imperialism, was relocated physically and intellectually by the state, demonstrating the power of the state to make history legible.

EXHIBITIONARY COMPLEX

Thus, by the middle of the century, an "exhibitionary complex" had transformed a problem of apparent social and political disorder into one of national culture.[96] The cultural historian Tony Bennett explains this construct:

> Instead, through the provision of object lessons in power—the power to command and arrange things and bodies for public display—they [state institutions] sought to allow the people, and *en masse* rather than individually, to know rather than be known, to become the subjects rather than the objects of knowledge. Yet, ideally, they sought also to allow the people to know and thence to regulate themselves; to become, in seeing themselves from the side of power, both the subjects and objects of knowledge, knowing power and what power knows, and knowing themselves as (ideally) known by power, interiorizing its gaze as a principle of self-surveillance and, hence, self-regulation.[97]

In these terms, the objects from the catalogue of the National Museum, the maps and atlases of the SMGE, and the landscape specialty of the academy were located within the historical and geographical coordinates that completed the reversal of the power of the ancien régime's monarchy-centered disciplinary and objectifying apparatuses as manifest in the itinerant equestrian statue of Carlos IV. In fact, the relocation of the statue to the Paseo Bucareli certifies the full power of a historical-geographical system that was no longer just emergent but actually functioning.

In this way, the confusion of relating dislocated pre-Columbian objects to emerging history and to the abstract territory of a map could be resolved through the visual rhetoric of cultural exhibition. By viewing exhibitions of pre-Columbian remains, viceregal art, Mexican landscape, and national maps, citizens could remember who they are, where they came from, and where they were to go through the power to locate themselves in historical and geographic coordinates. As a result, Mexico's exhibitionary system became self-referential too, manifesting the power "to organize and co-ordinate an order of things and to produce a place for the people in relation to that order."[98] Thus, by the middle of the nineteenth century, Mexico was in full production and circulation of a visual culture that emphasized ordered seeing and display that grew exponentially to exhibit Mexico through an ever-growing inventory of images of its people, its objects, its land, and its history.[99]

Following the Mexican-American War, the intensification of such exhibitionary modes is manifest in print culture. Some publications continued prewar themes and ideas, borrowing from earlier travelers.[100] For example, the *Álbum Pintoresco de la República Mexicana* was published by Julio Michaud between 1849 and 1852, possibly in France, for European viewing but directed toward a Mexican audience too through its sale at Michaud's Estampería, a print shop located near the post office.[101] The *Álbum Pintoresco* includes the image of *Las tortilleras* (figure 41) and *Las poblanas*, themes derived from Linati (figure 28, chapter 3, p. 88); the images themselves, however, are taken directly from Nebel's *Voyage pittoresque et archéologique* (figures 30 and 31 in chapter 3, pp. 91, 92). Michaud's *Álbum Pintoresco* also includes images from Gualdi as well as images illustrating scenes from the Mexican-American War. Although published after the war, Michaud's *Álbum Pintoresco* continued prewar interest in an exotic Mexico.

Postwar publications by Mexican editors strengthened the geohistorical narrative structures formed prior to the war, impelling new exhibitions of Mexican cultures. Thus, various forms of Mexican cultural production, including the visual arts, addressed formation of a genealogy. For example, the 1851 *Calendario de Ignacio Díaz Triujeque* used as its theme the conquest of Mexico, reformatting images from Prescotts's *Historia de la conquista de México* published by Cumplido, into small rectangular images on two pages. Díaz Triujeque selected images that display narrative images illustrating events of the conquest and, between these images are pre-Columbian objects, such as the Aztec Sun Stone. In this way, the calendar further popularized history.[102] Geographical publications were important too. The ten-

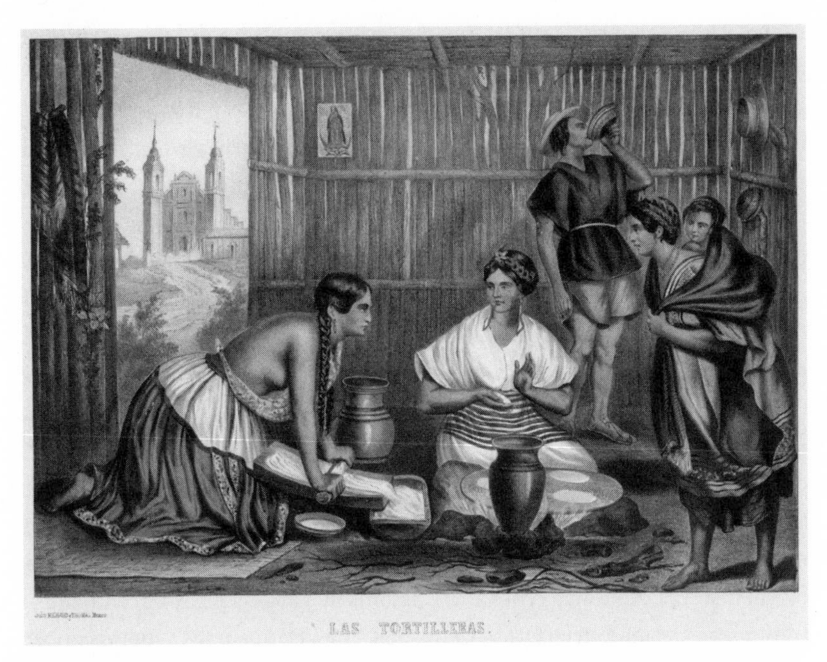

41 *Las Tortilleras*, from *Álbum Pintoresco de la República Mexicana* (circa. 1849–52). Courtesy of Yale University Library.

volume *Diccionario universal de historia y de geografía . . .* , part of which was edited by Manuel Orozco y Berra (1816–81), gathered information from diverse sources and gave greater detail to the previous outlines of the nation.[103]

Ignacio Cumplido published *El Álbum Mexicano* in 1849, which allowed Mexican writers to voice their anger over the recent loss of territory and identify Mexico's social defects and their improvement. Appearing weekly, *El Álbum* had an encyclopedic approach and, at the same time, vindicated the ideas of Cumplido's earlier publication, *El Museo Mexicano*. Importantly, to reinforce its nationalist focus, Cumplido did not include translations of foreign articles in this publication as was popular at the time.

After the *El Álbum Mexicano*, Cumplido launched *La Ilustración Mexicana* in 1851. Moving away from the localism of *El Álbum*, the editor contextualized Mexico in universal culture, locating its beautiful physical nature, its intellectual achievements, and articles about its patriots such as Hidalgo and Morelos. In addition, refining the theme of *tipos sociales* found in *El*

Museo Mexicana, articles in *La Ilustración Mexicana* also discussed national archetypes. In an 1851 article, for example, the *ranchero*, an important managerial figure in agricultural production, is described as a mixture of Spanish and Indian blood that formed the basic characteristics of this special national type. More importantly, associated with specific clothing including pants with a waist sash, a white shirt, a short jacket, and a large hat, the ranchero is identified as "una clase productiva, que bien dirigido podría ser útil a nuestra sociedad" (a productive class, which well directed, could be useful to our society).[104] This figure points to the nation's potential for national progress. *La Ilustración Mexicana* concluded in 1853, as the Plan de Ayutla of 1854 and the ensuing War of Reform (1856–61) brought major political reforms that were the focus of subsequent publications on political issues.[105]

After the Mexican-American War, changes in the publication of genre books of popular culture are demonstrated in *Los mexicanos pintados por sí mismos. Tipos y costumbres nacionales por varios authores* (Mexicans Painted by Themselves. National Types and Customs by Various Authors) with illustrations by Hesiquio Iriarte and published by Murguía in 1854. In this book, six Mexican authors narrate thirty-three national types who are identified primarily by their work or profession. Absent are the Linati or Nebel Indian types such as the tortilleras.[106] While lower-class individuals—servants, laundresses, venders, porters, and the like—appear across the pages, others who belong to the professional class, such as lawyers and ministers, are also described and depicted. Noticeably remaining is the *china*, derived from Linati's *Jeune ouvrière* and Nebel's *poblana* and repeated in Michaud. As described by José María Rivera, the china has been recast as "that daughter of Mexico as pretty as its blue sky; as fresh as its flowery gardens, and as smiling and happy as the delicious mornings of this land blessed by God and its saints."[107] Thus, the authors exhibit new national characters and characteristics of Mexico that focus away, for the most part, from the exotic indigenous types highlighted by Linati, Nebel, and Waldeck. In fact, in *Los mexicanos pintados por sí mismos*, indigenous groups have disappeared from Mexico!

Mexican publications also manifest the theme of travel, constructing Mexicans as travelers in their own land through production of albums and books about themselves. *México y sus alrededores. Coleccion de monumentos, trajes y paisajes* (Mexico City and Its Environs. A Collection of Views,

Monuments, Costumes and Landscapes), published in various editions between 1855 and 1864, is an example of a collaborative production by Mexican authors and lithographers.[108] It contains lithographic images by Casimiro Castro, J. Campillo, and Dacaen with accompanying essays by various authors.[109] As evident in the title, the emphasis of the publication is Mexico City, as a site of origin in its pre-Columbian identity of Tenochtitlán, and, at the same time, Mexico as a country. The images in *México y sus alrededores* depict scenes from in and around Mexico City, with all classes present in various scenes.

It is clear that Bustamante's compelling questions from before the Mexican-American War continue to haunt these postwar authors. Thus, in his description of La Fuente del Salto del Agua, a well-known public fountain, Francisco Zarco, newspaper editor, begins his description by writing:

> We are not Aztecs, we are not Spaniards; bastard race of the two, we have the indolence of the one, the arrogance of the other; but even though we do not constitute our own race, [we are] distinct from the rest, with peculiar qualities, good or bad. People of yesterday, without traditions, without great memories, our history of few years is the chronicle of inexperience, of insanity and of the discord and lacks more notable events that fascinating prestige of the distance that gives to men and things the numerous centuries that are interposed among the generations.[110]

While somewhat dark in its sentiment, Zarco concludes more positively, arguing for the continuing need to construct a history for Mexico and emphasizing that "without history there are no monuments, without events, there is nothing to retain in the memory." Thus, *México y sus alrededores* became an exhibition of constructed memory.

One of the more interesting prints included in the publication is *Trajes Mexicanos* (Mexican Costumes) by Casmiro Castro and J. Campillo, which depicts over thirty individuals in their various forms of dress, spread somewhat randomly across the page, with some shown in work activities (figure 42). Their general appearance and many of the characters are similar to the images of Iriarte in *Los mexicanos pintados por sí mismos*. The images continue to treat Mexicans as types, really representing class.

In another example, the print *Antigüedades mexicanas que existen en el Museo Nacional de México* (figure 43) illustrates archaeological objects, in-

42 C. Castro and J. Campillo, *Trajes Mexicanos*, from *México y sus alrededores* (1855–56). Brown University Library.

cluding monolithic as well as small sculptures and numerous other objects. These objects, many of which appeared as single objects in publications by León y Gama, Humboldt, Waldeck, and Nebel—as well as in Icaza's and Gondra's *Colección de las Antigüedades Mexicanas*—are now amassed on a single page, lacking systematic order. For the present-day viewer, the visual effects of Castro's *Antigüedades* are both profusion—the awe-inspiring diverse remnants of Mexico's past—and, at the same time, disorder—jumbled, unnarrated content. This early format specifies the functions of collecting and aggregating and is consistent with the contemporary functional notion of the museum as an inventory.[111]

Another fascinating example of this theme of travel is Marcos Arróniz's travel guide entitled *Manual del Viajero en Méjico*, written in Mexico City in 1857 and published in Paris in 1858. Marcos Arróniz was a well-known poet who also wrote an article for *México y sus Alrededores*.[112] In the introduc-

43 C. Castro and J. Campillo, *Antigüedades mexicanas*, from *México y sus alrededores* (1855–56). Brown University Library.

tion to this pocket-size book, Arróniz tells the reader that it is his intention to avoid reporting on Mexico's government institutions or systems. Instead, because much has been written about the negative aspects of Mexican culture, he wishes to present "á la vista del viajero todo lo que pudiese interesarle, y estuviera en relación con lo útil y pintoresco" (to the view of the traveler that which might be interesting to him and would be in relation to the useful and the picturesque).[113] His subsequent chapters combine history with sightseeing. The first chapter, reminiscent of Bustamante's *Mañanas de la Alameda de México*, provides a brief history of the ancient cultures of Mexico, avoiding any mention of the Spanish conquest. His following chapter reviews Mexico City's geography and information about its urban contents and resources. Recalling Gualdi's cityscape prints, Arróniz locates major sites such as the Cathedral, the Academy of San Carlos, and the statue of Carlos IV, as well as business establishments such as barbershops, bakeries, and laundries.[114] He also identifies major print shops—including those of Señores Cumplido, Murguía, and García Torres—and important lithographers—including Señores Dacaen, Salazar, and Murguía. He especially recommends Dacaen's recent publication *México y sus alrededores*.[115] The third chapter, "Trajes, Usos y Costumbres," includes descriptions of about eight types of people found in Mexico City, including the water carrier and a woman identified as *china poblana*.[116] While the fourth chapter reviews literary achievements of Mexican culture, the fifth and sixth chapters lead the traveler outside of Mexico to view important archaeological sites such as Teotihuacan and Chichen Itza, as well as geographic sites such as the Cacahuamilpa cave and Cascada de Regla, sites mentioned by Humboldt and others. Arróniz concludes his guide with the hope that his readers enjoyed his lecture and that it proves that, as he puts it, "our" country deserves to be visited and studied by enlightened and impartial travelers. And, as if directing comments to the anonymous author of *Conquest of Mexico!*, he reprimands those writers who have "pintado á Méjico como un desierto estéril, y á sus habitantes casi al nivel de las tribus bárbaras de la frontera" (painted Mexico as a sterile desert and its inhabitants as almost at the level of the barbarians of the northern frontier).[117] While addressed to foreigners, Arróniz's narrative also seems to be explaining Mexico to himself and other Mexicans.

Extending the travel theme, stereoscopy arrived in Mexico in the late 1850s. Stereoscopic photography required a special camera that produces a pair of images taken from slightly separated views.[118] The two images, stereographs, were paired on a card, then viewed through a stereoscope. The result is that the brain perceived the images as a single image with depth perception, that is, the objects in the image appeared to be three-dimensional.[119] Miguel Rodríguez set up a stereoscopy studio, promising "*estereoscópicas* con toda precisión (stereotypes [made] with all exactness).[120] Requiring the head to be held stationary in the stereoscope, these pictures resulted in a corporeal immersion in an image; that is, the perception of three-dimensionality caused the viewer to feel part of the place, scenes, or landscape depicted. Early stereoscopic images included landscapes, domestic interiors, and images of people. In 1857, possibly the first exhibition of stereographs was held which included portraits taken from history with the theme of independence. By 1861, one newspaper advertisement announcement offered 10,000 "vistas estero" from all over the world.[121] Daguerreotype and stereotype images produced at this time continued the themes found in lithographic prints, including portraits, landscapes, cityscapes, re-created historical scenes, and ethnographic images.

By the 1850s, photographic technology, which allowed multiple prints from a single negative, was established in Mexico. In fact, there was considerable competition as a number of photographic studios were founded utilizing photolithography, available by 1853. These studios both produced images and sold available images from Paris, New York, London, and other cities.[122] Foreigners traveled through Mexico, taking pictures and publishing them in various formats. The *Álbum fotográfico mexicano*, edited by Désiré de Charnay and published by Julio Michaud, consists of photographs of monuments and ruins of Mexico with texts by Manuel Orozco y Berra, a respected historian.[123] Competing with Charnay in the same year was the *Álbum fotográfico*, which contains twenty-four views of Mexico City and its surroundings. Mexican photographers were active as well during this time with ten studios functional by 1859.[124] Photographs formed another archive of images that repeated the imagery of lithographs of earlier decades and expanded, even more, the circulation of images about postwar Mexico for both Mexican and foreign audiences.

CONCLUSION

Recently, scholars of nineteenth-century Mexico have pointed out the growing importance of visuality and identified aspects of this visuality during this period.[125] In this chapter, I have traced the extensiveness and exchange within visual production. By the 1850s, visual culture—through prints, and lithographic and photographic technologies—supported and stimulated the coalescence of an imagined geography for Mexico. The power of display inherent in these technologies located objects, sites, and people within an emerging historical-geographical discourse about the nation-state, allowing individuals to be situated, as well as to situate themselves, in time and space.

Image production and its reproducibility fundamentally redefined the connection to materiality, experience, and truth. Photographs and lithographs promoted meanings to be inscribed and ascribed and dematerialized real objects, people, and places into types or constructs. This is to say that the image could represent a reality to such an extent that the original object seemed almost unnecessary, for the picture was more real than the object.[126] For example, rationalizing his fabricated history, Prescott certifies the authenticity of Bernal Díaz's writings as a historical source by comparing the verbal precision of his visual description to the exactness of daguerreotypes. Such visual certification also undergirds the addition of illustrative images to the Cumplido and García Torres' Spanish editions of *The History of the Conquest*. In another example, the mesmerizing images of Catherwood's camera-lucida-based lithographic images made the effect of Maya ruins more real than the actual remains, as the images transformed and dematerialized the actual dilapidated conditions of the ruins into a discourse about mysterious presences, lost worlds, and the desire for their recovery through exhibition.

At the same time, spaces for women were constructed through allegorical imagery as well as outlined in the construct of the educated "muger de casa." And how could the authenticity of *chinas poblanas* or *trajes mexicanos* or *antigüedades mexicanas* be doubted when their existence was repeatedly certified across diverse texts and images? Sights, whether people or place, became *sites* for and within the geographic imagination. These visualizations created and situated memory and memories in deep, evolving discourses about the nation manifested as an exhibition. Overall, by the late 1850s, Bustamante's foundational questions of 1834 had not really changed as Mexico and Mexicans still asked:"¿Quienes somos? ¿de donde venimos? ¿para donde caminamos?" What had changed were the formatting, repetitiveness, and intensity of the responses.

Antonio García Cubas initiated his cartographic work in the midst of this shifting understanding and use of visual images and as Mexico's geographical imagination emerged. Such recognition of Mexico's nationalist evolution in the broader developments of visual culture allows García Cubas's works to be dislodged from the restrictive categories of cartographic history, art history, nineteenth-century history, or a combination of those, and placed within this emerging scopic regime of the nineteenth century. Working within this understanding, his mappings of Mexico may be examined through analyses that connect his work to all of these fields of knowledge at once. As García Cubas reached adulthood in the 1850s, Mexican mapping practices—not only in the form of cartographic images, but also in the form of texts, print images, and exhibitions—emphasized a consolidation of the national boundaries and a restructuring of the nation's content to reflect an emerging state. As he began his work as a geographer and cartographer, Antonio García Cubas embedded himself and his work in the intensity and flux of this visuality.

5

FINDING MEXICO

The García Cubas Projects—1850–1880

Antonio García y Cubas entered the world on 24 July 1832, just as Bustamante posed his questions in *Mañanas de la Alameda de México* and grew up as passionate answers were proclaimed and tested. Born in Mexico City into a middle-class family, he was educated at the Colegio de San Gregorio.[1] In his memoirs, he states that he was not a child prodigy. In fact he preferred the "trompo y la pelota" (the top and the ball) to schoolwork; he did enjoy his musical studies.[2] As a fourteen-year-old boy, he witnessed the United States invasion and occupation of Mexico and the subsequent loss of territory. Describing this agonizing experience, he remembers, "Sentí oprimido el corazón y mis ojos se humedecieron. ¡Lágrimas puras vertidas por el amor de la patria!" (I felt my heart oppressed and my eyes were damp. Pure tears spilled for the love of the homeland).[3]

Having to support his widowed mother, Antonio attempted business ventures, then applied for and received an appointment in the Dirección General de Colonización e Industria (General Office for Colonization and Industry) in 1851, which would become the Secretaría de Fomento (Ministry of Development) in 1853. He was a tenacious civil servant too, working in the Secretaría de Fomento for much of his work life. His access to the rich library of this institution and that of private collectors stimulated Antonio's interest in geography and led to his part-time study of engineering at the Colegio Nacional de Minería, from which he graduated in 1865. He also furthered his education at the Academia de San Carlos, where he studied engraving with Luis G. Campa, a lithographer who turned his talents to photography in the 1860s. There he also met José María Velasco and Luis Coto, fellow students who would become important painters of landscape and history, respectively.[4]

García Cubas witnessed the diverse efforts at nation building: the dictatorship of Santa Anna, the rise of Benito Juárez, a push for political and social reform and an ensuing Guerra de Reforma (1858–60), the French occupation under Maximilian and its defeat (1863–67), and the return of Benito Juárez followed by the authoritarian government of Porfirio Díaz (1876–1911). These phases of Mexican political development were marked by continued experiments with government formats from liberal and conservative perspectives as various regimes attempted to rationalize and legitimize authority while searching for a unifying ideology and national identity.[5]

Mexican leadership of the 1840s and 1850s, García Cubas's formative years, was also marked by the rise to power of individuals who were neither from metropolitan locales nor of elite Spanish heritage but rather of mixed-blooded ancestry. In his study of this period, Richard Sinkin, a historian, identifies this leadership group as made up of individuals of mestizo, as well as poorer criollo background. A little less than half of the members of the Constitutional Congress were from Mexico City or its environs. Because these leaders had no roots in the elite social structure, they conceived of reform as a way to restructure society to include themselves.[6] As a result, the Constitution of 1857 called for a move from personalist politics to recognition of the supremacy of the law through a strong, unicameral legislature and universal suffrage.[7] Public education was an integral part of these proposed reforms. Thus, García Cubas became an active member of a generation of intellectuals such as Ignacio M. Altamirano, Manuel Larraínzar, Manuel Orozco y Berra, and José Fernando Ramírez, who grew up after the battles for independence from Spain were won, and as Mexico continued to struggle for its political and territorial integrity and to conceptualize Mexican citizenship.

The geographer's memoirs additionally confirm that he was an idealist and optimist as well. García Cubas relates that when he was appointed to the city council, he was elated as he believed this gave him a venue to realize certain projects including improving public health, destroying poverty, deterring drunkenness and prostitution, improving public instruction and criminal rehabilitation, turning orphans into good citizens, and alleviating the suffering of those suffering in hospitals. Recognizing this zealous idealism of his younger years, he writes, "¡Era yo tan inocente! (I was so innocent!!)"[8] Nevertheless, García Cubas retained throughout his life's work a fervent belief that he could and would contribute to Mexico's improvement. He accomplished this goal, in part because his politics were moderate and pragmatic, which allowed him to weather tumultuous times and retain his government appointments.[9]

Along with this social and political background, Antonio García Cubas initiated his cartographic work as the SMGE and Academy of San Carlos were in the process of collecting, ordering, displaying, and preserving the Mexican state as well as in the midst of the shifting understanding and use of visual images, which continued to expand in the 1860s and 1870s. Thus, his mappings of Mexico from the late 1850s to the late 1870s, the topic of this chapter, can be contextualized within the broad but threaded discourse and practices of nineteenth-century visual culture. Beginning with his 1858 atlas project, García Cubas's mappings incorporate this emerging discourse into his work. His subsequent work of the 1860s and 1870s, and especially after the French intervention, focus on pedagogical themes and placing Mexico and its culture in international contexts.

THE *ATLAS GEOGRÁFICO, ESTADÍSTICO É HISTÓRICO DE LA REPÚBLICA MEXICANA* PROJECT

García Cubas came to cartography and geography through avocation rather than vocation. In his twenties, while working at the Secretaría de Fomento, Antonio spent his free time expanding his geographic knowledge and studying mathematics. In his memoirs, nevertheless, he constructs a historical context for himself and his lifework. He laments the fact that at the time of the Treaty of Guadalupe Hidalgo, which ended the Mexican-American War, geography and cartography were so neglected that the treaty's boundary negotiations were based on a highly deficient map by John Disturnell (figure 37 in chapter 3, p. 105). García Cubas goes on to state emphatically that the geographic knowledge of Mexico at this time was so deplorable that it was best explained through a personification: "a girl, deformed and wasted away." And he explains, "Curing such an unfortunate being was a matter that offered serious difficulties, at the same time her nutrition and the arrangement of her dislocated members had to be attended to. To correct the damages of the girl and to cure her of her profound anemia, I directed all my efforts."[10] These efforts were made initially through cartography. In 1853, García Cubas launched what he calls his "obra magna": *Carta general de la República Mexicana* (General Map of the Mexican Republic), likely based on a map by García Conde.[11] He firmly recalls that when this map was presented to General Santa Anna, the extent of Mexico's loss of territory to the United States was realized for the first time. While dismayed, Santa Anna recognized the importance of García Cubas's work and ordered that the cartographer receive 100 pesos, a great fortune for Antonio at the time.[12]

García Cubas's career as a geographer-cartographer leapt forward because his 1857 map of Mexico, *Carta general en major escala*, was extremely well-received by members of the Sociedad Mexicana de Geografía y Estadística (SMGE). This map was completed using the compilation method based on comparing information of earlier maps and atlases—the oldest tradition of mapping—rather than field surveys.[13] The longitude and latitude of this map coordinated Mexico with the eastern corner of the Mexico City Cathedral, as prevalent in previous Spanish maps. At the same time, at the bottom of the map García Cubas included the coordinates of this Mexican meridian in relation to the Greenwich meridian, bringing Mexico into "cartographic consonance with what were then construed to be the icons of advanced civilization."[14] This is important because the coordination with the prime meridian signals both real and imagined dimensions of García Cubas's first national map: he places Mexican space in consonance with current European cartographic standards and, at the same time, this association indicates the nation of Mexico's breaking of its spatial links to Spain.[15]

García Cubas, however, saw his task as more than creating an accurate map; he wanted to provide nutrition for the geographic anemia of the nation. The next year, the *Carta general en major escala* was incorporated into a larger project, the *Atlas geográfico, estadístico é histórico de la República Mexicana*, published by J. M. Fernández de Lara. The *Atlas geográfico* was the first atlas published in Mexico, as the SMGE was unable to publish the *Atlas y Portulano de los Estados Unidos Mexicanos* of 1851. Previously, only Humboldt's 1811 *Atlas géographique et physique du Royaume de la Nouvelle-*

Espagne provided published comprehensive information about Mexico. García Cubas's first atlas project articulates a full projection of Mexico as both a real and imagined nation.

The *Atlas geográfico* is related to older cartographic and atlas traditions. García Cubas acknowledges this directly in the preface, stating he viewed maps of the Mexican states and the works of "los Sres [Señores]. Moral, Humboldt, García Conde, Teran, Rincon, Narvaez, Camargo, Lejarza, Orbegoso, Iberri, Harcort, Mora y Villamil, Robles, Clavijero, Prescot [*sic*], Alaman [*sic*], etc. etc."[16] At the same time, the *Atlas geográfico* is a product of nineteenth-century visual culture because not only does it incorporate circulating and mobile images to construct a coherent display of Mexico; it emphasizes visual strategies of display that carefully guide the viewer's consumption and understanding of the *Atlas*'s imagery.

The *Atlas geográfico*, consisting of text pages and thirty-two double-page maps reproduced through lithography, was published in 1858, ten years after the loss of one-half of Mexican territory to the United States, during the tumultuous time when major political reforms were promulgated by liberals and on the eve of civil strife in Mexico caused by the Reforma movement. Mexicanists of all disciplines have tended to ignore completely, generalize, or only look at specific sections of the *Atlas geográfico*.[17] In his excellent study, *Cartographic Mexico: A History of State Fixations and Fugitive Landscapes*, however, Raymond Craib persuasively explains that the *Atlas geográfico* supported an ascendant liberal regime's hope for colonization of Mexican land, capitalist development, and the disentailment of Church and Indian lands. Craib especially focuses on the *Atlas geográfico*'s *Carta general de la República Mexicana*, arguing that through it

> history and geography came together to compose Mexico as a coherent historical and geographic entity; that is, as a legitimate nation-state. In one sense the two disciplines came together in García Cubas's own conception of history, which he understood as a geographically descriptive enterprise aimed at discerning how the country literally took shape. His maps and atlases were genealogies of the territory, narrating a kind of property history in which the historical existence of the nation-state was taken as a given and a history of "its" territory was simply recounted.[18]

Craib's work is significant in its critical examination of the *Carta general* as more than cartographic achievement as he explores its historical and iconographic contexts. He cogently argues that the 1858 *Carta general* provided a place for history to be written.

It is, however, also critical to analyze the *Carta general*'s placement within and relationship to other sections of the *Atlas geográfico*, as well as to examine its imagery in the context of the circulating inventory of visual images that was well established by the mid-1850s. Such broad-based analysis expands Craib's reading of the *Carta general*, offering a holistic understanding of this particular page and its integration into the *Atlas*. The *Atlas*, then, is best examined as a whole, not only because it was conceived of as a single work, as García Cubas verifies in his introduction to the *Atlas*, but also because each of its sections focus on distinctive mappings of Mexico that coalesce into a panoramic narrative display of Mexico just past mid-century.

García Cubas envisioned the *Atlas* as consisting of five sections, four of which included splendid chromolithographs. In the introduction, he identifies the sources for his work and explains his objectives for the *Atlas*:

> The principal objective with which I have structured this present Atlas is to give recognition to this beautiful country so rich in natural products. The lack of maps and geographical information has been one of the obstacles for the realization of great projects. Well-known country, the projects of colonization, those of [building] roads, those of rich and abundant mines that we possess, those of agriculture and many others, will give the result of prosperity which the vows of all Mexicans ought to aspire to.[19]

For García Cubas, the *Atlas* becomes a way to display Mexico through its material aspects.

In the following description of the *Atlas geográfico*, I follow the order of the text and images as described by García Cubas in his introduction.[20] After the title page and introduction, the reader or viewer next opens to the *Cuadro geográfico y estadístico de la Republica Mexicana*, consisting of a general map of Mexico surrounded by extended text (figure 44). The *Cuadro geográfico* is followed by twenty-nine individual maps of each of the Mexican states and territories, such as *Oaxaca* (figure 45). These first two sections do not prepare the viewer for the superb *Carta general de la República Mexicana*, consisting of a map of Mexico surmounted by an elaborate title block (figure 46). The *Carta general* is followed by two reproductions of supposed indigenous maps entitled *Cuadro histórico-geroglífico de la pere-*

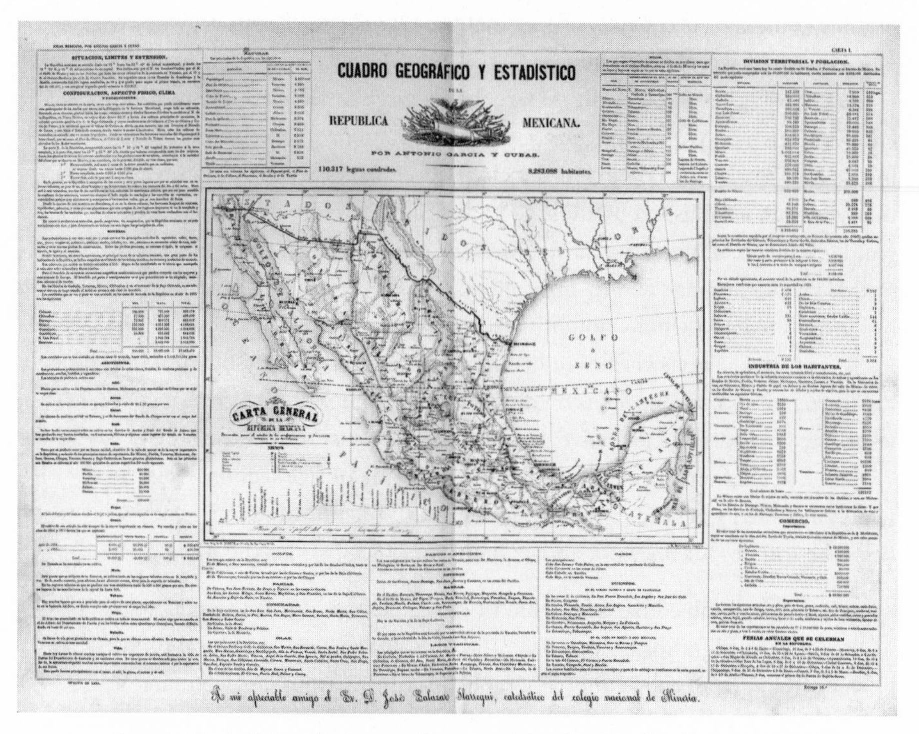

44 Antonio García Cubas, *Cuadro geográfico y estadístico de la Republica Mexicana*, from *Atlas geográfico, estadístico é histórico de la República Mexicana formado por Antonio García y Cubas* (1858). 55 × 37 cm. Courtesy of the Newberry Library, Chicago.

grinacion [peregrination] *de las tribus Aztecas*, enumerated as *número 1* and *número 2*, and surrounded by explanatory text (figures 49, p. 157, and 50, p. 158). The final section, *Noticia cronologica de los reyes anteriores de la conquista, y governantes que ha habido desde aquella epoca hasta nuestros dias* (Chronological Information of the Kings Prior to the Conquest and the Rulers from that Time to the Present Time), consists of a table listing all of the political leaders and governors of the territory of Mexico from 669 to 1858.

While clearly an overview of Mexico, its states, and history, the ordering of the atlas initially appears to lack coherence, seemingly just sets of contemporary and copied antique maps, associated data, and texts. A systematic and thorough examination of each section of the *Atlas geográfico* reveals more details about its content and allows a better understanding of

45 Antonio García Cubas, *Oaxaca*, from *Atlas geográfico* (1858). Courtesy of the Newberry Library, Chicago.

García Cubas's intertwining of maps, text, and images through the guided consumption strategies of nineteenth-century visual culture.

The *Cuadro geográfico y estadístico de la Republica Mexicana*, first section of the *Atlas* (figure 44), consists of a general map of Mexico surrounded by statistical data providing details about the physical territory, demographics, natural resources, and commerce of Mexico. As summarized by García Cubas, this page includes

> Geographic Section: Location, limits and extensions, physical aspects, climate and production. Gulfs, low lands, inlets, estuaries, peninsulas, islands, banks, or reefs, channels, headlands, ports, rivers, lakes, lagunas, mountains, volcanoes. Statistical Section. Territorial division and population. Manufacturing, agricultural, and mining industries. Information on minerals and ranches. Factories of cotton, wool silk, paper, glass, china. etc. Commerce.[21]

46 Antonio García Cubas, *Carta general de la República Mexicana*, from *Atlas geográfico* (1858). Courtesy of the Newberry Library, Chicago.

The *Cuadro*'s central map, titled *Carta general de la República de Mexico*, made by J. M. Muñozúren from H. Iriarte's workshop, shows Mexican boundaries, topography, rivers, mountains, cities, and towns. The map's cartouche located in the lower left corner profiles the land from Acapulco to Veracruz. This topographic profile is copied directly from Humboldt's print *Perfil del Camino de Acapulco á Mégico, y de Mégico á Veracruz* (figure 22, chapter 3, p. 76). Instead of placing the profile in a text, García Cubas positions it on the map page, making a graphic cartouche that replaces the allegorical cartouches found in maps of New Spain from the seventeenth and eighteenth centuries. The use of this topographic profile invokes the authority of Humboldt, allowing the mathematical data of the topographic calculation to be visualized. The text surrounding this page provides detailed information about specific agricultural products, commercial imports, and exports and major holidays celebrated in the Republic.

The *Carta general*'s cartouche of the west to east route from Acapulco to Veracruz suggests a travel itinerary. And thus, from this broad overview of Mexico, the viewer then proceeds in the next section of the *Atlas* to travel through Mexico, surveying extensive information about the states and territories of the Republic.[22] Each page consists of a state map surrounded by statistical data and geographical information. While the text surrounding these maps recalls the information of the *relaciones geográficas* and the *Theatro Americano*, their visual layout imitates the pages of Carey's *Atlas* pages of 1822 (figure 25, chapter 3, p. 80). Enumerated by Roman numerals that appear in the corner of the each page, the state and territory maps are ordered as follows in the *Atlas*'s table of contents:

2. Sonora
3. Chihuahua
4. Coahuila
5. Nuevo-Leon
6. Tamaulipas
7. San Luis Potosí
8. Zacatecas
9. Aguascalientes
10. Durango
11. Sinaloa
12. Jalisco
13. Guanajuato
14. Michoacan
15. Querétaro
16. México (estado de)
17. México (valle y Distrito de)
18. Puebla
19. Veracruz
20. Guerrero
21. Oaxaca
22. Chiapas
23. Tabasco
24. Yucatán

Territories:

25. Baja California
26. Sierra Gorda
27. Colima
28. Tlaxcala
29. Tehuantepec
30. Isla del Cármen

Through this ordering, the reader or viewer zigzags across Mexico; that is, beginning in the northwest corner of Mexico, the states were roughly ordered in west to east tiers, concluding with the Yucatán. Mexico's territories, however, are not integrated into this arrangement. In addition to a map of the state or territory showing its boundaries, topography, towns, and cities, each page displays relevant statistical data divided into the same categories as those used in the *Cuadro geográfico*. For example, from the Oaxaca page, one learns about the state's location and boundaries, physical details, climate, natural products, animal production, hunting resources, territorial division and population, industries, commerce, principal populations, public income, and mountains and rivers (figure 45). Continuing the theme of travel, these maps also note itineraries through the interior of states and territories that allow the viewer to travel through the places of each state to gain detailed information about kinds of people, where and how they live, the foods they eat, and their local industries. Through these separate maps, we move from the territorial space of Mexico, abstracted through the graticule of longitude and latitude, to particular Mexican places. The state maps emphasize the parts, not the whole of Mexico. This structuring of the state maps, however, is different from earlier centuries' division of New Spain, such as the maps of the Bourbon intendancies, because García Cubas's state maps are more than administrative divisions of space. They spatially identify and differentiate geographic and ethnographic characteristics. At the same time, the similar formatting of the pages and coloring of the maps themselves, as well as the common categories of textual data that surrounds each map, consistently connect the state maps to the whole. Put in political terms, the *Atlas*'s state maps demonstrate the constituent and constitutional parts of Mexico's confederated whole, reflecting the liberal administration's nominal federalist's policies.[23]

As envisioned by García Cubas, the *Carta general de la República Mexicana*, the next section of the *Atlas* is "Third [section]: General map in greater scale, which displays the principal populations of the Republic, its rivers, its mountains, roads, etc. etc."[24] The magnificent chromolithograph consists of a general map of Mexico similar to that of the *Cuadro geográfico* but with an updated northern border and without the surrounding text (figure 46).[25] The cartouche on the lower left of the page depicts graphs showing the lengths of the major rivers and heights of certain mountains, also borrowed from Humboldt's publication.

The *Carta general's* title block adds another dimension to this page and the *Atlas geográfico* as a whole. The image on the left of the title depicts an assemblage of well-known geographic locales, including the Organos de Actopan, Iztaccíhuatl volcano, Cofre de Perote, Popocatépetl volcano, Montañas de Jacal, Orizava, and Cascada de Regla (figure 47). The image on the right illustrates the archaeological sites of Palenque, Pirámide de Papantla [Tajin], Mitla (likely, Monte Alban), and Uxmal (figure 48). The landscape and archaeological imagery of these cartouches is derived directly from published lithographic images of travelers to New Spain / Mexico from

47 Antonio García Cubas, title block detail (from left side), *Carta general de la República Mexicana*, from *Atlas geográfico* (1858). Courtesy of the Newberry Library, Chicago.

earlier centuries. For example, four of the title block's landscape images are reproductions of prints from Humboldt's original *Vues des Cordillères*:

- Cascada de Regla image is a direct copy of plate XXII, *Rochers basaltiques et Cascade de Regla* (figure 20, chapter 3, p. 73);
- Montañas de Jacal image is a direct copy of plate LXV, *Montagnes de Porphyre Colonnaire de Jacal*;
- Cofre de Perote image is a direct copy of plate XXXIII, *Coffre de Perotte*; and,
- Organos de Actopan image is similar to Humboldt's Lámina LXIV, *Cúspide de la montaña de los Órganos de Actopan*.

These images also appear in both the original and the abridged versions of the *Vues des Cordillères*.

Similarly, images of archaeological sites of the right side of the title block are copied and edited from other nineteenth-century travel images (figure 48). For example, the *Carta general's* Palenque scene of corbel arches is derived from a lithographic print from John Lloyd Stephens's *Incidents of Travel in Central America, Chiapas and Yucatan* published in 1841 (figure 35, chapter 3, p. 97). Pirámide de Papantla is copied from plate 37 of Carl Nebel's 1836 publication *Voyage pittoresque et archéologique* (figure 32, chapter 3, p. 93), and the Uxmal image is a reproduction of plate 15, *Portion of the La Casa de las Monjas*, from Fredrick Catherwood's 1844, *Views of ancient monuments in Central America, Chiapas and Yucatan* (figure 34, chapter 3, p. 96).

The fourth section of the *Atlas* contains reproductions of two maps from

48 Antonio García Cubas, title block detail (from right side), *Carta general de la República Mexicana*, from *Atlas geográfico* (1858). Courtesy of the Newberry Library, Chicago.

supposed ancient times; both are entitled *Cuadro histórico-geroglífico de la peregrinacion de las tribus Aztecas que poblaron el Valle de México* and distinguished as *número 1* and *número 2* (figures 49 and 50). García Cubas's introduction explains these *Cuadros histórico-geroglífico*: "Fourth [section]. Map with information about the ancient history of Mexico. Origin, peregrinations, settlement, epoch of major splendor and destruction of the first inhabitants of this country. Celebrations and public games. Mexican mythology. Barbary and civilization, manifesting the first in their religion and the second in their works. Civil calendar, century, years and months. Famous wars. Mexican antiquities. Information about chronology."[26]

Cuadro histórico-geroglífico, número 1 is a copy of a map known as the *Mapa Sigüenza*, first published in the eighteenth century by Gemelli in his *Viaje a la Nueva España. Cuadro histórico-geroglífico, número 2*, also a copy, is known as the *Codex Boturini* or *Tira de la peregrinación*. Both images illustrate the migrations of the Aztec from Aztlán to Tenochtitlán using the narrative space formatting of *relaciones geográficas* maps such as the Amoltepec image of figure 12, chapter 2, p. 45. In addition, each *cuadro* is surrounded by the explanatory text by José Ramírez, head of the Museo Nacional, who elaborates on the details of the Aztec's search for a homeland in central Mexico. By inserting these maps, García Cubas repeats their use found in various narratives, including those of Humboldt and Prescott, as well as Doña Margarita's lectures in Bustamante's *Mañanas de la Alameda de México*.

The fifth and final section, *Noticia cronologica de los reyes . . .*, divides

49 Antonio García Cubas, *Cuadro histórico-geroglífico núm. 1*, from *Atlas geográfico* (1858). Courtesy of the Newberry Library, Chicago.

Mexican political history into the phases or periods with associated dates from 669 through 1858. These are Reyes Toltecas, Chichimeca, Aztecas, Dominación español, Regencia, Imperio, Gobierno Provisorio, República Federal, República Central, Dictadura, República Federal, and Dictadura. Names of indigenous kings, Spanish kings and viceroys, and the postindependence government leaders with their respective dates are listed under each division.

Analyzing the contents of the whole atlas, we see that the second section as well as the fourth and fifth contain a prolific amount of statistics, geographic data, historical information, and, of course, cartographic images. The third section, the *Carta general* with its elegant map and elaborate title block, seems to be out of order or misplaced as the middle section or center image of the *Atlas* because it does not document factual data in text format but strictly utilizes visual imagery. Why would García Cubas, a cartographer

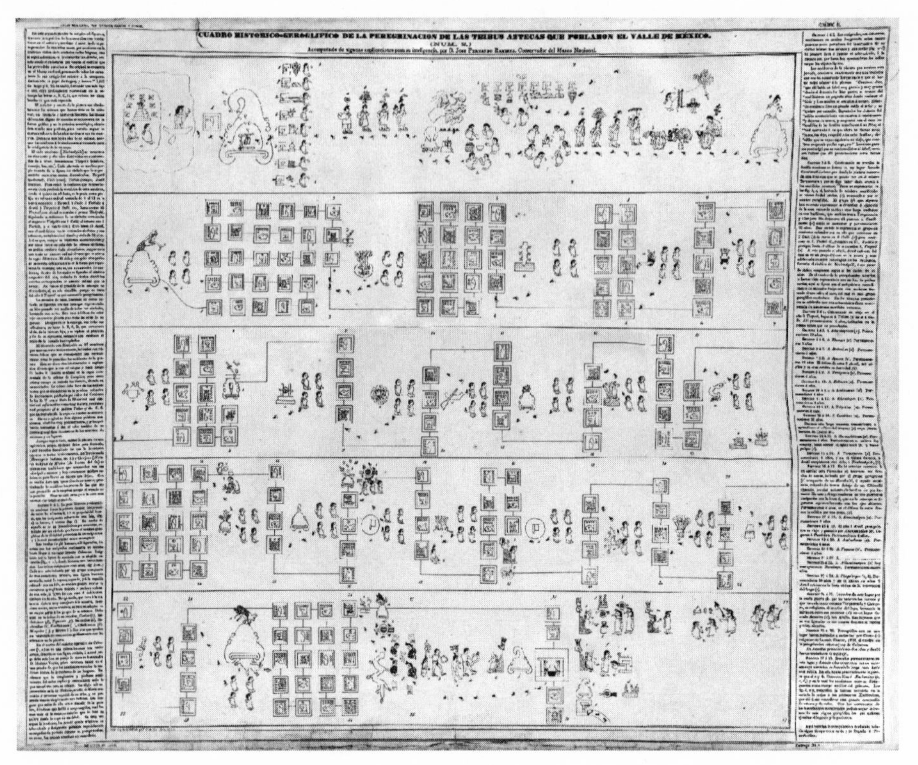

50 Antonio García Cubas, *Cuadro histórico-geroglífico núm.* 2, from *Atlas geográfico* (1858). Geography and Map Reading Room, Library of Congress.

concerned with order and linearity, stop the flow of statistical texts and insert an image-based map as the literal centerpiece of his *Atlas geográfico*? As a geographer, García Cubas would also be concerned with narrative flow; in this way, his placement of the *Carta general* provides focus and direction for the *Atlas* as a whole. Thus, the landscape imagery of the left cartouche with its emphasis on impressive geography broadly visualizes the abstract space of the *Cuadro geográfico* of the first section and the state and territory maps of the second. And the right cartouche, depicting important archaeological sites, points to the *Atlas*'s history-themed fourth and fifth sections, the *Cuadros histórico-geroglífico* and the *Noticia cronologica*. In this way, the title block imagery shifts the visualization of Mexico from space to time. Here, García Cubas links contemporary Mexican territory identified in the *Atlas*'s general and regional maps with the indigenous spaces and historical chronology. At the level of the reading structure of an atlas, the *Carta general* creates a bridge or segue for travel from a review of national and regional territory, as seen in the *Cuadro geográfico* and twenty-nine state and territory maps, to an examination of an emerging history through the ancient maps and information about the political history of Mexico.

Through the 1858 *Atlas geográfico*, García Cubas spatializes the history and geography that specifically emerged in the 1840s and 1850s through the Museo Nacional's catalogues of pre-Columbian objects, through the Academy of San Carlos's new specialization of landscape painting, and through the SMGE's statistical and geographic studies. At the same time, the *Atlas geográfico* parallels the translations of Prescott's *History of the Conquest* published ten years earlier by Vicente García Torres and Ignacio Cumplido. Through the title block images of the *Carta general*, with their visual references to travelers' views—the illustrations of Humboldt, Nebel, Stephens, and Catherwood—García Cubas also links and, at the same time, co-opts the geographies and prehistories conjured by foreigners to narrate Mexico through an exhibition of space and time. Thus, as the nexus of the *Atlas geográfico*, the map and cartouches of the *Carta general* fuse Mexican spaces and places with this emerging Mexican history through nineteenth-century visual strategies of display. Antonio García Cubas envisions Mexico for stationary observers—Mexican and foreign—as a journey through time and space; it is a journey that requires no travel.

The *Carta general*, with its title block, broadly structures the space and time flow of the whole *Atlas geográfico*. It interweaves the distinct sections of Antonio García Cubas's atlas into a comprehensive and panoramic story of Mexico, manifesting a growing interest in visually creating and displaying Mexico.[27] This is to say that through the *Atlas*'s *Cuadro geográfico y estadístico*, state maps, the *Carta general* with its map and title block images, two *Cuadros histórico-geroglífico*, and *Noticia cronologica*, García Cubas employs quintessential strategies of nineteenth-century visual culture: Mexico, derived from travelers' maps and imagery, becomes reproducible, collectible, and exchangeable and visualized as a great exhibition. The mobile and circulating reproductions of Mexican landscape and archaeological sites coalesce into a panoramic display of Mexico as place and history. Through the *Carta general*'s combination of map and cartouches, Mexico is formed in traditional cartographic practices as well as the emerging visual economy of mobile images and visual display. Stated concisely, Mexico is mapped through and into nineteenth-century visual culture strategies as the *Atlas geográfico* announces Mexicans as active consumers *and* producers of visual culture.

AFTER THE *ATLAS GEOGRÁFICO, ESTADÍSTICO É HISTÓRICO*

After the success of the *Carta general* and the *Atlas geográfico*, García Cubas claims, he was offered a position at the Escuela Nacional to teach geographic and topographic drawing. Aware of his educational deficiencies, however, he refused this appointment and continued to teach himself, studying in the early morning in the Alameda.[28] Despite the civil unrest that marked Mexican life, García Cubas actively produced new texts and cartographic works in the following decade. He published his lengthy statistical text, *Memoria para servir á la Carta General de la República Mexicano*, in 1861. García Cubas's work also came to be appreciated by art critics at this time. In February of 1862, a critical overview of the 1862 Exposition in the Academy of San Carlos cites the excellent work of young engravers, noting the work of "el señor García y Cubas, que ejecutó con precisión y paciencia unos signos geográficos y el Estado de Nuevo León" (Mr. García y Cubas, who executed with precision and patience geographic emblems and [a map of] the state of Nuevo León).[29] In March of the same year, he was awarded the copper medal for a print showing geographic symbols.[30]

As García Cubas's work appeared in the academy, Mexico continued to come to terms with its history of Spanish domination through the appropriation of viceregal art into national history. Following its mandate as an educational institution, the Academy of San Carlos addressed the issue of how to understand the art produced in New Spain between 1524 and 1810. As discussed previously, the Museo Nacional collected, recovered, and synthesized preconquest material remains; in this way pre-Columbian art could be silenced and narrated through comparisons to Egyptian art. In contrast, in the 1830s and 1840s, the art of New Spain, primarily religious in content and formally nonacademic, was viewed as marking the condition of Spanish domination, from which Mexicans were in the process of freeing themselves but had yet to come to terms fully with at mid-century. As Raymond Hernández-Durán, an art historian, writes, "The conflicted reaction to the Colonial material was thus born of the physical and symbolic confrontation between the material remains and the old Colonial system and a new political arena which was ideologically at odds with the original context that motivated the production of those works of art."[31] In this context, immediate acceptance of this art could signal identification with Spanish cultural supremacy.

Viceregal art, as a delimited historical and cultural category, was demarcated in the late 1850s and 1860s by the Academy of San Carlos, under the direction (and with the support of the government) of José Bernardo Couto, director of the institution between 1844 and 1862. By 1862, Couto created an exhibition space for displaying viceregal art identified as from the "Old Mexican School" and, thus, converted from the art of the Spanish domination to an early phase of history of Mexican national art.[32] This identification as art of the old Mexican School initiated a canon of Mexican art, it nationalized viceregal art, and it reinforced an emerging tripartite historical narrative (pre-European contact, European contact, and the blended Mexican culture).[33] In this way, like the SMGE, the Academy of San Carlos collected, aggregated, synthesized, and displayed another part of an emerging display of Mexico. These institutions' activities overlapped and reinforced each other's activities. For example, an article by Rafael Lucio on viceregal painting first appeared in 1863 in the *Boletín de la Sociedad Mexicana Geográfica y Estadística*.[34]

In 1861, the same year García Cubas's *Memoria* was published, the French initiated a military and political intervention into Mexico, ostensibly to recover debt payments owed by the Mexican government. Napoleon III initially sent troops to Mexico and then established a well-financed scientific mission in 1864, the *Commission scientifique du Mexique*. Premised on the belief that Mexican science was deficient, the *Commission scientifique*, sometimes working with the SMGE, attempted to produce an "accurate" map of Mexico, calling for more precise topography using the method of triangulation and possibly photographic survey. The teams of scientists also published information between 1868 and 1915 on archaeology, botany, geology, and zoology.[35]

Perhaps stimulated by the work of the *Commission scientifique*, García

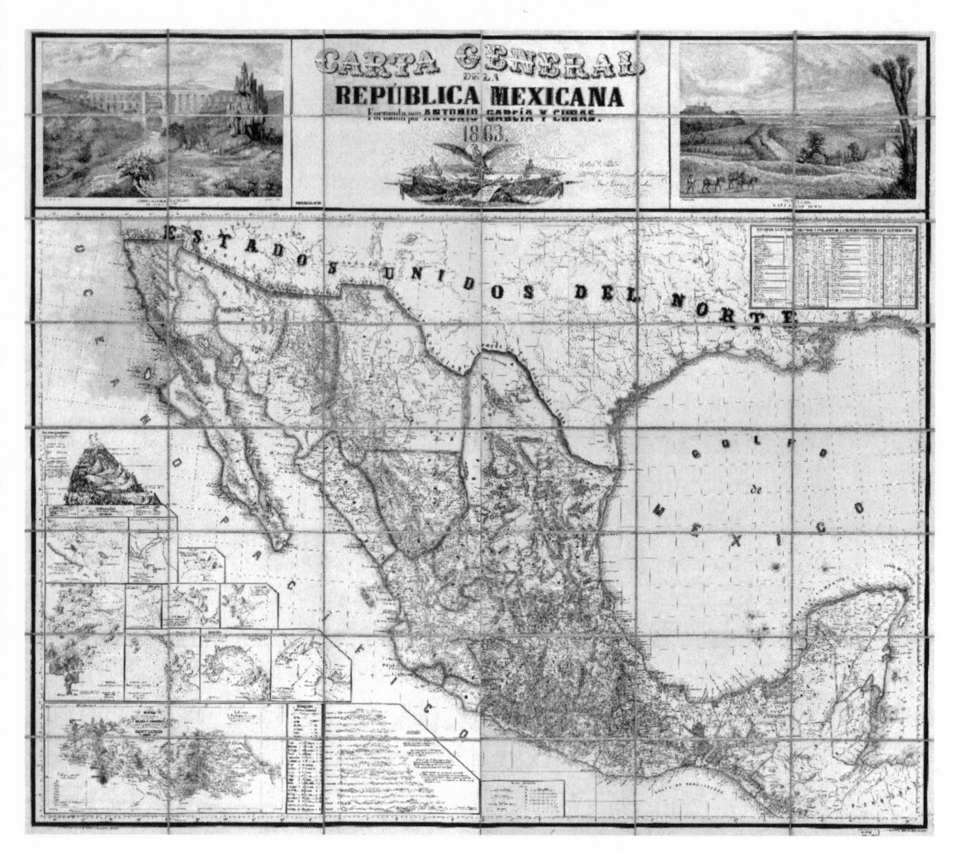

51 Antonio García Cubas, *Carta general de la República Mexicana formado por Antonio García y Cubas* (1863). Geography and Map Reading Room, Library of Congress.

Cubas's *Carta general de la República Mexicana* of 1863 attempted to correct some of the defects of the 1857 *Carta general* (figure 51).[36] Following the format of the *Carta general* of 1858, a central map was associated with cartouches in the lower left-hand corner and a title block with two images. The left image of the title block, titled *Acueducto de la hacienda de Matlala, Valle de Izucar, Estado de Puebla*, depicts a landscape scene with cactus plants, scrub brush, and four deer in the foreground. An aqueduct cuts across the scene and leads the eye to distant hills and a snow-covered mountain. E. Landesio, a teacher of landscape at the National Academy, is identified as the painter of the image and H. Iriarte as its lithographer. The image to

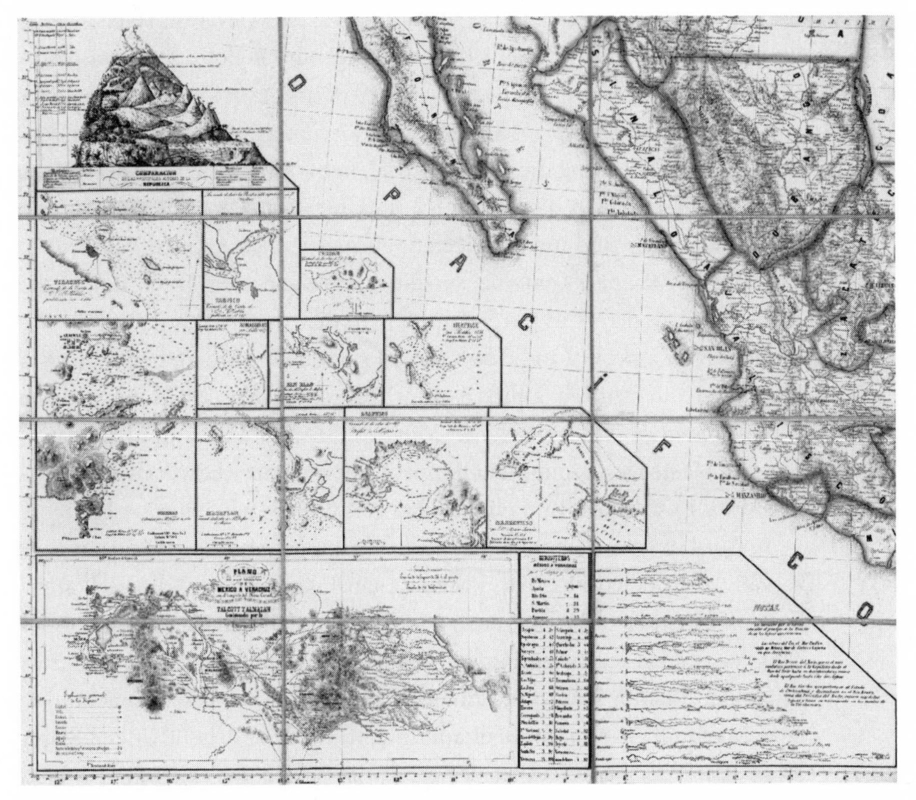

52 Antonio García Cubas, detail, *Carta general de la República Mexicana formado por Antonio García y Cubas* (1863).

the right of the title block, entitled *Valle de México tomado desde las lomas del Molino del Rey* (Valley of Mexico from the hills of Molino del Rey) and signed only by H. Iriarte, depicts a view of Valley of Mexico with the Chapultepec Palace to the left, Mexico City looming in the distance, and the volcanoes Iztaccíhuatl and Popocatépetl to the left. In the foreground, sheep and goat herders, muleteers and individuals walk toward the city. These images reference neither the ancient history nor the inventory of travel images as in the *Carta general* of 1858; rather, they seem to depict contemporary scenes of Puebla and the Valley of Mexico.

The interest in these landscape scenes, however, is explained through the cartouches located in the lower left of the map (figure 52). Stacked in rows, the first chart shows specific vegetation at various elevation levels, which is similar to Humboldt's depiction of the geography of equatorial plants (figure 19, chapter 3, p. 71). The next groups of images are maps of the ports, such as Veracruz, Tampico, Mazatlán, and Acapulco. In the lower section, we see a graph showing the comparative length of various rivers of Mexico. Next to this is a map displaying the area between Veracruz and Mexico City, indicating roads and a projected railroad service, entitled *Plano de los Caminos de México a Veracruz con el trayecto del Ferro Carril* (Map of the roads from Mexico [City] to Veracruz with the route of the Railroad). The title refers to a project to bring railroad service to Mexico, begun in 1837 and slowly and fitfully completed over the next five decades. In this way, the *Carta general de la República Mexicana* of 1863 represents an idyllic landscape of Mexico at an imminent moment when rail service will change communications significantly. Political changes were also arriving.

In April of 1864, Napoleon III, with the support of certain Mexican conservatives, installed Ferdinand Maximilian Joseph, prince imperial, archduke of Austria, prince royal of Hungary and Bohemia, as emperor of Mexico; and his wife, Charlotte of Belgium, known as Carlota, as empress of Mexico. Benito Juárez and his Reforma followers refused to acknowledge Maximilian's appointment and fought, with the help of United States arms, the French troops that protected the prince. Maximilian's ill-fated reign lasted only three years, when he was executed by order of Benito Juárez in June of 1867.

Maximilian undertook his duties with great zeal, attempting to bring progressive European ideas to Mexico, some of which actually reflected Reforma ideas. He had very focused ideas about the importance of visualizing space. In an effort to transform Mexico City, he envisioned the Calzada del Emperador, a boulevard that connected his Chapultepec Palace to the downtown area. The equestrian statue of Carlos IV was to be moved, again, to the entrance of the park that surrounded the palace.[37]

Further, understanding the need to legitimize his reign, Maximilian crafted an imperial genealogy that included the indigenous princes of Anahuac and heroes of the independence such as Hidalgo and Morelos. Thus, as the historian Robert Duncan points out, "Far from either ending or challenging the tradition of celebrating independence, Maximilian tried instead to expropriate it from the republic."[38] Moving away from Prescott's version of Mexican history that emphasized Cortés and the Spanish conquest, Maximilian endorsed a cult of heroes of the independence, even calling Hidalgo's Grito de Dolores "the anniversary of our nationality."[39] Maximilian also aided this formation of nationalism through undertaking various visual projects. He proposed a monument to independence that included bronze statues of Hidalgo, Iturbide, Morelos, and Guerrero crowned with a representation of the nation and with an inscription commemorating Maximilian as also a liberator of Mexico.[40]

EXPANDING VISUALITY: PHOTOGRAPHY

The 1860s and 1870s were marked by a significant expansion of photography, also precipitated in part by the French intervention. Photographs became important sources for García Cubas's work of the next decades. Photographic studios increased from ten in 1859 to twenty-two in 1866. In 1862, there were approximately forty-four Mexican-born and European-born professional photographers working in Mexico City and another twenty in the principal cities.[41] Along with images of landscapes, cityscapes, and ruins, *tarjetas de visita* proliferated (as discussed in chapter 4). Following the *cartes de visite* format found in France and the United States, tarjetas, small portrait photographs of elite and nonelite personages, proliferated. Maximilian, recognizing the importance of public image and propaganda, allowed the production of photographic images of himself, Carlota, and members of their court.[42] One of the more important foreign photographers during this period was François Aubert, a French photographer, who arrived in Mexico in 1864 and unofficially chronicled the life and times of the Emperor's brief reign. Aubert produced tarjetas de visita images of Maximilian and Carlota individually and together, often in full royal dress. Aubert's production also included studio photographs of *tipos mexicanos* (Mexican social types), among them elite and nonelite individuals and couples that bring *cuadro de casta* images to mind. His photograph of *Tortilleras* is a striking image (figure 53). Photographing a type whose origins date back to Linati and Nebel, Aubert photographed the tortilla makers in a staged tarjeta de visita studio format, depicting one woman grinding the maize while the second forms a tortilla over her cooking stove, and two children sit next to them. Perhaps reflecting an allusion to Indian character, a picture of the

53 François Aubert, *Tortilleras* (1860s). Courtesy Royal Military Museum, Brussels.

Virgin of Guadalupe is placed on an easel in the background. These images continued the inventory of Mexican types and sites developed by foreign travelers in contemporary popular formats and downplayed in publications such as *Los mexicanos pintados por sí mismos*. Exceptions to these standard photographic forms were Aubert's dramatic images associated with the execution of Maximilian, including an image of the embalmed prince in his casket. Aubert's images confirmed the monarch's demise for the Mexican public and added images to a growing catalogue of images.

The Sociedad Cruces y Campa, a partnership between the two artist-photographers Antíoco Cruces and Luis G. Campa, was also very active during the 1860s and into the 1870s.[43] The photographers, both of whom had some training at the Academy of San Carlos, produced high-quality tarjeta de visita images for commercial purposes. They published a book containing images of political figures of Mexico, pointedly titled *Galería de personas que han ejercido el mando supremo de México, con título legal o por medio de la usurpación* (Gallery of Persons Who Have Exercised Supreme Control in Mexico, through Legal Title or by Means of Usurpation). Cruces y Campa's

54 Sociedad Cruces y Campa, *Trajineros* (1860s). Courtesy of the Latin American Library, Tulane University.

images of Mexican types, published as *Retratos fotográficos de tipos mexicanos*, were particularly noteworthy.[44] Like Aubert, the photographers did not go into the streets of the city or countryside to photograph their subjects; instead, for their *tipos mexicanos*, they staged scenes to control the composition, light, and the sitters' position. For example, figure 54, *Trajineros* (Canoe guides), depicts a staged scene in which an Indian woman is seated in a canoe, which is filled with vegetables and being pushed by two standing men; in another photograph, the same woman appears in the canoe without her companions. Also coming to life from the pages of Linati and Nebel publications are Cruces y Campa's photos of numerous other characters from daily life in Mexico. Like the lithographic images of 1840s, these photographs portray imagery that travelers would have seen in Mexico City and its environs but which were re-created in a studio.[45]

More importantly, instead of being part of limited editions of books, single photographs were readily available to private collectors through the numerous photo studios of Mexico City. Cruces y Campa's work won a bronze medal at the 1876 Philadelphia Exposition, indicating the high quality of their work as well as the proliferation of images of Mexico in the international exchange of commercial photographs. The importance of photographic technology during this and the subsequent decade cannot be underestimated. As Rosa Casanova, an art historian, points out, "The circulation of images during the Imperial period signifies a qualitative leap in the way Mexicans accustomed themselves to thinking about and using photo images."[46] Through this circulation of photographs, Mexicans saw themselves and they also saw other parts of the world. As one 1865 Mexican commentator on stereoscopic images asks, "¿Quién no se decidirá a emprender tan cómodo viaje por el orbe, para ver cuanto éste encierra, cuando el trayecto de ida y vuelta no cuesta más que un real para los adultos y medio real para los niños?" (Who will not decide to embark upon such a comfortable trip around the world, to see how much this includes, when the roundtrip journey does not cost more than a real for the adults and half a real for the children?)[47]

Further, Deborah Dorotinsky, a historian, argues that tarjeta de visita images established visual archetypes of elite Mexicans, consolidating membership in the upper class.[48] Obviously, they also certified the existence of tipos mexicanos. At the same time, through this medium, elite women were depicted elegantly dressed in parlorlike domestic settings, often sitting or standing next to their husbands. In contrast, Indian and lower-class women would be located in the streets of Mexico as vendors selling food and other goods. Photographs also functioned as a mechanism for vigilance over the nonelite population. Photo albums of known criminals were produced, and such collections were also used to locate particular kinds of women: *mujeres públicas* (prostitutes). The use of photographs to identify criminals and prostitutes was popularized in Europe in the 1850s. Under Maximilian, *El registro de mujeres públicas conforme al reglamento expedido por S.M. el emperador* (Registry of Public Women, Conforming to the Regulation Issued by His Majesty, the Emperor) was introduced. It contains images of supposed prostitutes, associated with personal data such as name, age, address, and place of origin. The register also identifies whether the woman worked independently or for a madam, and categorizes the individual as either a first-, second-, or third-class prostitute, seemingly determined by her skin color; generally, mestizo and Indian women constituted the second and third class.[49] Curiously, the images of these women are not "mug shots," as might be expected. These mujeres públicas, who were forced to register with the government, consciously presented studio images of themselves, set in elegantly staged parlors. Here, through photography, the women presented themselves with self-respect within the confines of state surveillance. *El registro de mujeres públicas* manifests a kind of mapping of spaces for women, a theme discussed more fully later in this chapter.

AFTER MAXIMILIAN: *THE REPUBLIC OF MEXICO IN 1876*

Perhaps spurred by Maximilian's interest in display of the nation, photography would grow and come to identify people and spaces of the nation. It was possible to view elite and nonelite spaces and people; it was possible to view political figures, Maximilian and Carlota; and it was possible to view historic moments, such as Maximilian in death. Finally, a subset of all of this seeing is surveillance practices: people could be watched and observed through photography as well. Richard Sinkin, a historian, claims that as part of their legacy, the French showed Mexicans that they were Mexicans, that is, they formed a social, political, and cultural community.[50] Perhaps overstated, the French intervention may better be understood to have expanded the growth of visual culture of earlier decades and intensified its production and consumption. As a result, the exploration of *mexicanidad* (Mexicanness) by Mexicans is a theme of their visual culture after the demise of Maximilian, as Benito Juárez resumed leadership and again attempted to stabilize the Republic until his death in 1871.[51] Thus, in the late 1860s and 1870s, there was a surge in nationalist themes and topics in various publications and imagery. Under Juárez, the Academy of San Carlos became the Escuela Nacional de Bellas Artes in 1867 and institutionalized ways of looking at *mexicanidad*. Stacie Widdifield's research on the E scuela at this time shows that the call for national history painting resulted in works such as José Obregon's *Discovery of Pulque*, and Petronilo Monroy's *Constitución de 1857*, which visually narrate Mexico through historical recountings as well as allegorical images.[52]

Particularly noteworthy at this time was the expansion of landscape painting, especially that of José María Velasco. Born in 1840 in a small rural village, Velasco moved to Mexico City in 1846 and began to study at the Academy of San Carlos in 1858. Velasco's drawing talent was considerable and, under the tutelage of Landesio, he became the premiere painter of Mexico's landscape.[53] Velasco's landscapes depict the vastness and openness of a land that dwarfs people and towns. In his painting, the Mexican landscape becomes heroic. Widdifield further points out that landscape painting provided what history painting could not.[54] This is to say that history painting was based on the artist's active imagining of a particular nationalist historical moment or event. In contrast, landscape was not understood as invented by an artist; rather, it actively promoted the illusion of reproducing an eyewitness moment of a specific physical place within the nation that, in turn, could be related to a growing international interest in landscape imagery. Through this expansion of landscape painting, the Escuela de Bellas Artes continued to develop the mapping practices of its predecessor institution.

And while national self-display is clear in García Cubas's *Atlas geográfico*, his works after the French intervention intensified self-reflection through education in international contexts, specifically the interest in geographic encyclopedias.[55] For example, in 1869 he published a world geography text of 131 brief lessons, fully titled *Curso elemental de geografía universal dispuesto con arreglo á un Nuevo método que facilite su enseñanza en los establecimientos de instrucción de la República* (Elementary Course in Universal Geography Arranged by a New Method that Facilitates Its Teaching in Establishments of Instruction of the Republic). With notes for the instructor, García Cubas begins his lectures with a history that summarizes the contributions of cosmographers and geographers going back to Ptolemy, then proceeds to provide introductory material on geometry, astronomy, and the principles of physical geography.

The next section contains "Geografía descriptiva" in which García Cubas defines basic terms such as *mapa* or *carta*, the "representation on a plane surface, of a country or terrain by means of conventional signs that represent mountains, rivers, populations, etc."[56] Continuing, he explains the differences among regional, general, and world maps. He also identifies thematic maps, amplifying how they provide political, administrative, orographic, communication, hydrographic, maritime, military, and scientific information, notations that curiously foreshadow the structure of his next major project, the *Atlas pintoresco é histórico de los Estados Unidos Mexicanos por Antonio García Cubas*, of 1885. These explanations are followed by geographic descriptions of Europe, Asia, Africa, Oceania, and America, including the United States, Mexico, and South America.

In this universal geography, Mexico takes up about 30 percent of the text's description, with South America at 8 percent, Europe 9 percent, Asia 6 percent, Africa 3 percent, and Oceania and the United States 1 percent. García Cubas openly articulates the nationalist purposes of the text, writing that while other geography texts are available, they are written for foreign nations, and are not appropriate for the teaching of geography in Mexico.[57] Explaining his reasons for writing the book, he writes that it comes from a "conviction and the desire to contribute to the progress of public instruction, foundation of our happy future, and the wish to make myself useful to my compatriots."[58] In this way, García Cubas clearly announces his educational focus, one of the themes from the decades of the Reforma and after the French intervention.

The 1870s saw a growing emphasis on the cult and personality of a heroic pantheon—heroes of independence, Indian heroes, and heroes of the Reforma—an expansion of a cult of heroes that had begun in the 1840s. More public monuments depicting these heroes were proposed, potentially in an effort to provide moral and civic lessons, and usurp the ubiquitous presence of churches.[59] The name of the Calzada del Emperador was changed to the Avenida de la Reforma. And in 1870, Vicente Riva Palacio and Manuel Payno, historians, published *El Libro Rojo 1520–1867*, an illustrated hagiography of heroes of the Independence.[60] This was followed by Manuel Rivera Cambas, a mining engineer and self-instructed historian, writing *Los gobernantes de México* (1872–73), which contains biographies of Mexican leaders from the first viceroy to Benito Juárez, with lithographic images by Murguía.[61]

From these evolving visual and social contexts, Antonio García Cubas published *The Republic of Mexico in 1876*.[62] With an English text translated by George F. Henderson, the book is divided into three sections: the "Political Part," "Historical Part," and "Ethnographical and Descriptive Part."[63] The publication's eight chromolithographs, each divided into three registers for a total of twenty-four vignettes, indicate clear awareness of previous surveys of tipos mexicanos and set out an emphatic agenda to rectify the misrepresentation of Mexico and Mexicans by foreigners. In the introduction to the book, García Cubas writes:

> This book has been written with the view of removing the wrong impressions that may have been left on the minds of the readers of those works which, with evil intent or with the desire of acquiring notoriety

as novelists, have been composed and published by different foreigners in regard to the Mexican nation. The impressions received during the rapid excursion of pure amusement, without making any longer stay in the various towns, than the time required to repack their valise and continue on a journey of useless results; the isolated facts that are observed in every society in contradiction to general rules, and a disposition to judge events without proper examination and careful study, are not sufficient to obtain a complete knowledge of any class of people, and much less to authorize such impressions through the medium of the press. The works of similar writers, in misleading the conceptions of the public, conspire against the real utility of general information, as their ideas (in direct opposite to those given to the world by such profound observers as Humboldt, Burkart, Sartorius, Jourdonet,) cannot convey any instruction to our intelligence, but only dispose the mind to receive the impressions produced by the novel.[64]

Thus, García Cubas decries the fact that travel writing had conjured a fictitious image of Mexico and Mexicans in the public mind. In a subsequent section of this text, he specifically assails the work of Louis Figuier, a French science writer popular at the time.

> Mr. Figuier, in his works of "The human races," reproduces an engraving from another European publication, which represents the type of a woman of the people, and not that of one of the principal ladies of Mexico, as he supposes. This type is taken from a correct photograph by Mr. Jules Michaud. . . . On comparing Mr. Michaud's photograph with the engraving shown in the work referred to, I have formed the conviction that there has been bad faith in its reproduction. In the photograph, which has been transferred in the number referred to, a woman is seen of an agreeable and lovely figure, and not with the characteristics of the negro race, as she appears in the adulterated plate of Mr. Figuier's work.[65]

Here, García Cubas explicates his full awareness of the potential of appropriated and circulating images to be manipulated in order to misrepresent Mexico and her people.

García Cubas lays out another larger issue when he concludes his criticism of Figuier's writing with the comment "Moreover, this gentleman, who, doubtless, has produced many recommendable works, would have done well in abandoning the routine of classifying the Mexican nation among the red skins."[66] To rectify this misrepresentation and misclassification of Mexico and Mexicans by foreigners, he lays out his plan for a new kind of understanding of Mexico, stating that

> it is requisite to make known those vital elements and fountains of wealth that yet remain unexplored, and with this purpose, the present work only leads the way to a series of publications destined for the information of those abroad and written by Mexicans devoted to the prosperity of the Republic, and which will doubtless contribute to the development of so wished-for a result.[67]

García Cubas develops this new program for describing Mexico by organizing his narrative not around personal travel observation or opinions but in the language of scientific concepts and formats. Words such as *data*, *population*, and *statistics* float through his text: "The more recent data, as a natural result of the advancement of the people and of more reliable statistics."[68] Returning to a Humboldtian format, the fourth page of the "Political Part" of *The Republic of Mexico in 1876*, entitled "Political Division and Population," begins with a table showing the population of each state in the Republic. Along with data, statistics, and charts, García Cubas includes a thematic map entitled *Carta etnográfica de México*, showing the geographic location of twenty-nine ethnic groups identified linguistically (figure 55).[69] While the faint outlines of the Mexican states of the *Carta general* are apparent, the map—produced by the lithographer Salazar—divides Mexico up into the territory associated with Indian groups. For example, we see the Maya occupying the Yucatán peninsula and parts of Veracruz; the Othomi in central Mexico; and the Mixteca and Zapotec in southern Mexico.

In addition, the text is illustrated by eight chromolithographs, which García Cubas employs to visually convey the cartographic, statistical, and descriptive data on the Mexican population. Introducing the populations of Mexico, he explains, "The differences of dress, customs and language in the Mexican Republic, make known the heterogeneousness of its population, which may be divided into three principal groups: viz the white race, and the more direct descendants of the Spaniards, the mixed race and the Indian race."[70] He goes on to explain the characteristics and economic role of each group. "The habits and customs of the individuals who compose the first

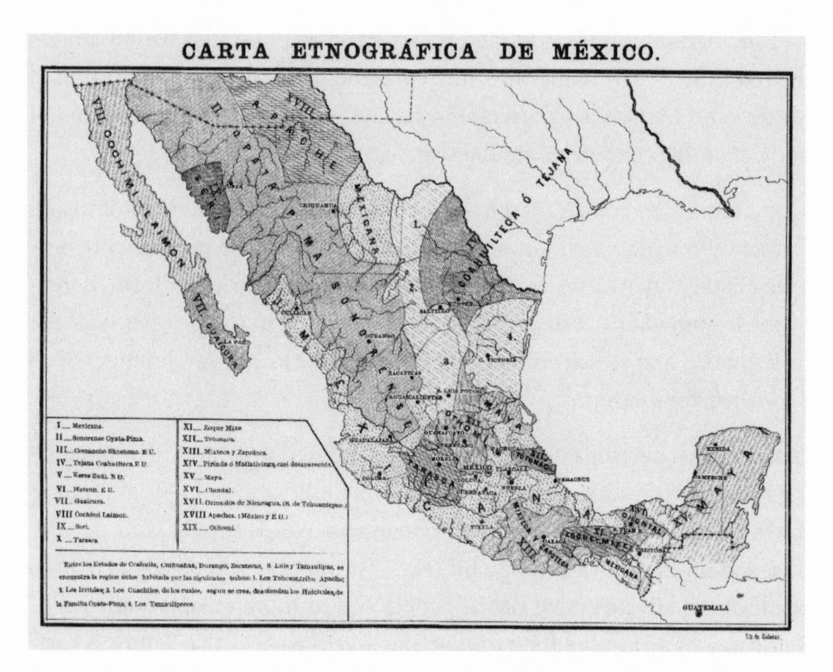

55 Antonio García Cubas, *Carta etnográfica de México*, from *The Republic of Mexico in 1876*. Author's collection.

division [whites], conform in general to European civilization, and particularly to the fashions of the French with reminiscences of the Spaniards."[71] Concentrated in urban areas and speaking Spanish and some French, English, Italian, or German, whites are associated with various professions. This segment of the population is illustrated with four images showing leisure activities of this group, including ladies and gentlemen going to church, a coach trip, a boat ride in the Papaloapan springs, and an elegant dance (figure 56, and the upper register of figure 57).[72]

Describing the regional diversity, he explains that mixed-race people constitute the working population of Mexico, speak Spanish with various dialects and accents derived from Indian languages, and are "Sagacious, intelligent and with a special gift for imitation, this race is remarkable for the taste and perfection of its manufactures. . . ."[73] The next sections of this chapter review the social and economic activities of this group, emphasizing the people's potential for economic expansion. This mixed race is illus-

56 Plate 1 from *The Republic of Mexico in 1876*. Author's collection.

57 Plate 2 from *The Republic of Mexico in 1876*. Author's collection.

trated with plates showing servants in *poblana* dress style with a male, identified as a "Guard with bullion from Real de Monte," followed by traditional dancers of the *jarabe*, and once again, a tortillera (figure 57, middle and lower registers).

Over half of this 129-page book, and more than half of the vignettes, are devoted to the Indian race under the section heading of "Ethnographical and Descriptive Part." García Cubas's initial indignation over Figuier's misclassification is elaborated upon: "There is much to be said in regard to the indigenous race, numerous and extended as it is throughout the territory of the Mexican Republic: its habits and inveterate customs, diametrically opposed to those of the white and mixed races."[74] In his overview, he writes that the indigenous race exhibits tendencies towards "idolatry and blind fanaticism" and are "imbued in their ancient habits, they preserve their customs, dress and dialect."[75] Their characteristics include "misconfidence, dissimulation, cunning, obstinacy and inclination for spirituous drinks, and other general characteristics of the Indian, although he is brave, daring and long suffering." Moderating his generally negative description, however, García Cubas continues, "Many circumstances show that the degradation of the Indian race is not derived from their original nature, but from their customs and mode of living."[76] And, in general, he concludes, "If we make a careful examination of the state of the population in different parts of the Republic, we shall find the fact confirmed and our assertions corroborated, when stating that the indigenous race is gradually approaching towards its complete extinction."[77] García Cubas, then, exhibits a cultural racism that is spreading across contemporary texts and that is focused on the Indian population as posing a problem for the prosperity of Mexico. For García Cubas, however, hope for this situation is found in statistics.

Eighteen of *The Republic of Mexico in 1876*'s images depict Indian races divided by their language group or ethnicity. For example, the third register of plate 3 illustrates the "Mexican natives of Santa Anita and Ixtacalco to the South of the capital" (figure 58). The image, showing a woman seated in a canoe with two males standing, is a colorized version of Cruces y Campa's tarjeta de visita of *Trajineros*, a scene staged in a studio (lower register of figure 58, and figure 54). Background vegetation and the outlines of volcanoes have been added to make the scene appear more authentic. Other vignettes from this plate depict the indigenous groups from Veracruz, each

58 Plate 3 from *The Republic of Mexico in 1876*. Author's collection.

image displaying details of ethnic costume and customs. García Cubas's lithographic illustrations recall the travel images of Nebel and Linati as well as tipo mexicano lithographs of *Los mexicanos pintados por sí mismos*. The reappearance of the poblana, tortillera, and other figures marks a continuing attempt to relocate these types into a national narrative about the nation's potential for economic growth, with population as a contributing element.

García Cubas concludes this work with a comparison of data from the censuses from 1810 and 1875. Claiming that the white and mixed-race groups have increased, he concludes, "*the Indian race has decreased and continues on the road of its decline*, unless civilization and other unforeseen causes should modify these lamentable results, converting them into others of a more favorable character. That manifest destiny is successively observed from North to South [of Mexico]."[78] This statement allows García Cubas to claim the diminishment of what was perceived to be a significant impediment for the continuing prosperity of Mexico—the Indian population. At the same time, the reference to the term *manifest destiny*, coming from mid-century expansionist rhetoric of the United States (perhaps the term assigned by the translator), nevertheless articulates a goal of *The Republic of Mexico in 1876*: it manifests a hoped for-destiny for Mexico. Consequently, by the third quarter of the nineteenth century, García Cubas's publication constructed a comprehensive archive of specimen citizens marked by their dress and habits derived not from colonial sources but in response to travel literature and its associated illustrations. Through *The Republic of Mexico in 1876*, the national characters of earlier travel literature, and to a certain extent tipos mexicanos, are subsumed into a discourse on demographics and economic projection.

The Republic of Mexico in 1876 demonstrates that Mexico can be mapped as more than a landmass articulated in the graticule of longitude and latitude. García Cubas's work employs a process of erasure and fabrication: the sign system of dress and costume associated with travel images of Mexican national characters was renarrated into a comprehensive nationalist story about specimen citizens, their geographic location as well as their contribution to economic and social prosperity and a hoped-for manifest destiny. This erasure of images of the viceregal subject associated with travel literature and its illustration was a critical step in Mexico's evolving nationalist narrative and reflects a reevaluation of race and ethnicity. Further, as legal designation of race disappeared, cultural racism appeared. Physiological distinctions were replaced by strong cultural inflections embedded in wider structures of domination tied to national sentiment.[79] Race was a key to the fundamental change from romantic nationalism of the earlier nineteenth century to the state-oriented nationalism of these decades.[80]

The next year, García Cubas collaborated on a project with Casmiro Castro fully titled *Álbum del ferrocarril Mexicano* (*Album of the Mexican Railway*. Its extended text in both Spanish and English, included with a series of images, describes the landscape along the line illustrated by twenty-four print images. The *Álbum* includes a map, *Plano orográfico de la zona recorrida [*covered*] por el ferro-carril mexicano de Veracruz a México*, and associated cartouche showing a profile of the railway by García Cubas, depicting the path of the railroad from Veracruz to Mexico City. The map is a copy of the *Plano de los Caminos de México a Veracruz con el trayecto del Ferro Carril* found in the cartouches of the 1863 *Carta general de la República Mexicana* with color and greater topographic detail added (figure 52). Recalling the prose of Prescott and landscape paintings of Velasco, García Cubas's text is full of exuberant visual narrative.

> In advancing along the road, the vegetation displays itself in all the loveliness of tropical life: gigantic fig-trees, arborescent ferns, elegant mimosas, tall bamboos and flowery vines, that wind themselves around the gnarled trunks and outspreading branches of lofty trees, all combine to make these woods impenetrable. Orquideas and other parasites, each more beautiful than the other, in their rich variety, completely cover the trees of many years standing, and replace their primitive foliage by their floating and florid clusters, enlivened by the most brilliant colors.[81]

Casmiro Castro's images also emphasize the dramatic and diverse terrain, always illustrating either railroad tracks or a train cutting through the scene. The people remain inconsequential, overwhelmed by the landscape as well as the train, always shown leading to the next part of the journey. García Cubas concludes the text by ringing his hope for the railway: "The Mexican Railway, which has been built by dint of the tenacious struggle with natural difficulties and at the cost of immense sacrifices on the part of the nation and the company, now constitutes the mightiest artery of our social being, from whence others will branch off, which being ramified throughout the whole of the Republic, will diffuse the fruitful seed of national prosperity."[82] The *Álbum del ferrocarril Mexicano*, like *The Republic of Mexico in 1867*, highlights Mexico's potential for material progress.[83]

In sum, the *Atlas geográfico, estadístico é histórico de la República Mexicana* was García Cubas's magnum opus of the early part of his career. It coalesced distinct earlier mappings of Mexico, displaying Mexico for Mexicans. As he matured and continued to educate himself, García Cubas shifted his focus and considerable energy from holistic panoramas of Mexico to developing more educational and nationalist works, such as the *Curso elemental de geografía universal*, *The Republic of Mexico in 1876*, and the *Álbum del ferrocarril Mexicano*. In the next decades, the cartographer-geographer returned to larger-format works.

We cannot leave this early work by García Cubas without addressing its significant silences. Cartographic silences are absences that are as important as presences in maps and are "active performances in terms of their social and political impact and their effect on consciousness."[84] García Cubas's silencing of contemporary indigenous viewpoints is apparent in his cartography. In the *Atlas geográfico*, although he refers to Indians in demographic statistics, García Cubas visualizes only dead Indians through the *Carta general*'s title block images of ancient ruins and unoccupied, open land and the assimilation of the two *Cuadros histórico-geroglífico*. In *The Republic of Mexico in 1876*, framed through the supposed scientific perspectives of ethnography and demography, Indian groups are incorporated into the nation through the thematic map as well as their identification as one of the three groups or races of Mexico.

While scholars have studied the absence of indigenous groups in Mexican cartographic history, another silenced group has yet to be considered: women and their distinctive spaces are not mapped in García Cubas's cartographic or geographic images of Mexico. Women do not appear as a separate category in any of his demographic statistics, and they are subsumed into the racial and ethnographic descriptions of *Mexico in 1876*. But women are mapped through other practices, such as those of the *El Semanario de las Señoritas Mejicanas*, discussed in the previous chapter, which intended to locate the private places and roles for women in the Mexican republic, and of the *El registro de mujeres públicas*. This need to certify the role of women was amplified in the 1860s and 1870s.

Alberto Bribiesca's 1879 painting, *Educación moral. Una madre conduce*

59 Alberto Bribiesca, *Educación moral. Una madre conduce a su hija a socorrer a un menesteroso* (1879). Museo de Querétaro, INAH. Reproduction authorized by INAH.

a su hija a socorrer a un menesteroso (Moral Education. A Mother Guides her Daughter to Aid a Beggar), presents a mapping of private spaces of the previous decades (figure 59). Here, with her left hand gently touching her daughter's head and her right hand gesturing toward a poor man, an elite mother gently guides the young girl to share her wealth with this man who extends his hat in supplication. The beggar's position in the doorway indicates that he comes from the exterior, public world; while the mother and daughter are firmly located in the bounded, interior space of a home, a light-colored curtain marking the separation between the two spaces. Their moral action, assisting those in need, improves not just the mendicant but also the nation. In depicting this moralizing scene, Bribiesca charts an important role for elite and middle-class women that was resolutely promoted in the second half of the century: domesticity. Domesticity included management of the house as well as education of the young.

Such education goes beyond moral acts, however. A map of Mexico hangs prominently on the back wall along with the books stacked on the desk, likely indicating that along with having lessons in reading, writing, and math, young citizens must also learn geography.[85] This map may also be read in another way: its presence insists that domestic space and political space are both part of the ongoing construction of nation. In this way, women's presence in domestic spaces rationalizes their seeming absence in cartographic spaces.

6

TRAVELING FROM NEW SPAIN TO MEXICO—1880–1911

PORFIRIATO SPACES AND HISTORIES

Maximilian's execution marked the rout of the French as well as the defeat of Mexican conservative groups and the triumph of the liberal agenda of reform. Between 1867 and 1876, however, the restored Republic—the governments of Benito Juárez and Sebastián Lerdo de Tejado—were unable to turn the liberal victory into sustained political and economic stability, much less a consolidated nation. As a result, negative images of Mexico's political and economic status persisted in the minds of foreign government officials as well as investors through the late 1870s.[1] These circumstances became an opportunity for José de la Cruz Porfirio Díaz Mori to usurp political power and initiate an era between 1876 and 1911 often identified as the Porfiriato.[2]

Ultimately achieving the rank of general, Porfirio Díaz climbed to power through his military acumen in fighting Santa Anna, supporting Juárez in the Guerra de Reforma, and battling against the French intervention. As the leader of Mexico for over thirty years, Díaz is a historical figure who has drawn both admiration and disdain in scholarly as well as popular writings.[3] On the one hand, writers associate his authoritarian government with the modernization of Mexico, referring to the expansion of industrialization, improvement of foreign diplomatic relations, the immersion of Mexico into international capitalism, and further expansion of print and other mass media.[4] On the other hand, Díaz often accomplished such modernization through unscrupulous, illegal, and sometimes violent means, leading to the Revolution of 1911, and his demise and exile. Drawing on his strong personality and strategist skills, Díaz's dictum "Order and Progress" asserted definitive responses to Bustamante's questions: "¿Quienes somos? ¿de donde venimos? ¿para donde caminamos?" Political, economic, and social order and progress assured the convergence of an alleged chaotic past into an integrated, peaceful present and promising future.

The Porfiriato was an era of state projects: projects of industrialization, investment, land redivision, architecture, and renovation. The rhetoric of visual power embodied in the exhibitionary complex sustained these projects conveying Porfiriato ideology through national and international displays. In a tradition going back to Bullock in the 1820s through Maximilian in the 1860s, Díaz was also a great showman, who used multiple media to exhibit a Mexico of order and progress to the nation as well as the world. Display became so intense, in fact, that the historian Eric Van Young associates the Porfiriato with the "commodification of historical meaning."

> Commodification occurred both in the sense that [cultural] signs were increasingly manufactured, consciously manipulated, and broadly diffused by power holders to naturalize their authority, to legitimate it and provide it with a genealogy, as well as in the sense that representations of forms of community came to center on the images and ownership of things, or to be imbricated with them.[5]

Further, as commodification presumes consumption, it may also be suggested that during Díaz's regime the production and consumption of visual culture, which had intensified during the reign of Maximilian, not only continued but also greatly expanded.[6] This predilection for commodification and consumption was also exemplified in the Porfiriato's insistence on participation in late-century world's fairs that was fomented by the desire for and anxiety about Mexico's presence on the international stage.[7]

Like earlier state projects, the Porfiriato's undertakings helped Mexicans recognize themselves and remember what could not be remembered. And increasingly, media and new institutions further prompted Mexicans to consume more images of themselves—real and imagined.[8] For example, 1880s newspapers and periodicals were filled with advertisements for all the things that bourgeois Mexicans needed to purchase to be a modern Mexican: trips through the countryside and abroad, European couture and perfume, French brandy, photographs, encyclopedic books, and so on (figure 60). In fact, the Porfiriato era saw the rise of that surest sign of mass consumption, the shopping arcade. The department stores—El Centro Mercantil, El Puerto de Veracruz, El Palacio de Hierro—all opened in the 1880s.[9]

Porfirio Díaz was also keenly aware of space, all kinds of spaces. National space would be conceived of as a space for development, with the railroads connecting raw resources to commerce centers and transporting people across Mexico's vast territory. The space of public lands was surveyed for immigration purposes. The communal spaces of indigenous peoples were deemed "empty," and converted to privately owned spaces. The public spaces of Mexico City were revamped to display a history that expanded the hagiography developed by Maximilian and implicitly predestined the achievements of Porfirio Díaz and the Porfiriato. Urban space was reorganized to separate commerce from fashionable residences, commoners from elites.[10] Finally, domestic space was concisely restricted as feminine. All of these spaces required reformed mappings.

Díaz also recognized that the cartographic knowledge of Mexico was highly problematic because, despite the efforts of the SMGE, there were significant errors and misinformation in existing maps of Mexico. In 1877, the Comisión Geográfico-Exploradora (CGE) was established to create an accurate map of Mexico. Military personnel with surveying experience administered this new institution. The CGE maps eschewed the assemblage methods of cartography based on existing maps used earlier in the century, instead utilizing surveying fieldwork. As Craib summarizes, "The final products would render obsolete existing vernacular, unreliable, and presumably inferior maps of the territory, standardizing their information into a new, unifying structure."[11]

García Cubas did not work directly on this new cartographic effort. Instead, his work focused on geography and was part of an exponential increase in the 1880s of publications that focused on the exposition of the diverse spaces of Mexico, making them visible and available for visiting and investment. These were published by private publishing houses as well as government print shops. The Secretaría de Fomento purchased the best typographic equipment available and published works such as the *Primer*

La mujer juzgada por una mujer.—Se halla próxima á agotarse la tercera edicion de este interesante libro debido á la pluma de la conocida escritora *Concepcion Gimeno de Flaquer.* Hállanse todavía ejemplares en las librerías de Murguía, Buxó, Chavez y Ballescá, al precio de **$1.**

GRAN FABRICA DE TABACOS Y CIGARROS

EL BORREGO

Fabrica, Libertad núm. 2 } MÉXICO { Despacho central, Capuchinas, 12

La gran acogida que el público ha dispensado siempre á los famosos PUROS y CIGARROS de esta marca, nos obliga á emplear en nuestras manufacturas las mejores ramas de San Andrés, Jaltipan y Acayúcan: papel catalan, marca «España y México,» que es sin disputa alguna el mejor que se ha importado, por su blancura y buen arder.

Tambien recomendamos á nuestros constantes favorecedores la elegante marca LA ASTURIANA, cuyos cigarrillos de hebra pueden competir con los mejores en su clase.

MÉXICO Á TRAVÉS DE LOS SIGLOS.

HISTORIA GENERAL
DEL DESENVOLVIMIENTO SOCIAL, POLITICO, RELIGIOSO, MILITAR, ARTISTICO, CIENTIFICO Y LITERARIO DE MEXICO,
DESDE LA EDAD MAS REMOTA HASTA LA EPOCA ACTUAL.

Obra única en su género imparcial y concienzudamente escrita por una reunion de distinguidos literatos mexicanos.

OBRA DE INUSITADO LUJO.

La obra más monumental de todas las publicadas hasta el presente. Adornarán este monumento histórico, científico, artístico y literario de México, un considerable número de magníficas cromolitografías, láminas de gran tamaño y grabados intercalados, representando sucesos históricos, hechos de armas, objetos de arte, vistas, trajes, monumentos, planos, ciudades, retratos, moviliarios, autógrafos, numismática, grandes cuadros de pintores mexicanos, y todo cuanto pueda contribuir á dar idea cabal del adelanto y cultura de la nacion mexicana.

Ya han salido los veinte primeros cuadernos. ***J. Ballescá y Comp.***

LA CIUDAD DE MÉXICO.

TIENDA Y ALMACEN DE ABARROTES
ESQUINA de la 2ª de SAN FRANCISCO y COLISEO

Ventas por mayor y menor.

En este Establecimiento se encuentra el más completo surtido de vinos de mesa, tanto tintos como blancos; riquísimos licores; cognac de las más acreditadas marcas; encurtidos en vinagre y mostaza; frutas secas y en su jugo; té perla y negro; estearina de la Estrella y alemana, y galletitas de exquisito gusto.

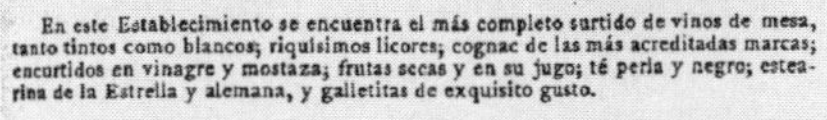

Se garantiza por la casa la legitimidad de las mercancías.

FRANCISCO ZEPEDA.

GUANTERIA Y PELUQUERIA

DEL BUEN TONO

OBJETOS DE LUJO Y DE GUSTO

GUANTES DE CABRITILLA

P. MICOLO
ESQUINA DEL ESPIRITU SANTO Y 3ª DE SAN FRANCISCO

4 LA DROGUERIA PINTADA DE AZUL 4

Calle de la Profesa 4 — **ANTIGUA DROGUERIA DE LA PALMA** — CARLOS FELIX Y COMPAÑIA — Calle de la Profesa 4

CALLE de la PROFESA

El ALMACEN MEJOR SURTIDO y MAS ACREDITADO de la CAPITAL

Surtido completo de todos los artículos del ramo.
EFECTOS DE TLAPALERIA.
DEPOSITO DE TODAS LAS ESPECIALIDADES anunciadas en los periódicos.
Aparatos Hidroterápicos de Walter Lécuyer.

APARATOS NEVERAS DE DELPY.
VINOS TINTOS PUROS Y JEREZ, garantizados.
COGNAC Y CHAMPAGNES superiores.

PERFUMERIA EXTRAFINA de Pinaud, Coudray, Roger y Gallet, Lubin, Legrand (Oriza), Atkinson, Gosnell, Rieger, Colgate y Comp., Lundberg, etc., etc.
Veloutina CH. FAY, legítima.

Vinagre de Benjuí, de Gelle, la preparacion más higiénica para el cútis.
Gran surtido de CAJAS y NECESERES para obsequios.
Polveras y Borlas.—Cepillos de todas clases.—Adornos para tocador.

4 LA DROGUERIA PINTADA DE AZUL 4

«EL ECO DE LA MODA» y «EL GLOBO.»

GRAN CAJON DE ROPA Y SOMBRERERIA
POR MAYOR Y MENOR
De Valera, Elizondo y Compañía.

SURTIDO COMPLETO de OBJETOS de LUJO, MEDIO LUJO y CORRIE[illegible]
Ajuares austriacos, Espejos para salon, Pianos, Máquinas Singer
CASA COMISIONISTA
para toda clase de compras, ventas y reexpedicion de mercan[illegible]
AGUASCALIENTES.—Portal de Aldama núms. 12, 13 y 17

IMPRENTA Y LITOGRAFÍA

DE E. DUBLAN Y COMPAÑIA

MEXICO—2ª DE PLATEROS Nº 3—MEXICO

En esta casa, montada á la altura de las mejores de su género en el país, y que cuenta con un gran número de prensas movidas por vapor, se desempeñan toda clase de trabajos de Tipografía y Litografía, ya sean

Libros, Esquelas, Tarjetas,
Folletos, Periódicos, Brevetes, Estados, Libranzas, etc.

Ferrocarril Interoceánico de ACAPU[illegible] MORELOS, MEXICO, IROLO y VERACRUZ.

Itinerario de los trenes desde el di[illegible] Setiembre de 1884, hasta nueva disposicion.

LINEA DE MORELOS.

TREN MIXTO DE MEXICO A OZUMBA		TREN MIXTO DE MEXICO A YAUTEPEC		ESTACIONES.	TREN MIXTO DE YAUTEPEC A MEXICO.		TREN MIXTO DE OZUMBA A MEXICO	
Llega	Sale	Llega	Sale		Llega	Sale	Llega	Sale

LINEA DE IROLO.

Tren de Pulque de México á la Luz		Tren Mixto de México á Calpulalpam		ESTACIONES.	Tren Mixto de Calpulalpam á México		Tren de Pulque de la Luz á México	
Llega	Sale	Llega	Sale		Llega	Sale	Llega	Sale
„	8,00 A.M	„	6,30 A M	México (S. Lázaro).	[illegible]		[illegible]	

60 Advertisements from *El Álbum de la mujer* (1885). General Collection, Library of Congress.

almanaque histórico, artístico y monumental de la República Mexicana (1883), actually highlighting mining, finance, and the railroad network; *Noticia histórica de la riqueza minera de México y de su estado de explotación* by Santiago Ramírez (1884); and *Mexico: its trade, industries and resources* (1893) by García Cubas. Many publications were translated into English, French, or both to assure a broad audience that not only included Europeans but Mexico's northern neighbor, whom Díaz recognized as potential investors.[12]

In turn, foreign publications about Mexico multiplied, and the titles of these works indicate that Mexico's open-door investment policies were being noticed. Such publications included travel books such as *A Trip to Mexico: being notes of a journey from Lake Erie to Lake Tezcuco* (Becher, 1880); *Where to spend the winter months. A Birdseye View of a Trip to Mexico* (Wineburgh, 1880); and *Voyage au Mexique, Un parisien au Mexique . . .* (Leclercq, 1885–86). Also appearing were resources guides, such as *Mexico and her Resources* (Dunn, 1888) and *Mexique à la portée des industriels, des capitalistes, des negociants* [traders] *importateurs et exportateurs et des travailleurs* (Bianconi, 1889). Some of these works were part of Mexico's extensive publications for world's fairs and commercial expositions, which increased during the 1880s.[13] This travel literature remains distinct from that of previous decades because it was less interested in identifying Mexico's national character than in promoting its national wealth. The cumulative impact of these 1880s displays of Mexico was the expansion of national pride as is evident in an editorial published on 15 June 1885 in *El Siglo XIX*. It stated that Mexico's representation in the New Orleans exposition was brilliant and successful and, as a result, "there has been reserved for Mexico a major space that is destined for well-known advanced nations in the industry and in the arts; . . . Among those peoples, it remains already shown that Mexico is worthy of having an honorable place and first in the United States and later in Europe."[14]

Finally, these 1880s cultural, economic, and commercial exposés were supplemented with comprehensive and revised mappings of national narratives, one of which was Antonio García Cubas's most visually powerful work, the *Atlas pintoresco é historico de los Estados Unidos Mexicanos*, published by Debray in 1885. To understand the context and distinction of this atlas, the focus of this chapter, I begin with a brief analysis of two coeval examples of nationalist narrative: *México pintoresco, artistico y monumen-*

tal (1880–83) by Manuel Rivera Cambas, and *México a través de los siglos* (Mexico through the Centuries) edited by Vicente Riva Palacio (1884–90). Both multivolume works were highly illustrated, integrating a plethora of images from previous decades with extended texts in an attempt to form (or imagine) a nationalist community. While Rivera Cambas's and Riva Palacio's works were quite divergent in their respective approaches and intentions, both works manifested a complex set of themes, ideas, and images that were shared with García Cubas's *Atlas pintoresco é historico de los Estados Unidos Mexicanos*.

MÉXICO PINTORESCO, ARTISTICO Y MONUMENTAL

The expansion of Mexican intellectuals' production of comprehensive histories is exemplified by Manuel Rivera Cambas's *México pintoresco, artistico y monumental: Vistas, descripción, anecdotas, y episodios de los lugares mas notables de la capital y de los estados, aun de las poblaciones cortas, pero de importancia geográfico ó histórico* . . . (Picturesque, Artistic and Monumental Mexico: Views, Descriptions, Anecdotes and Episodes of the Most Notable Locales of the Capital and of the States, although the Population Is Limited, but Having Geographic and Historic Importance), consisting of three volumes published in 1880, 1882, and 1883. As this section of the extended title indicates, the texts repeat the encyclopedia-entry format found in Rivera Cambas's *Los gobernantes [Rulers] de México*, published in the early 1870s. Further, *México pintoresco* is an updated and expanded version of Marcos Arróniz's 1858 *Manual de historia y cronología de Méjico*. Both are structured as tours through Mexico with visits to specific sites and monuments associated with events of Mexican history or culture.

A portion of the subtitle of Rivera Cambas's work, *Obra ilustrada con gran número de hermosas litografías* . . . (Work Illustrated with a Great Number of Beautiful Lithographs), justifies the intense visuality of the production. Many of the numerous prints are initialed with "L.G.," likely Luis Garcés, who illustrated Rivera Cambas's earlier works, as the artist with Murguía signing as the lithographer. The images included by Rivera Cambas refer back to those of Sartorius, Nebel, Gualdi, and other artists and, in some cases, are almost directly copied. Other illustrations adapt imagery from early nineteenth-century publications. For example, figure 61, *Varios*

61 *Varios objetos de los mas notables que se encuentran en el Museo Nacional*, from *México pintoresco, artistico y monumental*, tomo 1 (1880), p. 175.

objetos de los mas notables que se encuentran en el Museo Nacional, shows a jumble of archaeological objects from the Museo Nacional along with a few colonial items. The print references objects from Casmiro Castro's *Antigüedades mexicanas* of *México y sus Alrededores* published by Decaen in 1855 (figure 43, chapter 4, p. 139).

México pintoresco, artistico y monumental's frontispieces introduce the content and theme of each volume. Volume 1's print depicts a well-dressed woman and her daughter looking at what appear to be architectural remains: fragments of columns, a pedestal with a column or possibly the lower section of a viceregal cross, a rectangular stone with an abraded inscription, and pre-Columbian objects (figure 62). The pair stands in a landscape that draws from the inventory of images of Mexico's geography; desert and tropical plants frame the picture while a church and volcano (Orizaba?) form a backdrop. Here, the nation's present and future, the woman and child, peer at the remains of the past, the topic of this volume. The entries in the vol-

62 Frontispiece, tomo 1 (1880), from *México pintoresco, artistico y monumental.*

ume cover New Spanish and Mexican history from the 1520s to the 1870s, particularly focusing on the identification of sites and monuments and their role in historical events. For example, in Rivera Cambas's recitation on the Palacio Nacional, the visitor is told of its use as a palace by Moctezoma and subsequently by the Spanish viceroys, the attacks on the building during various wars, its occupation by the U.S. general Scott during the Mexican-American War, and its return to use by Juárez and Maximilian. Similar histories of particular sites and monuments located in Mexico City are elucidated; these include the El Parián, a marketplace; the Zócalo, the central plaza; the Alameda; the central train station; the National Theater; and the Sun Stone.

Rivera Cambas's volume 2 completes the historical tour of Mexico City's spaces and places, then moves to its immediate *alrededores* (environs), describing and illustrating numerous churches and sites from towns surrounding Mexico City. The frontispiece from this volume represents a scene with a great gazebo holding people who are standing and likely listening to music. This is less of a city site than a scene and place found in the central plazas of towns and villages.

The frontispiece image of the final volume recollects the imagery found in García Cubas's and Casmiro Castro's *Álbum del ferrocarril Mexicano* of 1877 (figure 63). Two male figures—one dressed in traditional indigenous clothing and one dressed in a modern, Western suit—stand on the railroad tracks that curve around a rock outcrop. Here, two traditions, contemporary and ancient, await the train. The railroad theme is pertinent here because Rivera Cambas's tour continues from the outskirts of Mexico City to the states surrounding the city that are connected by railway: Mexico, Hidalgo, Morelos, Guerrero, Michoacán, and Colima, with a brief mention of the territory of Baja California. Descriptions and images cite major towns and cities, and geographical sites and archaeological ruins including the pyramids of Teotihuacan and Xochicalco. In these pages, we also find depictions of numerous natural sites such as the cave of Cacahuamilpa and, of course, the Cascada de Regla, an image copied from Humboldt and reappearing in the title block of García Cubas's *Carta General* of 1858.

Rivera Cambas's distinctive approach to his description of Mexico is made clear by one of the promises of the other section of his subtitle: *señalar el grado de nuestro adelanto y el aspecto fisico, moral é intellectual de la*

63 Frontispiece, tomo 3 (1883), from *México pintoresco, artistico y monumental.*

República (Indicating the Degree of Our Advancement in Physical, Moral and Intellectual Aspects of the Republic). His entries and lists are premised by the idea that monuments, events, and people manifest the progress of history. This is to say, that by viewing a monument, the Palacio Nacional, for example, one can associate it with specific events and people. Rivera Cambas's work amplifies the call made by Francisco Zarco (cited in chapter 4) in the 1860s work *México y sus Alrededores*: "Without history there are no monuments, without events, there is nothing to retain in the memory."

MÉXICO A TRAVÉS DE LOS SIGLOS

In 1883 the first volume was published of *México a través de los siglos: Historia general y completa del desenvolvimiento social, politico, religioso, military, artistic, científico y literario de México desde la antigüedad más remota hasta la época actual* (Mexico Across the Centuries: General and Complete History of the Social, Political, Religious, Military, Artistic, Scientific and Literary Development [unfolding] of Mexico from the most Remote Antiquity to the Present). When completed in 1889, this five-volume project, conceived and edited by Vicente Riva Palacio, author of *El Libro Rojo* (1870), constituted a comprehensive narrative of Mexican history and culture. Its audience—made up of 7,000 subscribers in 1882—was the domestic reader, evidenced by the advertisement located in the upper right corner of the page from a weekly periodical (figure 60). It announces the work as "La obra más monumental de todas las publicadas hasta el presente" (the most monumental work of all those published to date). *México a través de los siglos* was Riva Palacio's response to an 1880 call by Manuel González—a figurehead who was Mexico's president from 1880 to 1884 and whom Díaz put in office until he returned in 1884—to write a history of the French intervention and empire.[15]

In his meticulous examination of the production of *México a través de los siglos*, the historian José Ortiz Monasterio reviews a prospectus for the work found in the archives of Vicente Riva Palacio, and likely written by the author, which outlines not only the content but also the approach of the work.[16]

> The intellectual movement of Mexico and its notable advancement, in sciences and in literature, that each day takes greater flight, makes indis-

> pensable the publication of a complete History that would bring together all the conditions of the works of this kind, in an epoch like the present, in which neither taste nor good judgment now conform within the average or incomplete works.
>
> Also, the need to give light to the general History of Mexico from its most remote times is tied to the great standards that are known everywhere in the civilized world the progressive march and fecund in the achievements of a people whose origin, development and culture are unknown in the most enlightened countries of Europe and even in the American continent itself.[17]

The work that Riva Palacio proposes goes significantly beyond a simple history of the French intervention: it continues the search for a feasible history of the previous four decades. At the same time, *México a través de los siglos* diverged from the 1850s history written by Lucas Alamán, which saw the Spanish conquest as the beginning of Mexican history and the viceregal period as a golden age. In the breadth of its conception, it was also distinct from the histories written in the 1870s because it combines liberal rhetoric with the professional language fostered by a positivist scientific approach.[18]

Unlike Rivera Cambas's encyclopedia-like texts that enumerated places and events of history, *México a través de los siglos* emphasizes a "new," more narrative history that also traces intellectual history. The prospectus specifically mentions that previous histories emphasized war and violence, often obscuring the intellectual and moral development of nations.

> In this new History, the authors, separating themselves from that method, do not give secondary place to the means that have served for the progress of the intelligence and of the morals in Mexico, following the foundation, development and function of the well-founded Universities, the main Colegios, the schools of arts, the Academies, museums, and literary societies that have flourished or flourish at present in the Republic, of the houses of Charity, hospitals, institutes of public instruction, of the writers, poets, scientific men, publicists and artists that have distinguished themselves and making themselves known in the world of letters, of the sciences and of the arts.[19]

México a través de los siglos was to be an imposing work that attempted to display the intellectual workings of the social, economic, political, intellectual, and moral history of Mexico for "todo el mundo civilizado," not merely a chronology of events, monuments, and people.

In its viewpoint, *México a través de los siglos* was to be a liberal history. It begins with the earliest human occupation, echoing Villaseñor's writing, and concludes with the defeat of the French. In this way, like all nationalist history, the whole work is premised by its conclusion: history foreshadows and culminates in the present. Thus, it is a history written from a present that sought to elucidate the evolution of Mexico's rich past towards its imminent potential for modern prosperity, that is, Porfirian Mexico, the successor of the Reforma.

This progressive national history is outlined in the subtitles of each volume of *México a través de los siglos*. The first volume, *Historia Antigua y de la Conquista*, written by Alfredo Chavero, an archaeologist and historian, summarizes the early occupation of the Americas and the indigenous groups of Mexico.[20] Vicente Riva Palacio himself wrote the second volume, *El Virreinato, Historia de la dominación español en México desde 1521–1808*, which overviews the period of Spanish domination. The next two volumes divide the last part of Mexico's history. *La Guerra de Independencia*, the third volume, written by Julio Zárate, summarizes the movement to independence from 1808 to 1815. The fourth volume, written by Enrique Olavaria y Ferrari, continues this history of independence under the title *México Independiente, 1821–1855*. This great history concludes with the fifth volume, written by José María Vigil, director of the Biblioteca Nacional from 1879 to 1909, and simply entitled *La Reforma*. It summarizes historical events from the 1858 triumph of Ayutla to the expulsion of the French in 1867. Thus, through its texts and supplementary images, *México a través de los siglos* is a wide-ranging retelling of a now four-part history formulated by the end of the 1880s.

Further, this national history is visually articulated in the frontispieces of *México a través de los siglos*, which encapsulate the content of each book.[21] For example, the frontispiece of the first volume portrays two Indians seated on the Stone of Tizoc, an Aztec-Mexica monument, and on either side of a zoomorphic brazier (figure 64). In front of the massive sculpture are the shields, bow and arrow, and armor previously associated with the statuesque Indian warriors and the conquistador. Also, a version of the central image of the Aztec Sun Stone placed on a wall decorated with references to pre-

64 Frontispiece, tomo 1 (1883). *Historia Antigua y de la Conquista* of *México a través de los siglos*. Rare Books, Library of Congress.

Columbian-like architectural elements serves as a backdrop for the seated figures. Here, by combining archaeological objects with imaginary people, the originating time and space of the Mexican nation are elaborated. Woven through the written narrative of this volume are numerous images of pre-Columbian objects and archaeological sites, also found in 1840s and 1850s publications that validate Mexico's origin not in the moment of conquest, as argued by Prescott and Alamán, but in the pristine epoch of pre-Columbian civilizations suggested by Villaseñor, Bustamante, and others.[22]

In the second volume, *El Virreinato, Historia de la dominación español en México desde 1521–1808*, the frontispiece depicts a tonsured friar possibly shown in the process of baptizing two babies. The scene takes place in front of an arch engraved with two dates used to delimit the viceregal period: 13 August, the date attributed to Cortés's conquest of Tenochtitlán in 1521, and 27 September, marking Mexico's declaration of independence from Spain in 1821. The Iztaccíhuatl and Popocatépetl volcanoes, deep-rooted imagery in the recounting of the history of the conquest, loom in the background. An indigenous mother presents a child, and a black mother (or possibly a servant) holds a second child. A conquistador figure stands in the background on the left, while an indigenous male who holds a shield oversees the scene. While the viewer cannot determine the precise maternal or paternal association of the figures, the identity of the child as an offspring of a mixed-blood union is clear. Notably, this image, which initiates the discussion of viceregal history, emphasizes a moment of Mexico's history when the mixing of Spanish, Indian, and black blood produces the Mexican race, rather than the moment of Spanish conquest associated with Cortés.

The frontispiece of the next volume, *La Guerra de Independencia*, summarizes the movement to independence. In this image, a seated winged female figure wearing a long white tunic, marked by a flame on her forehead and a quill in her right hand, holds a folio with its open page bearing the heading "Independencia 1810." This figure raises her head to peer at a second figure, a hovering young woman, dressed in a short white tunic. This figure appears to be rising from the smoke of an intense battle scene below. In her left hand, she brandishes a sword, while in the right hand, she holds a gathered flag whose slightly unfurled edge reveals a sliver of the image of the Virgin of Guadalupe. Here, two women metaphorically represent the events that liberate Mexico from Spanish domination: the Grito de Dolores, denoted by the image of Guadalupe, resulting in the war of independence; and the 1810 Treaty of Córdoba. The two allegorical figures represent the nation in battle, the nation rising in victory, and the nation claiming its independence.

The fourth volume continues this history of independence, under the title *México Independiente, 1821–1855*. The frontispiece of this volume depicts the tumultuous period after independence. Another metaphorical image of nation rises above a dark and bleak landscape marked by a burning city, broken trees, and a grave marker etched with the name Guerrero, a hero of the Independence movement and Mexico's second president. She holds a knife in her right hand and a torch in her left. In this context, the figure is menacing, yet the rainbow in the background and her torch may allude to hope.

México a través de los siglos concludes with the fifth volume, *La Reforma*, and another female figure introduces this volume (figure 65). This frontispiece is a reproduction of the allegorical image of *Constitución de 1857* by Pertronilo Monroy and exhibited circa 1869. As a symbol of Mexico's reformist constitution, the figure was understood as an apparition of Mexico. In January of 1869, one critic described the Monroy's figure in rather seductive terms: her face, open to thought and love; her black eyes in which sleep the rays of passion; and her mouth trembling, with promises, caresses. He concludes, "Es México, es la patria querida, es la glorificación de la razón" (She is Mexico, the beloved motherland, the glorification of reason).[23] As appropriated for this volume, the image does not so much narrate historical moments as the previous frontispiece figures but is a metahistorical continuation of the visual configuration of a deep discourse on sovereignty and the corporeal imagery of the nation as a body, which seduces and enamors citizens. Her allegorical form as the Constitution of 1857 indicates the continuity of the female body as marking the shifting site of sovereignty.

Along with these frontispieces, the text of *México a través de los siglos* is intertwined with the inventory of images that had accumulated by the 1880s. In fact, each volume has over 100 images, many of which are copied from previous publications.[24] This vast array of visual images includes portraits of historical characters, archaeological sites, and maps, as well as facsimiles of documents such as the Spanish Inquisition's 1810 indictment against Miguel Hidalgo and the 1836 peace treaty with Spain. Overall, unlike Alamán's history—which was written with less attention to illustrations—and unlike

65 Frontispiece, (after Petronilo Monroy's *Constitución de 1857*) tomo 5 (1889). *La Reforma*, of *México a través de los siglos*. Rare Books, Library of Congress.

Rivera Cambas's historical digests—which are told with a catalogue of images but without a specific narrative thread—Riva Palacio uses visual imagery to tell a comprehensive and holistic story of Mexico through the centuries.

ATLAS PINTORESCO É HISTORICO DE LOS ESTADOS UNIDOS MEXICANOS

Between 1884 and 1889, Antonio García Cubas enthusiastically contributed four major publications to this move to develop comprehensive overviews of Mexico as part of his work for the Secretaría de Fomento. The *Cuadro geográfico, estadístico, descriptivo é histórico del los Estados Unidos Mexicanos* (1884) was produced for the Universal Exhibition in New Orleans. This information was republished in the *Cuadro geográfico, estadístico, descriptivo é histórico del los Estados Unidos Mexicanos, obra que sirve de texto al Atlas pintoresco* de *Antonio Garcia Cubas* (1885). These works contain similar general information and statistical data, including I Location and political division; II Ethnographic data; III Ecclesiastic divisions; IV Transportation; V Public education; VI Mountains; VII Hydrography; VIII Agriculture and Industrialization; IX Mining; X Geographic description of Mexico; XI City of Mexico; XII Historical Review, part 1; and XIII Historical Review, part 2. The information in these studies was updated and translated into French and English resulting in the *Étude géographique, statistique, descriptive et historique des États Unis Mexicains* in 1889, appearing at the Paris World's Fair in the same year, and in 1893 as *Mexico: its trade, industries and resources.*[25]

In 1885, García Cubas also published a related fifth text, the *Atlas pintoresco é historico de los Estados Unidos Mexicanos*. Each of the thirteen pages is composed of a thematic map of Mexico associated with statistical information in the form of charts and graphs and framed by small images. Repeating the basic content of the 1884 *Cuadro geográfico*, the titles and subject matter of the thirteen atlas pages are

I. *Carta política*	Past and present political figures of Mexico and municipal buildings
II. *Carta etnográfica*	Indigenous groups of Mexico
III. *Carta eclesiástica*	Major churches and dioceses of Mexico

IV.	*Vías de comunicación y movimiento marítimo*	Water and overland transportation resources
V.	*Instrucción pública*	Institutions of public education and major intellectual figures
VI.	*Carta orográfica*	Mountain systems of Mexico
VII.	*Carta hidrográfica*	Water resources
VIII.	*Carta agrícola*	Agricultural resources
IX.	*Carta minera*	Mineral resources
X.	*Carta histórica y arqueológica*	Archaeological monuments and sites
XI.	*Reyno de la Nueva España*	Major figures of the viceregal period
	Valle de México	Views of the Valley of Mexico
XIII.	*México y sus cercanías*	Mexico City and its environs

The *Atlas pintoresco* did not include a written text; however, García Cubas's 1885 *Cuadro geográfico, estadístico, descriptivo é histórico del los Estados Unidos Mexicanos* was written to elucidate the atlas pages.[26] In the preface to this text, García Cubas summarizes the *Atlas pintoresco* project, emphasizing that it offers the public the latest information on geography, statistics, and history of the República Mexicana and that the maps express graphically the principal statistic evidence and important comparative information.[27]

Turning these pages, we are presented with important historical, social, and political data as well as information about physical features and resources of Mexico. The *Atlas*'s imagery—graphs, charts, and vignettes—derives from images of the eighteenth century and travel images of previous decades of the nineteenth century, with some new representations. In seamlessly integrating this diverse imagery with maps, García Cubas displays his skillfulness with cartographic and geographic bricolage.

The combination of images with a map in the *Atlas pintoresco* pages expands the itinerary and exploratory travel format found in García Cubas's 1858 *Carta general* and the 1863 *Carta general* but in a 1880s context. The *Atlas pintoresco* was published just after the publication of Rivera Cambas's *México pintoresco* and about the same time that the first volumes of *México a través de los siglos* appeared. The three publications portray Mexico holistically as a nation with a common interest in visual display but with distinctive approaches. As a result, the García Cubas 1885 atlas is in visual exchange with both of these works. Specifically, all three publications access an inventory of images now firmly used to map the content of Mexico, including archaeological sites, pre-Columbian objects, regional churches, important historical figures, railroad images, and types of people and landscapes. Likewise, all three publications retrieve images from depictions by travelers from earlier in the century. For example, both *México a través de los siglos* and the *Atlas pintoresco* use Nebel's images of the pyramid of Papantla (or Tajin) and Catherwood's depictions of Mayan sites. And they borrow illustrations from each other: the map from *VIII Carta agrícola* appears in tomo 5 of *México a través de los siglos*; while the *Atlas pintoresco's* images of churches in the *III Carta eclesiástica* page come from the *México pintoresco*.

There are critical differences among the publications, too. Rivera Cambas's *México pintoresco* is structured like an encyclopedia, with discrete citations and images used as illustrations. In contrast, Riva Palacio's *México a través de los siglos* and García Cubas's *Atlas pintoresco* utilize a tightly integrated narrative structure to describe Mexico's history, cultural progress, and natural abundance. Riva Palacio accomplishes this through extended written narratives by authors considered to be experts in their topics, employing images to supplement the dense text.[28]

In turn, García Cubas's *Atlas pintoresco* reiterates and extends Riva Palacio's sociopolitical and historical emphasis through cartographic and geographic perspectives that structure a strong *visual* narrative. Through the central placement of a map on each page, the designated theme is spatialized. The adjacent small image narrates the theme by referencing both an inventory of circulating images and imagery from previous decades as well as contemporary publications. Here, images are not illustrations or supplements but primary means to narrate Mexico's past, present, and future. For example, in *I Carta política* the map identifies the political boundaries of Mexico and provides a sense of population density of the country and its territory (figure 66).[29]

Repeating Humboldt's concept of using graphs to relate dense data, the cartouche in the lower left corner of the map graphically compares the territorial areas of the various states as well as their populations. Above the map, in a long rectangular block, are fifty-five portrait images of important leaders of Mexico, beginning with Iturbide and concluding with Manuel

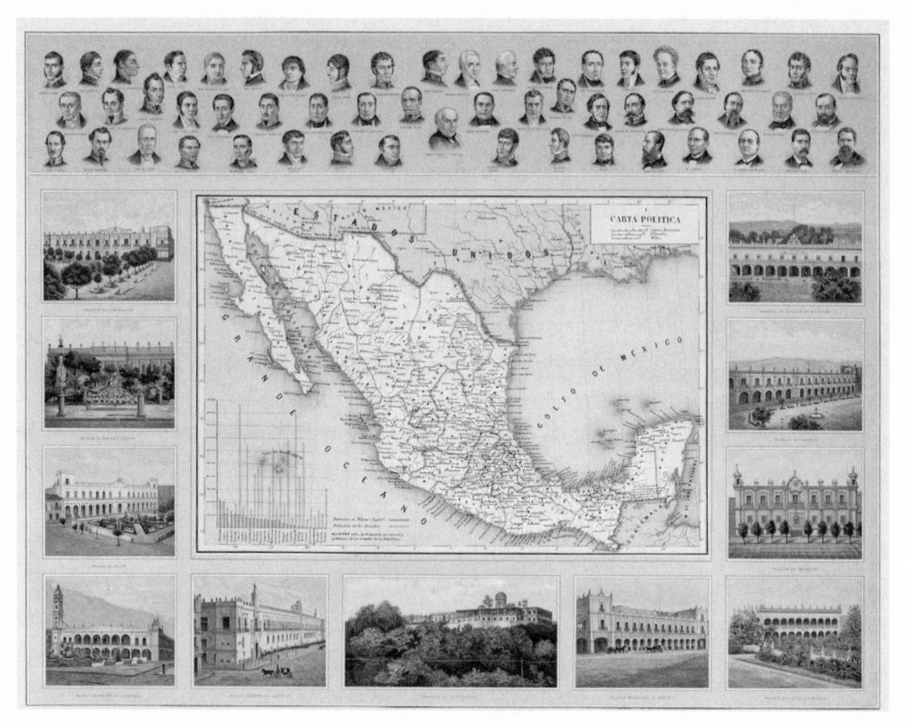

66 Antonio García Cubas, *I Carta política*, from *Atlas pintoresco é historico de los Estados Unidos Mexicanos*, 63 × 80 cm (1885). Geography and Map Reading Room, Library of Congress.

González, president of Mexico at the time. The portraits have a *tarjeta de visita* format (see chapter 5) and are similar to those found in Rivera Cambas's *Los gobernantes de México*. Surrounding the map on three sides are images of municipal buildings from eight states, along with Mexico City's government buildings, such as the Palacio Nacional de México and Palacio de Chapultepec. The *I Carta política*'s map shows Mexico's completed (present) form and does not represent Mexico's changing shape under various pictured leaders.

In another example, the *II Carta etnográfica* is one of the most visually engaging pages (figure 67). The page's map is a colorized reproduction of the *Carta etnográfica de México* from García Cubas's *Mexico in 1876* and shows the geographic location of twenty-nine ethnic groups identified linguistically (figure 55, chapter 5, p. 174). Three graphs comparatively visualize the population by *razas*: the white race, indigenous race, mixed race; total

67 Antonio García Cubas, *II Carta etnográfica*, from *Atlas pintoresco é historico de los Estados Unidos Mexicanos* (1885). Geography and Map Reading Room, Library of Congress.

Indian population in relation to the general population of each state; and comparison of the demographic statistics of fourteen distinct indigenous groups.

This statistical data comes to life through twenty-six small illustrations depicting the major ethnic groups. In a format much like that of Blaeus' atlas page (figure 3, chapter 1, p. 27) and Alzate's *Nuevo mapa geográfico de la América septentrional española* in 1767 (figure 16, chapter 2, p. 60), vignettes framing this map depict the kinds of people who make up the nation of Mexico. As should now be clear, such images of indigenous groups come from a long tradition of cultural description going back to the sixteenth century and continuing to the nineteenth century. All but five of the images are derived from the chromolithographic plates of *The Republic of Mexico in 1876*. Slight changes are made to the images; for example, figures 68 and 69 compare the illustration of *Zapotecas de Coatecas* from the two

68 *Zapotecas de Coatecas Altas y de los alrededores de Oaxaca*, detail, *II Carta etnográfica*, from *Atlas pintoresco é historico de los Estados Unidos Mexicanos* (1885). Geography and Map Reading Room, Library of Congress.

69 *Zapotecas Indians from Coatecas Altas, Oaxaca*, detail no. 3, *The Republic of Mexico in 1876*, plate 5. Author's collection.

de Veracruz y el de Coyoacán, que se decía de México, pues no es indudable que Segura de la Frontera tuviera ayuntamiento. Cortés, con una gran actividad, hizo que se fundase en esos momentos una villa con el nombre de Medellín, para tener un ayuntamiento y un procurador más, y envió los nombramientos de alcaldes y regidores y el de procurador de esa villa, que concurrir debía á la junta de Zempoala, para cuyo cargo fué designado Andrés de Monjarás, el mismo á quien Gonzalo de Sandoval había dejado encargada la gente conque se pacificaba Tuxtepec.

Reuniéronse en Zempoala con Cristóbal de Tapia, Pedro de Alvarado, como procurador del ayuntamiento de México; Cristóbal Corral por el de Segura de la Frontera ó Tepeaca; Bernardino Vázquez de Tapia por el de la Villa Rica de Veracruz, y Andrés de Monjarás por el de la villa de Medellín, y como procuradores ó representantes de Cortés, Gonzalo de Sandoval, Diego de Soto, Diego de Valdenebro y fray Pedro Melgarejo de Urrea.

Insistía Tapia en presentarse á Cortés y reclamar la gobernación de la Nueva España; hacíanle presente los procuradores todos los peligros y perturbaciones que de esto se podían originar; urgía él, oponíanse ellos, el uno alegando las órdenes reales, los otros protestando obedecer pero no cumplir los dichos mandamientos, y manifestando apelar ante el emperador. Por fin Tapia,

India zapoteca.—Oaxaca. (Tipo actual)

accediendo mal de su grado, convino en volverse á la Villa Rica para embarcarse; procurando, sin embargo, sacar el mayor provecho de su situación y del empeño de sus adversarios, contrató con los procuradores de Cortés que se le comprarían por éste algunas de las cosas que él había traído de España.

Señaló Tapia precios exorbitantes como era de suponerse; enviáronse correos á Cortés en demanda de oro que tardó poco en llegar, y Tapia vendió un navío, unos negros esclavos y tres caballos.

Volvió Tapia á Veracruz para embarcarse; pero allí, bien porque sintiera abandonar un gobierno al que creía tener tanto derecho; bien, y es lo más probable, por cubrir las apariencias, disimular que había sido cohechado con el pretexto de las ventas, y preparar una vuelta menos vergonzosa á Cuba y Santo Domingo, fingió nueva y más obstinada resistencia para embarcarse.

No quisieron los amigos de Cortés usar de la fuerza para obligar á Tapia, sin dar antes á la providencia un barniz de legalidad, y, como en todos estos casos acontecía, el poder municipal vino en ayuda de sus propósitos.

Era á la sazón alcalde de la Villa Rica Francisco Alvarez Chico, parcial de Cortés y amigo de los procuradores, y de acuerdo con ellos dió orden para que Tapia se embarcase, y de no hacerlo de grado le compeliese á ello por la fuerza el alguacil mayor, que era nada menos que Gonzalo de Sandoval, el más leal, el más inteligente y el más osado de los amigos del Conquistador.

Como Tapia podía suplicar de la providencia ante el

70 Page from *México a través de los siglos*, tomo 3 (ca. 1885), p. 45. Brown University Library.

publications. In the *Carta*'s vignette, titled *Zapotecas de Coatecas Altas y de los alrededores de Oaxaca* (figure 68), the figure of the woman on a donkey is copied almost exactly from the third register of plate 5 of *Mexico in 1876*, titled *Zapotecas Indians from Coatecas Altas, Oaxaca* (figure 69); while the position of the seated woman and standing man that face her are changed. More importantly, the *Carta etnográfica* group emphasizes spatial specificity by placing the group in front of a building, instead of a vague, nondescript background. The woman on a donkey image was also published in tomo 3 of *México a través de los siglos* (figure 70). Likewise, the image, *Mexicanos de Santa Anita*, of figure 71, is reproduced almost exactly from plate 3 of *Mexico in 1876* (figure 72), which was a colorized reproduction of Cruces y Campa's 1860s staged photograph of the *Trajineros* (figure 54, chapter 5, page 167). The significant circulation of images from the mid- to late nineteenth century is exemplified here.

The *X Carta histórica y arqueológica* serves as final example of the complex assemblage of the Atlas pages (figure 73). The central map and inset overlay information from the ancient maps of *Cuadros histórico-geroglífico núm. 1* and *núm. 2* of the *Atlas geográfico* (figures 49 and 50, chapter 5, pp. 157, 158) onto a contemporary cartographic semblance of Mexico. Red lines crisscross the map, marking the indigenous migration routes of the Nahuatlacas (Nahuatl) and Mexica groups. The inset identifies the location of specific indigenous groups and marks the arrival of the Mexica in the Valley of Mexico.

The panel image across the top of the page depicts archaeological objects from the Museo Nacional.[30] The objects repeat those found in Casmiro Castro's *Antigüedades mexicanas* from *México y sus Alrededores* (figure 43, chapter 4, p. 139), as well as the adaptation by Rivera Cambas's *Varios objetos . . .* print (figure 61). In the *X Carta histórica y arqueológica* image, however, the jumbled positioning of the objects of the early versions has given way to a more orderly presentation. The antiquity of this material is further certified by views of eight archaeological sites that surround the map. These images are copied from prints from Nebel, Waldeck, and Stephens, as well as Catherwood, all discussed in chapter 3 and appear in the introductory pages of tomo 1 of *México a través de los siglos* (figure 64).

The remaining *Atlas pintoresco* pages elaborate their themes through similar tight compositional structure and cross-referencing of imagery.

71 *Mexicanos de Santa Anita*, detail from *II Carta etnográfica*, from *Atlas pintoresco é historico de los Estados Unidos Mexicanos* (1885). Geography and Map Reading Room, Library of Congress.

72 *Mexican Natives of Santa Anita and Ixtacalco to the South of the Capital*, detail no. 3, from *The Republic of Mexico in 1876*, plate 3. Author's collection.

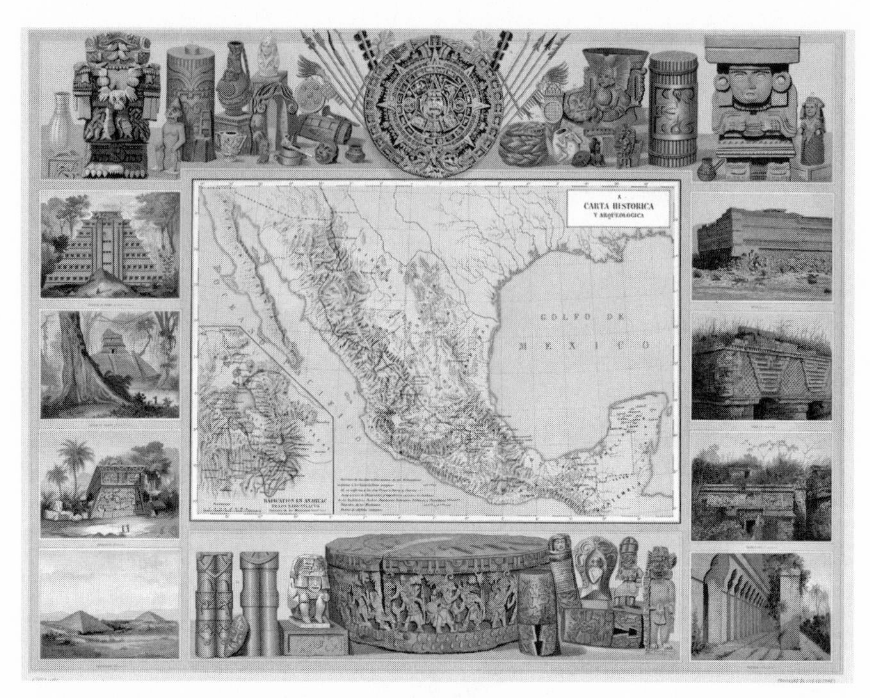

73 Antonio García Cubas, *X Carta histórica y arqueológica*, from *Atlas pintoresco é historico de los Estados Unidos Mexicanos* (1885). Geography and Map Reading Room, Library of Congress.

The detailed examination of these three pages demonstrates how the *Atlas pintoresco é historico de los Estados Unidos Mexicanos* pages become visually multilayered, with the depictions of people, places, objects, and activities emphatically linking the cartographically imagined various *Mexicos* of the *Atlas*. The content and composition of all of its pages reveal a highly structured visual program that not only scaffolds the *Atlas*'s thematic content but also sustains García Cubas's argument for a holistic display of the nation.

Scholars have cursorily assessed the *Atlas pintoresco* as a propagandistic and commercial project.[31] Indeed, reflection on how such themes are embedded and elaborated reveals that the atlas pages do display Mexico's investment promise for potential immigrants. In nineteenth-century political terms, nationhood required governmental stability and deep history: two supposed Mexican shortcomings consistently pointed out in the commentaries of foreigners. Thus, propaganda, as publicity and promotion, is a broad theme and intention of García Cubas's *Atlas pintoresco*. For example, in *I Carta política*, García Cubas subsumes Mexico's constant political struggles and disequilibrium into a thematic map whose cartographic certainty verifies the nation's predestined and stable political existence. The images that surround the map further confirm political continuity. They illustrate the genealogy of its political leaders through the tarjeta de visita portraits and attest to a stable federalism that connects regional governments, illustrated by the state municipal buildings, to the central government, signified by the Palacio Nacional and Palacio de Chapultepec.

Supporting the theme of *I Carta política*, the *II Carta etnográfica* demonstrates the control and management of another potential detriment to economic expansion—the indigenous population. Each image emphasizes the costume and activities associated with a particular regional or ethnic group. In this way, the group's identity within the nation is reified through specific features, languages, and customs, and is able to be located on the map and through graphs. The associated map emphasizes the linguistic spaces, not physical spaces, occupied by indigenous groups. This is concordant with the Porfiriato view that indigenous land, especially communal land, was empty and available. The *II Carta etnográfica*'s depictions of ethnic groups visualize inhabitants of the nation and demonstrate population stability. At the same time, in presenting indigenous peoples as Mexico's "others," the imagery suggests nineteenth-century anthropological perspectives and their colonialist premises.[32] Here, Mexico's indigenous populations are named, contained, and controlled, and pose no problem for political or economic development.

The subsequent pages of the *Atlas* visually elaborate Mexico's potential for economic and commercial development. Transportation and communication within Mexico is explicated in *IV Vías de comunicación y movimiento marítimo* (figure 74). The map displays Mexican railroad systems (constructed and projected), telegraph lines, roadways, and the boat and steamship routes between Mexican ports and those of the United States and Europe. An inset in the lower left corner profiles the train line between the cities of Veracruz and Mexico. The viewer is further presented with a summary of the water resources of Mexico through the *VII Carta hidrográfica* (figure 75). Its map locates numerous rivers, indicating the area of their

74 Antonio García Cubas, *IV Vías de comunicación y movimiento marítimo*, from *Atlas pintoresco é historico de los Estados Unidos Mexicanos* (1885). Geography and Map Reading Room, Library of Congress.

75 Antonio García Cubas, *VII Carta hidrográfica*, from *Atlas pintoresco é historico de los Estados Unidos Mexicanos* (1885). Geography and Map Reading Room, Library of Congress.

76 Antonio García Cubas, *VIII Carta agrícola*, from *Atlas pintoresco é historico de los Estados Unidos Mexicanos* (1885). Geography and Map Reading Room, Library of Congress.

approximate watersheds, and images framing the map—some of the most beautiful of the *Atlas*—depict these various rivers and waterfalls. Through these pages, investors are assured that Mexico has the transportation infrastructure needed for production and industrialization.

The *VIII Carta agrícola* and *IX Carta minera* visually and statistically demonstrate the immense natural resources Mexico has to offer. The *VIII Carta agrícola*'s central map locates the major crops grown in various parts of Mexico and identifies woods and tropical rain forests, as well as uncultivable land (figure 76). To the left of the map, the distinct vegetation at elevations from sea level to the snow-covered volcano is shown. This cartouche is derived from Humboldt's geography of equatorial plants print (figure 19, chapter 3, p. 71) and directly copied from García Cubas' *Carta general de la República Mexicana* of 1863 (figure 51, chapter 5, p. 162). On either side of the map, there are two beautiful natural history prints of plants from the warmer regions of the country. Above and below the map are scenes of agricultural production.

Turning to *IX Carta minera*, García Cubas displays the country's rich

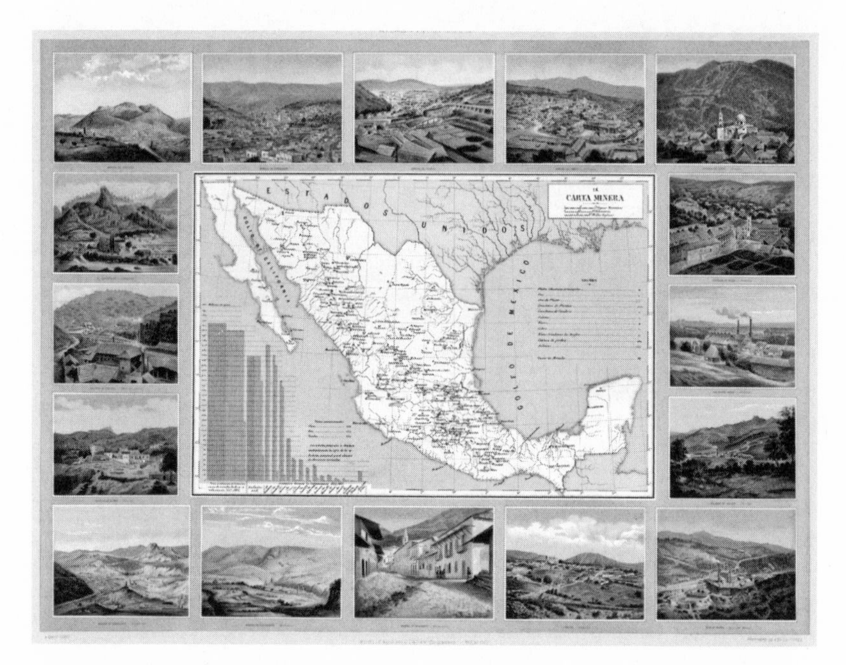

77 Antonio García Cubas, *IX Carta minera*, from *Atlas pintoresco é historico de los Estados Unidos Mexicanos* (1855). Geography and Map Reading Room, Library of Congress.

endowment of mineral resources (figure 77). The history of these resources are presented on the map, which shows the location of mineral resources—including silver, gold, iron, copper, and salt—and various mining towns. A bar graph in the lower left records the total minting of silver and gold coins, separately and together, in millions of pesos and a second graph illustrates the production of coins from 1822 to 1882. These graphs recall Humboldt's *Cantidad de Oro y Plata sacado de las Minas de Mégico*, which related the amount of gold and silver mined in Mexico (figure 18, chapter 3, p. 70). The images bordering the map identify various mines and their operations.

V Instrucción pública assures the presence of cultural and intellectual capital (figure 78). The central map verifies the growth of public education using dark to light shading to show high to low education rates in each state. A register of about ninety small portrait images of the major scientists, writers, and artists going back to viceregal times, listed by name and profession or area of study is placed above the map. Alexander von Hum-

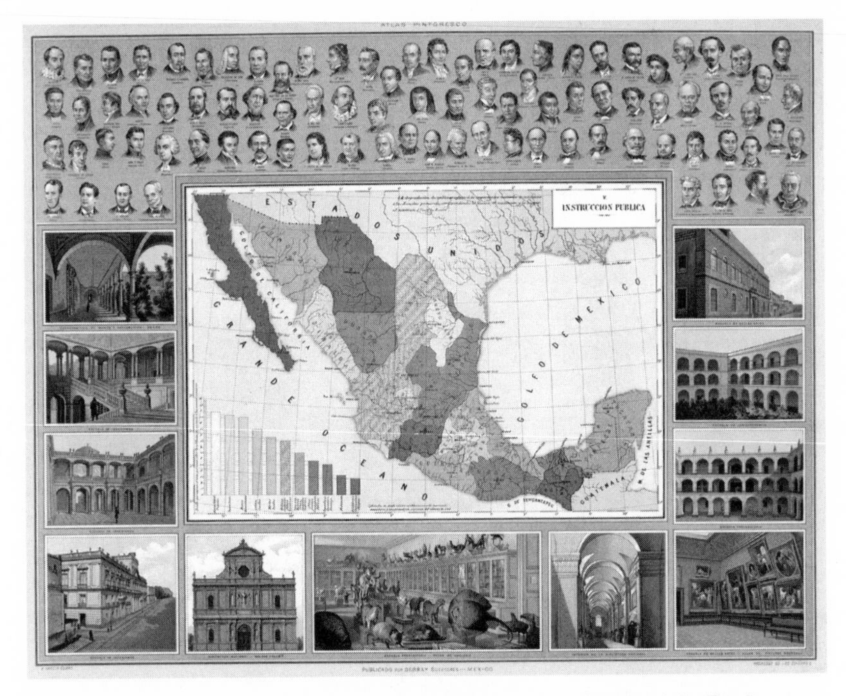

78 Antonio García Cubas, *V Instrucción publica*, from *Atlas pintoresco é historico de los Estados Unidos Mexicanos* (1855). Geography and Map Reading Room, Library of Congress.

boldt appears among these esteemed Mexican intellectuals, with the words "Fundador de la Est. Mex." (Founder of Mexican Statistics). The portraits confirm the depth of Mexico's intellectual heritage and achievements. *V Instrucción publica* also reiterates the stability of Mexico's intellectual achievements through exhibiting state institutions that shape and mold the nation's citizens and expand Mexico's intellectual capital. Here, Mexico can claim to have institutions—museums, libraries, professional schools, and a conservatory—comparable to those of Europe.

Nationalist propaganda also requires a grand history to retell, and in the *Atlas pintoresco* such a history is revealed as deep and continuous. This begins with the *X Carta histórica y arqueológica*, where sculptures and architecture from the pre-Columbian past are exhibited as precursors to contemporary Mexican culture, again ordained by the map. This history is continued in the next page of the *Atlas*, the *XI Reyno de la Nueva España*

79 Antonio García Cubas, *XI Reyno de la Nueva España a principios del siglo XIX*, from *Atlas pintoresco é historico de los Estados Unidos Mexicanos* (1885). Geography and Map Reading Room, Library of Congress.

a principios [at the Beginning of] del siglo XIX (figure 79). The central map traces New Spain roughly at the beginning of the nineteenth century, divided into intendancies, the Bourbon administrative units established in the eighteenth century. The graph to the left shows the territorial extension and population of each intendancy, based on information originally summarized by Humboldt. Here, we see Mexico prior to the loss of its northern lands to the United States. Below the map, García Cubas includes a copy of the 1797 print representing the equestrian statue of Carlos IV in the Plaza Mayor and signifying the viceregal period. The Hapsburg and Bourbon coat of arms flank the image, while above the map floats the coat of arms of New Spain, showing both Bourbon and indigenous emblems.[33] This viceregal theme is broadened with portrait images of important conquest and viceregal period personages that surround the map, such as Christopher

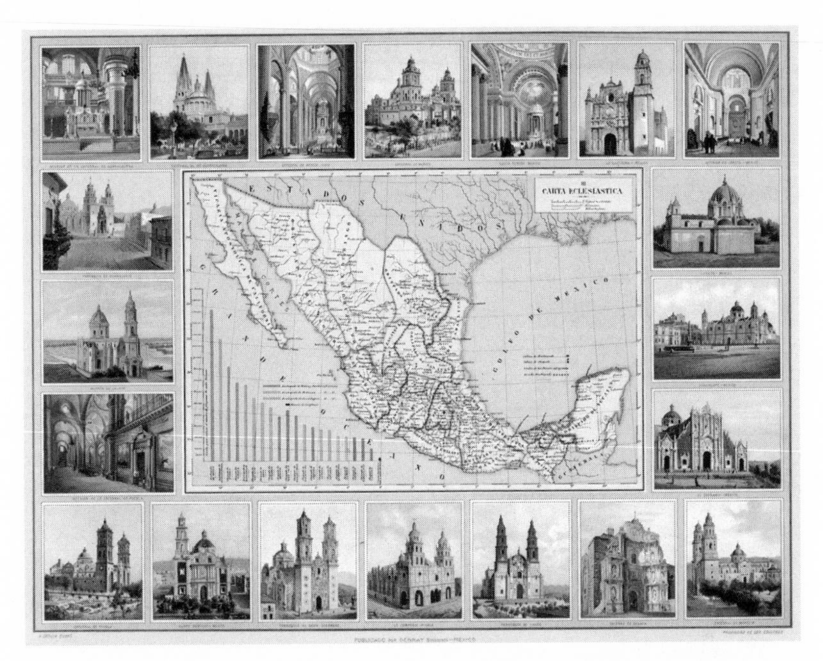

80 Antonio García Cubas, *III Carta eclesiástica*, from *Atlas pintoresco é historico de los Estados Unidos Mexicanos* (1885). Geography and Map Reading Room, Library of Congress.

Columbus, Queen Isabel, King Ferdinand, and Cortés. Flanking the map are seventy-one small images of Hapsburg and Bourbon kings of Spain and viceroys of New Spain, many derived from Rivera Cambas's *Los gobernantes de México* of the 1870s. *XI Reyno* confirms that Mexican history has regal origins, predestined to traverse from Aztec to European nobility. More importantly, the period of Spanish domination becomes a contained episode within the historical narrative of Mexico.

Two atlas pages seem to connect indirectly to the themes of propaganda and commercial expansion. The *III Carta eclesiástica* represents major churches and maps the twenty-one Roman Catholic dioceses (figure 80). The Porfiriato saw the churches as functioning like museums and honoring Mexico's past.[34] This theme, not found in García Cubas's previous works, may refer to an incipient rapprochement with the church in the late 1880s after the nationalization of ecclesiastic goods during the Reforma movement.[35] Likewise, *VI Carta orográfica*'s map displays the topography of

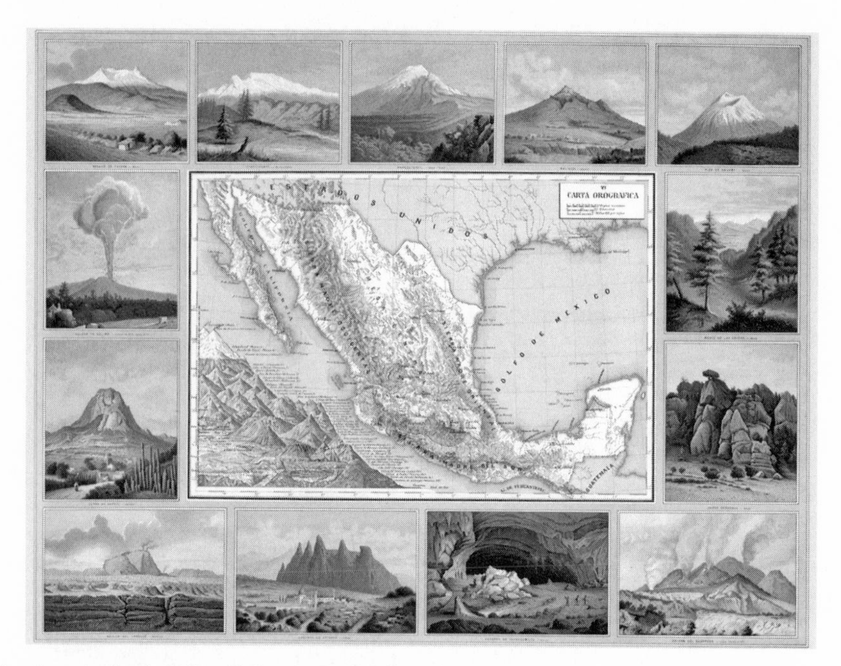

81 Antonio García Cubas, *VI Carta orográfica*, from *Atlas pintoresco é historico de los Estados Unidos Mexicanos* (1885). Geography and Map Reading Room, Library of Congress.

Mexico, noting the western, eastern, and southern Sierra Madre mountain ranges (figure 81). To show the comparative elevations of various mountains, with the volcano, García Cubas provides a graph, which depicts Popocatépetl being the highest peak and continues down to sea level, a format used by Humboldt in his *Géographie des plantes Équinoxiales* (figure 19, chapter 3, p. 71). The page does not so much comment on specific commercial resources as represent the visually dramatic, and available, landscape of Mexico.

The *Atlas*'s final pages illustrate the quintessential sign of the modern nineteenth-century Europe: the city. In *Valle de México*, an unnumbered page, the central map shows the topography of the Valley of Mexico (figure 82). Here, Mexico City is located in a web of train routes and roadways that weave around mountains and Chalco lakes. The surrounding images include various views of the Valle de México region and sites from the immediate area. Lastly, *XIII México y sus cercanías* depicts sites of Mexico City (figure

82 Antonio García Cubas, *Valle de México*, from *Atlas pintoresco é historico de los Estados Unidos Mexicanos* (1885). Geography and Map Reading Room, Library of Congress.

83). In the middle of the map, we see the city grid at the center with a web of roads and railroads leading in and out. The surrounding vignettes provide a panorama of the city, perhaps reflecting a mid-1880s renewed interest in the city and its maintenance.[36] Through the images of this page, potential immigrants are indulged with the panoramic views of the city from the east and south as well as scenic city spaces, the Bosque de Chapultepec, and industrial locales, the Belén paper factory. Likewise, *XIII México y sus cercanías* illustrates the spectacle of the inner spaces of the city: beautiful parks; striking architecture; civic monuments; wide, clean streets; and orderly movement of people as pedestrians, as well as riders of trams and railroads. It is a place to see and be seen.

Thus, the *Atlas pintoresco* may be read as an advertising project for foreign immigration and investment, and it is not difficult to conjecture that propaganda and commercial themes are intentionally elaborated in it. But

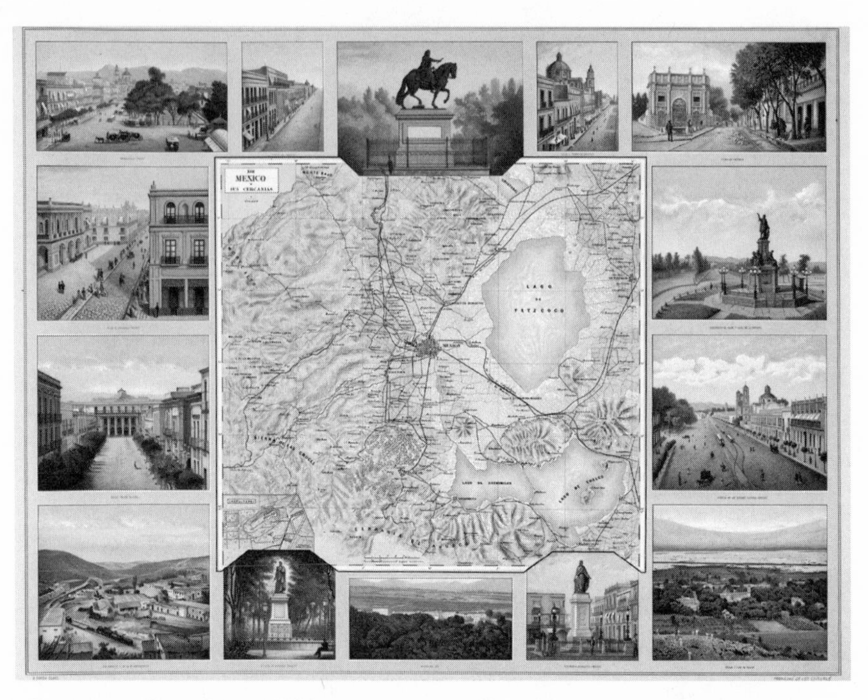

83 Antonio García Cubas, *XIII México y sus cercanías*, from *Atlas pintoresco é historico de los Estados Unidos Mexicanos*, n.p. Geography and Map Reading Room, Library of Congress.

could an unwieldy folio, twenty-five inches high and thirty-two inches wide, be an effective marketing tool? This is, of course, a facetious question that I use to point out that while the *Atlas pintoresco* may have had propaganda and commercialization aspirations, it could not be easily disseminated to its supposed target audience. Further, few foreigners would have understood some of the imagery included on the atlas pages, such as the identity and contribution of the intellectuals of the *V Instrucción pública* page. Having addressed themes of propaganda and commercialization as intended aspects of the *Atlas*, we now may return to the images and text to reflect on other levels of readings of this work.

Analysis of the *Atlas pintoresco* must also keep in mind that García Cubas chose to use the atlas format to form an overview of Mexico, which indicates a clear awareness of the atlas tradition as a gathering of vast and diverse cartographic and geographic data. The associated graphed statistical data on

84 Antonio García Cubas, *Carta general de los Estados Unidos Mexicanos*, from *Atlas geográfico, estadístico é histórico de los Estados Unidos Mexicanos* (1887). Geography and Map Reading Room, Library of Congress.

each page illustrates García Cubas's interest and belief in statistics, indicative of the Porfiriato's power to gather information about the nation and its people.[37] The atlas format, however, connotes movement through space, be it a journey, a route, a tour, a passage, or an excursion—all planned ways to move through a territory to construct meaning for the viewer-traveler. But García Cubas did not classify this project as a geographic atlas as he did his work of *Atlas geográfico of* 1858. In fact, he actually produced a new *Atlas geográfico y estadístico* in 1887. Its pages use the new printing technique of cerography, a wax engraving printing technique, to update the maps of his earlier atlas, such as the *Carta general de los Estados Unidos Mexicanos* (figure 84; cf. figure 46, chapter 5, p. 152).

In contrast, in the *Atlas pintoresco*, Mexico is displayed as picturesque and historic. The viewer is not presented with pages that are overwhelmed with texts of data, cartographic computations, or statistical tables, as in the *Cuadro geográfico y estadístico de la Republica Mexica* (figure 44, chapter 5,

p. 150). Instead, through the *Atlas*'s images, thematic maps, pictures, charts, and graphs, viewers are presented with various possible itineraries through Mexico: from the historical spaces of ancient cultures to that of New Spain; from bountiful land to plentiful mineral resources; from a valley of *nahua* migrations to a modern city of splendid spectacles; and from pristine but disparate landscapes to a landscape whose railroad network portends an interconnected territory. This theme of itinerary was a metaphor used by Porfirio Díaz himself. In 1882, he stated, "I am happy that today Mexico walks in hurried steps along the path of material and intellectual progress."[38] Thus, distinct from Riva Palacio's *México a través de los siglos* and Rivera Cambas's *México pintoresco*, the *Atlas pintoresco é historico* plots itineraries through memorable, real, and imagined places and spaces. These allow the viewer-traveler to move across and through space as well as backward and forward in time.

The intense visuality of the *Atlas pintoresco* allows García Cubas to exhibit not only the content but the creation of the nation as well.[39] This is accomplished by the persistent placement of the cartographic image of Mexico at the center of each page, and associating it with visual culture images derived from drawings, paintings, photographs, and prints, some of which were recent, many of which came from an inventory of images that emerged in the earlier part of the century. Various images were shared with *México pintoresco* and *México a través de los siglos*. This atlas structure makes connections between previous visual constructions of Mexico and contemporary visualizations, which seem stabilized and naturalized by the chimerical truth of cartography. Consequently, through the *Atlas pintoresco*, García Cubas stabilizes the circulation of images of Mexico that began to emerge in the 1840s and expanded in the 1860s through the expansion of international visual culture. He effectively connects and narrates images and imagery through various itineraries.

The *I Carta etnográfica* page again serves as an example of these complex workings of the *Atlas pintoresco* pages. Through its images, derived from *The Republic of Mexico in 1876*, some of which are shared with *México a través de los siglos*, the viewer can travel through Mexico ethnographically, gaining familiarity with the kinds of people who occupy Mexico's linguistic spaces. The theme of peoples of Mexico recalls the imagery of eighteenth-century *cuadro de casta* paintings of New Spain as well as the national types by travelers such as Bullock, Linati, and Nebel, which depict Mexico as a physical place inhabited by diverse kinds of bodies. Unlike these images, however, the lines of longitude and latitude and tables of the *Carta*'s map allege a scientific legitimacy for the *Carta etnográfica* groups and replace the speculative taxonomy of the casta panels and the exotica premises of travel images. More importantly, in the *Atlas* images, Mexicans are not identified as the products of miscegenation or tantalizing unfamiliarity; rather, they are the manifestation of race and ethnicity that define the nation. Consequently, the *Carta etnográfica* images, casta panels, and exotic images assemble space and people to form a sociopolitical place in very distinct ways. In the casta panels, mixed-blooded people of New Spain dominate and constitute place, while the Nebel and Linati depictions of national types represent the emerging nation. In contrast, in the *II Carta etnográfica*, Mexican national space dominates people.

Simultaneously, some of the *Carta etnográfica* images link to more contemporary constructions of Indians, as seen in the *Mexicanos de Santa Anita* of the *Atlas* derived from Cruces y Campa's staged studio portrait *Trajineros*. The sleight-of-hand of photography specifically, and images generally, comes into play here: how can the authenticity and veracity of the ethnic groups (or commercial resources, history, and scenic beauty, for that matter) be denied when it can be both photographically pictured and located on a map?

In another example, I have already traced how pre-Columbian sculptures and ancient ruins of the *X Carta histórica y arqueológica* were produced in the 1830s and associated in the 1840s with the translations of Prescott's *The Conquest of Mexico*, the 1851 *Calendario de Ignacio Díaz Triujeque*, and Castro's *Antigüedades mexicanas* print. In tomo 1 of *México a través de los siglos* published in 1883, the author weaves these pieces into an epic story about cultures prior to Spanish contact, reconstituting this past as a critical element of Mexican nationhood.[40] García Cubas spatializes this archaeology, locating sites on the central map of *X Carta histórica y arqueológica*. Conspicuously, in the *Carta*'s map, García Cubas also overlays the indigenous migration routes represented in the *Cuadro histórico-geroglífico de la peregrinacion de las Tribus Aztecas número 1* and *número 2* of the *Atlas geográfico* onto a landmass (Mexico) that could never have been imagined by the ancient indigenous cultures. But that is not the point, because neither nahua migrations nor archaeological sites and objects are to be related as indigenous history. They are subsumed into a display of a feasible history of Mexico that may be extolled nationally and internationally.

This historical excursion continues in the next page of the *Atlas*, the *XI Reyno de la Nueva España a principios del siglo XIX*. Viceregal history, the third part of Mexico's quadripartite history, is compressed into visualizations of territorial extension and biography. The map of New Spain's intendancies associated with the Habsburg and Bourbon coats of arms, and the equestrian statue of Carlos IV, represent Spain's three-century territorial and political control. This tour of viceregal history is proscribed, however. It is not a Spanish history; it is a Mexican history that subsumes viceregal history as an episode after the indigenous history of the previous atlas page and the contemporary Mexicos of the other pages. Importantly, the control of this viceregal history is firmly reiterated in the *XIII México y sus cercanías* page of the *Atlas*, where the infamous equestrian sculpture, whose move from the central plaza of Mexico City, to University patio, to the Avenida de la Reforma, marking Mexico's rethinking of history, has now become a full tourist site, another city spectacle.[41]

Other *Atlas* pages also point backward to move toward the future. The profile of the road from Veracruz to Mexico of *IV Vías de comunicación* references Humboldt's *Perfil del Camino de Acapulco á Mégico, y de Mégico á Veracruz* and, simultaneously, recalling the epic aura assigned to this route by Prescott in his description of Cortés's march to Tenochtitlán and elaborated upon by others. The mythic references of this path are overlaid with a new one: that of the railroad's promise for yet another age of economic discovery. And as a final example, Mexico's ecological diversity is emphasized through the botanical prints, recalling those of the eighteenth century, that decorate the side panels of the *VIII Carta agrícola* and the Humboldt-derived chart differentiating climatic environments. This ecological diversity points to Mexico's present and future agricultural productivity displayed in the vignettes above and below the map.

At its most fundamental level, *Atlas pintoresco é historico de los Estados Unidos Mexicanos* is a travel exhibit, employing the rhetorical strategies of travel writing. In this way, like displays of Mexico of previous centuries, the *Atlas* offers the viewer-traveler—foreign or Mexican—whatever he or she needs to locate. Foreign viewers may see an incomparable cornucopia of investment potential. And through travel, citizens—elite and bourgeois—may remember various *Mexicos*—political Mexico, ethnographic Mexico, archaeological Mexico, scenic Mexico, intellectual Mexico, historic Mexico, resource-rich Mexico. And, as it is critical for the mapping of any journey, the unfamiliar becomes familiar and, thus, it is possible to find oneself in some or all of these Mexicos.

The *Atlas pintoresco é historico*, in some ways, culminates García Cubas's work of the previous decades, as he maps the spaces of Porfiriato Mexico. The *Atlas* presents diverse Mexicos to inspire citizens, as well as to export to the international stage. For Mexicans, remembering is possible through the atlas and seeing is believing as we *view* answers to Bustamante's questions: "¿Quienes somos? ¿de donde venimos? ¿para donde caminamos?" Further, through the *Atlas*'s fabrication and consumption of the Mexican nation-state, Mexico would contribute to and participate in international visual culture production. In fact, the Library of Congress's copy of the *Atlas pintoresco*, signed by García Cubas himself, was dedicated to the Smithsonian Institution, an organization that at the time was forming itself as a nineteenth-century institution of collection and display.

García Cubas's 1885 *Atlas pintoresco é historico de los Estados Unidos Mexicanos* unified and framed the fragments of real and imagined Mexico that had emerged in the mappings of New Spain and were expanded upon and mobilized in later travel writing and its illustration by the mid-nineteenth century. As a result, through the *Atlas pintoresco*, Mexico exhibits the ongoing move toward cultural modernity discussed in the introduction, attempting to fabricate identity and destiny for itself. The *Atlas*'s inherent dialectic between similarity and difference, backward and forward in time, points to Mexico's construction of alternative modernities.[42] In this way, nationalist visual display does not exhibit new forms of Mexico but reconfigures the old and familiar forms. This display of modernity is incomplete and, thus, manifests Mexico in the process of becoming.

AFTER THE *ATLAS PINTORESCO É HISTORICO*

After publishing the *Atlas*, García Cubas continued to gather data into comprehensive texts about Mexico. Between 1888 and 1898, he published the *Diccionario geográfico, histórico, y biográfico de los Estados Unidos Mexi-*

85 Photograph of Antonio García Cubas in his study, from *El libro de mis recuerdos*. Courtesy of the John Carter Brown Library at Brown University.

canos and *Mexico: its trade, industries and resources* appeared in 1893. These publications contain massive amounts data about Mexico, its people, geography, and history. In 1909, Porfirio Díaz awarded García Cubas a Medal of Honor from Sociedad de Geografía y Estadística de la República Mexicana for fifty years of scientific work. Despite his advanced age and declining eyesight, he continued his geographic work. Antonio García Cubas died on 13 February 1912, at eighty years. At the time, he was compiling another comprehensive work, *Desarrollo de la civilización de México* (Development of Mexican Civilization).[43]

An undated photograph appearing on the introductory page of *El libro de mis recuerdos* immortalizes this tenacious cartographer and geographer (figure 85). It depicts the distinguished García Cubas seated behind a desk, presumably in his study. His forehead rests on his left hand, his right hand poised to write as he peers into the camera. Testifying to his erudition, numerous books surround him: some in wall bookcases, some in a small free-standing case, some on a table in front of his desk, and some on his desk. This library is filled with other items: a globe is placed to the left of his desk; pictures of religious images and landscapes are arranged on the back wall; and small sculptures, busts, and statuettes perch atop almost every flat surface in the room. A miniature copy of the equestrian statue of Carlos IV prances on the shelf directly behind the seated García Cubas.

Displaying Mexico, marking the passage of Mexico from the spaces of New Spain, was the life work of Antonio Garcia Cubas. Yet, notably, a map of Mexico is not visible on the walls presented to the viewer. While it might seem odd that the cartographer does not have a map displayed within such a photographic self-portrait, perhaps it is not peculiar at all. The photo demonstrates the process of mapping practice rather than its products. Through the quite sedentary activity of assemblage, García Cubas constructed his beloved Mexico from the fragmentary materials like those that surround him—texts, images, and objects. With pen in hand, Antonio García Cubas assembled Mexico as a place in-between the texts of the books and the imagery of the statuettes, paintings, and prints. As the title of his memoirs certifies, it was a place of recollection and memory.

CARTOGRAPHIC SILENCES AND THE MAPPINGS OF GENDERED SPACES

This chapter cannot be concluded without reiterating the fact that the late geographic and cartographic works of Antonio García Cubas, like his earlier works, are filled with profound silences. In the *Atlas pintoresco*, women appear among the vignettes of the *II Carta etnográfica* and only two women appear among the intellectuals depicted in *V Instrucción pública* pages. Thus, García Cubas's mappings display the highly gendered spaces of male activities—politics, commerce, agricultural production, and so forth. It appears that women's spaces become silenced and not part of the cartographic nation. In fact, while cartographic silences filled the *Atlas pintoresco*, gendered mappings continued to proliferate in other Porfiriato texts.

As discussed in the previous chapter, diverse kinds of women were visually mapped across and into distinct public and private spaces. With the Reforma movement, women's social and political roles were mapped through domesticity as seen in Alberto Bribiesca's *Educación moral. Una madre conduce a su hija a socorrer a un menesteroso*, of 1879 (figure 59, chapter 5, p. 182). Other spaces for different kinds of women were demarked in lithographs and photographs such as the figures of *china poblanas* and *mujeres públicas*. In this way, women were located in the nation through alternative mappings that located them in spaces deemed appropriate.

By the late 1880s, women's place and space in the nation continue to be located and certified. There were, however, conflicting views on women's specific role in society as illuminated in a written exchange in the journal, *El Correo de las Señoras* (The Ladies's Courier) of 1884. It begins with José María Vigil, author of tomo 5 of *México a través de los siglos*, suggesting that women's education needed to go beyond primary and secondary education, or training in the fine arts, and into fields that can assure honorable work and modest subsistence. Such education, he argues, is vital for the progress of the nation.[44] In his response, Luis Ruiz poses questions to Sr. Vigil: "Do you want her to practice medicine, to practice law, that she represent the people and assume an important role in government?" Ruiz answers his own questions, stating emphatically that "each part of humanity has its own well defined objective. The role of women is so fundamentally important in the sanctuary of the home, that whatever other activity she might want to devote herself to (however important she may suppose), would be small and miserable in comparison to her grand domestic duties."[45] In this exchange, it is clear that the importance of women's education was no longer an arguing point as it had been earlier in the century; the end purpose of this education was.

El Álbum de la Mujer, an important 1880s serial for women, would tread between the Ruiz and Vigil positions. *El Álbum* was initiated, owned, and edited by María de Concepción Gimeno de Flaquer, a Spanish writer who emigrated to Mexico in the late 1870s.[46] *El Álbum de la Mujer* continued the tradition of the earlier *El Semanario de las Señoritas Mejicanas* with emphasis on more serious topics, although there were a few articles on women's European fashions and travel adventures.

As editor of *El Álbum de la Mujer*, Gimeno de Flaquer shaped the weekly publication to stress the importance of women as intellectual achievers and, thus, active contributors to the nation. The cover page of *El Álbum de la Mujer* always includes the image of an important contemporary or historical

86 Title block, from *El Álbum de la Mujer*, 27 July 1884. General Collection, Library of Congress.

woman whose contributions, often self-sacrificing in nature, were the topic of one of the issue's essays. Other pages include short stories, descriptions of European fashions, poetry, and advertisements. The title block of *El Álbum*'s cover page is pertinent as it reveals the viewpoint of this publication (figure 86). In the center of this 27 July 1884 example, we see Knowledge, depicted as a winged figure with a flame on her head, instructing a young boy and girl from a book; a globe, more books, and a lute rest next to the figure. Behind this central figures are images of three seated women. The figure to the immediate left depicts a woman reading a book while seated next to a baby who rests in a cradle; a second figure behind the first plays the piano. In contrast to the interior placement of these figures but remaining in a private space, a third woman to the right paints on a canvas while seated on a balcony that overlooks a church, possibly the Mexico City Cathedral. The imagery of *El Álbum*'s title block displays ideal Porfiriato women as existing in enclosed spaces where enlightened motherhood includes knowledge of geography, literature, history, and the arts.

The 19 June 1887 issue of *El Álbum de la Mujer*, second edition of this publication, changed its focus.[47] The imagery of the title block is eliminated and the cover displays a housewife, wearing an apron and holding knitting in her left hand, standing in front of a mantel and slipping money into a box marked "Savings Bank" (in English). Like the elite woman of Alberto Bribiesca's painting, this cover image demonstrates women's role in overseeing the stability of the household with emphasis here on the management of the domestic economy rather than moral guidance.

This issue continues with an essay by Sofía Taritilan entitled "La educación de la mujer." In it, she states that great philosophers, writers, and historians agree on the necessity of educating women, the "cornerstone on which rests the base of the society and the family."[48] It disparages those who think women are too frivolous to take up serious topics as well as utopians who believe that society can exist without women. It lauds women's education as critical for the family because the "family is society, concentrated, reduced to its most delicate essence, and that man is born to live in society is unquestionable."[49] It concludes that woman's management of the home assures the progress of society. The issue also contains the biography of the poet Dolores Guerrero, short articles, the conclusion to a short story, updates on European fashions, poetry, and a short section of quotes by famous men and women. The final page of the issue displays a page of advertisements. Overall, issues of *El Álbum de la Mujer* provided its readers with reviews of current themes, ideas and trends, expanding women's minds and firmly placing them in the nation through their place in the home. Thus, while not mapped cartographically, the spaces of women were visualized in other nineteenth-century mapping practices.

While contemporary women were mapped through and into domestic spaces of Mexico, they were also mapped allegorically to represent the nation. The use of female imagery to represent the Americas / New Spain goes back to the sixteenth century. This continued to be elaborated in subsequent centuries, as I traced in chapters 1 and 2. With independence, allegorical figures such as the 1834 *Alegoría de la Independencia* (figure 40, chapter 4, p. 117) appeared, while in the late 1860s, Monroy's *Constitución of 1857* (figure 65) marked the Reforma movement. In the 1880s, allegorical figures reappear in *México a través de los siglos* to visualize history in the second frontispieces added to each *tomo*; through this imagery, Mexico's history is conveyed (figures 64 and 65). The Mexican nation continued to be conceived as an allegorical female figure. Mexican history and politics would be male spaces.

This allegorical notion of the nation underlies García Cubas's conceptualization of his own work. This is clearly laid out in his memoirs when he states that prior to his 1850s work, geography in Mexico was like a girl, "contrahecho y desmedrada" (twisted and wasted away). Manifesting paternal emotion, he further writes that his conscientious care rescued this abandoned daughter, bringing her into young adulthood and ultimately making her "una señora poderosa que fomenta los principales ramos de la riqueza publica" (a powerful woman who supports the principal branches of public wealth), meaning that geography is critical for the economic progress of the nation.[50] García Cubas also envisions the nation allegorically as a female figure. His cartographic and geographic works provided form and context for this beloved figure of *la patria*.

7

PERFORMING THE NATION

In the late evening of 15 September 1910, an estimated crowd of 100,000 people of all classes and ages assembled impatiently in the Plaza de la Constitución, where in viceregal times the great equestrian statue of Carlos IV had once stood. At precisely eleven o'clock, Porfirio Díaz stepped out onto a balcony of the Palacio Nacional, and solemnly pronounced, "¡Viva la Libertad! ¡Viva la Independencia! ¡Vivan los héroes de la Patria! ¡Viva la República! ¡Viva el Pueblo Mexicano! (Long live Liberty! Long live Independence! Long live heroes of the nation! Long live the Republic! Long live the Mexican people!). With the national anthem playing, fireworks exploding, and the crowd roaring, Díaz's rousing patriotic chant, remembering and reenacting Hidalgo's Grito de Dolores of 1810, was part of the monthlong Centenario de la Independencia de México, a celebration marking the 100th anniversary of the commencement of Mexico's independence movement.

In true Porfiriato style, the Centenario de la Independencia was orchestrated as an extravaganza, an acclamation of Mexico's past and present achievements and the promise of its future for a Mexican as well as international audience. The dizzying number of ceremonies and events that took place in September of 1910 are well documented with narrative texts and numerous photographs in the *Cronica oficial de las fiestas* (celebrations) *del primer* (first) *Centenario de la Independencia de México*, authored by Genaro García—director of the Museo Nacional de Arqueología, Historia y Etnología and one of the official historians of the Centenario—and published by the Museo Nacional in 1911.[1] Perusing the *Cronica oficial*'s pages, however, there is a sense of déjà vu: the themes, the ideas, as well as the people and events displayed in the centenary activities are apparitions from the inventory of images and imagery circulating in numerous publications of the mid- and late nineteenth century, such as *México y sus Alrededores* (1855); *Los mexicanos pintados por sí mismos* (1854); Rivera Cambas's *México pintoresco, artistico y monumental* (1880); Riva Palacio's *El Libro Rojo 1520–1867* (1870), as well as his *México a través de los siglos* (1882); and, even, Prescott's *The Conquest of Mexico* (1843). Further, García Cubas's *Atlas pintoresco é historico de los Estados Unidos Mexicanos* (1885) almost seems to have served as the conceptual and thematic blueprint for the Centenario celebration. Pages from the *Atlas pintoresco* were transformed into dioramas and performances during this centennial celebration. This nationalist visual display, like García Cubas's work, did not just exhibit new forms of Mexico. It too reconfigured old and familiar forms as real and imagined manifestations of Mexico as a nation. Although taking place at the beginning of the twentieth century, the Centennial's exhibitionary visuality and display attest that the great spectacle was rooted in the exhibitionary visuality and display of nineteenth-century visual culture. Through the numerous events, activities, and performances of this grand celebration, the Mexicos envisioned by Antonio García Cubas came to life.

The Centenario celebration was a synthetic moment. Through an examination of its content and structure, we may observe the impact of the mapping practices of the nineteenth century. This is to say, the Centenario celebration provides summation of Mexico's nineteenth-century visual culture traced in this study. Mexican participants in the Centenario de la Independencia played the dual roles that underlie the production and consumption of nineteenth-century visuality: they were both *observed in* the displays of Mexico and, at the same time, *observers of* this grand exposition of Mexico. Young students from various schools were marshaled to the Plaza de la Constitución to participate in a swearing of patriotic allegiance pageant. Factory workers, teachers, military officers and soldiers, Indians, and members of various other groups marched in parades or were exhibited as part of themed events. At the same time, transformed from the observed to observers, these diverse social groups would witness Mexico through specific thematic foci. For example, spectators could view Mexico's advances in public health during the inauguration of the Manicomio General (Mental Hospital) by the president and his wife, Carmen Romero Rubio de Díaz, in early September. They could also tour national schools for the blind, deaf and mute, and orphans. Manifesting educational development, institutions such as the Escuela Nacional Primaria Industrial para Niñas (Girls) and the Escuela Normal para Maestros (Teachers) were opened.[2] The Universidad Nacional de México was inaugurated with much fanfare in the presence of diplomats as well as representatives from various international universities.[3] Also calling to mind the *V Instrucción pública* and *VIII Carta agrícola* pages of the *Atlas pintoresco é historico*, intellectual advancement was displayed through showcases of Mexican scientific, medical, hygienic, and agricultural accomplishments. Mexico was further exhibited through expositions of geology, visual art, and industrial technology.[4]

Along with these material demonstrations, Centenario spectators also witnessed allegorical manifestations of Mexico. Independence was symbolized as *El Ángel de la Independencia*, a 115-foot column surmounted by a Winged Victory figure and located on the Paseo de la Reforma along with older monuments including the equestrian statue of Carlos IV. At the same time, celebrants were visually immersed in Mexican history. Mexico City streets were dedicated to important figures in Mexico's historical narrative such as Isabela la Católica, who supported Columbus's voyages. Numerous historical monuments were dedicated, such as those to the Niños Héroes de 1847, the young cadets who stood up to the invading U.S. Army, and a massive marble monument to Benito Juárez. Concurrently, historic memorabilia was enshrined and displayed.[5] On 2 September the baptismal font used by Hidalgo was transported from the town of Cuitzeo de Abasolo to Mexico City (figure 87). It was carried from the train station where it had arrived to its final destination, the Museo Nacional de Arqueología, Historia y Etnología on an elaborately decorated carriage. Described in the *Cronica oficial* as symbolizing the waters that initiated the emancipation of Mexico, the font was surrounded by the image of the sun, flag, and Phrygian cap.[6] This museum proudly displayed its pre-Columbian collections, too (figure 88). A photograph showing its exhibition of ancient monuments mirrors Casimiro Castro's *Antigüedades mexicanas* page of *México y sus Alrededores* (figure 43, chapter 4, p. 139), the *Varios objetos* print of tomo 1 from Rivera

87 "La fuente bautismal de Hidalgo," from Genaro García, *Cronica oficial de las fiestas del primer Centenario de la Independencia de México* (1911). Courtesy of the John Carter Brown Library at Brown University.

Cambas's *México pintoresco* (figure 61, chapter 6, p. 190), and, especially, the more ordered images associated with the *X Carta histórica y arqueológica* (figure 73, chapter 6, p. 210).

So intense was the desire for popular inclusiveness of this celebration that even those who lost the right to be spectators participated: on 16 September, the planning committee arranged for a choral group to perform for the prisoners in the central hall of the Penitenciaría del Distrito Federal. From their cells, convicts listened to patriotic songs, including the "Himno Nacional."

88 "Vista del gran salón de monolitos," from Genaro García, *Cronica oficial de las fiestas del primer Centenario de la Independencia de México* (1911). Courtesy of the John Carter Brown Library at Brown University.

Inmates of the city jail and the House of Correction for Men and Women enjoyed a modest banquet on that day as well.[7]

Dignitaries and officials from the Americas, Europe, and Asia also participated in the Centenario activities and honored Mexico with gifts. Mexico received a statue of Louis Pasteur from France and a facsimile of Donatello's *Saint George* from Italy. Some foreign governments also acknowledged their prominent role in and impact on Mexican history by donating and, in some cases, returning symbolic objects. For example, Mexico received a statue of Alexander von Humboldt from Germany. Spain returned personal garments belonging to José María Morelos, a leader of the war of independence who was captured and executed by the Spanish Army, which had been on exhibit at the Museo de Madrid, (figure 89). France restored the symbolic keys to Mexico City, acknowledging the French intervention. The United

89 "Prendas de Morales . . ." (Clothing of [José María] Morelos), from Genaro García, *Cronica oficial de las fiestas del primer Centenario de la Independencia de México* (1911). Courtesy of the John Carter Brown Library at Brown University.

States was silent about its invasion and annexation of Mexican territory during the Mexican-American War. Instead, the United States acknowledged Mexico's centennial with a statue of its first president, George Washington. Thus, Mexico's independence was celebrated by representatives of foreign governments as observers of a great panorama of Mexico; at the same time, through their gifts and associated ceremonies, they became incorporated into this panorama.

Perhaps the most revealing events of the Centenario were three parades: Desfile de carros alegóricos del comercio (Parade of Floats with Commercial Allegory) taking place on 4 September; the Gran procesión cívica formada por todos los elementos de la sociedad Mexicana (Great Civil Procession Formed of All Elements of Mexican Society) on 14 September; and the Desfile histórica (Historical Parade) on 15 September. Like other parts of the centennial celebration, these pageants incorporated real and imagined aspects of Mexico. But unlike the static displays of grand openings, inaugurations of monuments, or assorted exhibitions, the diverse aspects of Mexico were *performed*.

The Desfile de carros alegóricos del comercio consisted of a series of highly decorated floats focused on the themes of Mexico's physical resources and economic potential. The lead-off float was sponsored by El Centro Mercantil, a department store, and was pulled by horses with groomsmen who wore costumes in the style of Louis XV. It included the flags of Mexico and France and a portrait bust of Hidalgo being crowned by a woman who represented the nation. Busts of Juárez and Díaz placed on either side of Hidalgo's portrait were also crowned by women who respectively were allegorical representations of Justice and Peace. A coach carrying representatives of Empleados de Comercio (Employees of Commerce) and the military band of the Seventh Battalion followed this float.

The next float, sponsored by the department store El Palacio de Hierro, was also drawn by horses and preceded by elaborately costumed attendants. It too contained an allegorical figure of Mexico, recalling the frontispiece figures of the *México a través de los siglos*, who stood in front of an image of the sun, with four richly dressed women attending her. A carriage filled with employees, three automobiles owned by the Palacio de Hierro, and the military band of the Eleventh Battalion followed this float.[8] These elaborate moving displays tied commercial interests to patriotism.

The third float represented the agricultural wealth of Mexico. In a replay of the cartouche of the *II Carta etnográfica* and the *VIII Carta agrícola* pages, the float was in the form of a mountaintop upon which one could see rustic gods and rural scenes with individuals wearing the regional costumes of Tehuantepec, Jalisco, and Yucatán. They were associated with products from the various agricultural zones of the Republic, including grapevines, palm trees, cactus, wheat, and sugar cane. Pulled by pairs of oxen led by drivers, the float preceded a group of rancheros in leather outfits riding their horses.[9]

A subsequent float, "Carro alegórico de la Minería," allegorically displayed the mineral resources of Mexico, the theme of the *IX Carta minera*. It depicted a mountain crowned with a nopal cactus and eagle, with a *matrona* (matron) personifying the mining industry seated in front of the mountain and attended by two women representing Gold and Silver. Groups of miners with mining equipment walked alongside this float. The "Carro alegórico de la Industria" portrayed the industrial progress of Mexico with a matrona who represented industry and was seated on a throne with an image of the sun behind her head. She was surrounded by four females, who represented Ciencia (science), Trabajo (work), Exactitud (precision), and Fuerza (energy), critically elements of the expansion of industry.[10]

The floats of the Desfile de carros alegóricos del comercio recall the imagery of wonder and abundance combined with allegorical figures that go back to the earliest visualizations of the Americas / New Spain, such as the seventeenth-century engraving *America* by Jan van der Straet or the *Continent of America* ceramic figurine of the eighteenth century (figures 4 and 7, chapter 1, pp. 28, 31). Beginning with Bourbon reforms and continuing with early nineteenth-century travelers to Mexico, these themes were reformatted to emphasize Mexico's economic potential. Here, Mexico's substantial commercial potential was paraded for all to see and, hopefully, believe.

While Mexico's physical content and potential were clearly laid out in the first parade, the commission in charge of planning sought to develop a second parade as a "civic manifestation in honor of the liberators," which was officially designated the Gran procesión cívica formada por todos los elementos de la sociedad Mexicana. Led by members of the Comisión Nacional de Centenario de la Independencia, it commenced at nine o'clock in the morning on 14 September and, according to the *Cronica oficial*, participants in this parade exceeded 20,000.[11] These included individuals from every segment of Mexican society: representatives from banking, commerce, industry, and mining; representatives of states and territories; artisans and laborers; managers of factories; government officials; representatives of scientific and literary academies and institutions of higher education; and, veterans of the Mexican-American War. The *Cronica oficial*'s description captures the excitement of that morning: "The streets were filled with people; the buildings were tastefully adorned with decorations, and in the balconies and doorways of the commercial houses, a multitude crammed to witness the civic parade."[12] The parade groups marched to the Cathedral to participate in ceremonies honoring the Heroes of the Independence. This pageant focused on Mexico's human capital, that is, citizens' role in the formation and expansion of the nation.

With the content and citizens of Mexico visualized, the performance of Mexico turned to illuminating national history. Although smaller in scale than the Gran procesión cívica, the Desfile histórica was laboriously planned and executed. Meticulous preparations took place for the historical scenes reenacted by hundreds of participants, with extensive costuming and staging. On 15 September 1910, about 50,000 to 70,000 people saw the three great epochs of national history pass before their eyes: the conquest, Spanish domination, and independence.[13] Along with marching the length of the designated path, characters from each of these epochs reenacted a particular event that symbolized the designated era.

The conquest was marked by a parade of about 200 indigenous people dressed as the ancient tribes of central Mexico along with the emperor Moctezoma and his court, who received loud applause from onlookers (figure 90). They were followed by numerous individuals garbed as Spanish soldiers on horses and on foot. Then came Cortés with Doña Marina (or Malintzin), his translator and consort; and the Tlaxcalteca warriors and nobles who had conspired with Cortés against Moctezoma and his warriors. Arriving below the balcony of the Palacio Nacional, the first encounter between Cortés and Moctezoma was reenacted, a description going back to writers of the seventeenth and eighteenth centuries and a critical event of Prescott's epic tale.

Next, the era of Spanish domination was marked by the appearance of 288 persons dressed as individuals from the viceregal period. These included members of the Ayuntamiento, an oversight body; civil and military functionaries; and, of course, a viceroy (figure 91). In this section of the Desfile histórica, El Paseo de la Pendón (The March of the Standard) depicted these Spaniards carrying a black banner inscribed with a red cross.[14] It also re-created viceregal ceremonies: the Hapsburg's Jura del Rey (a swearing of allegiance to the king) and the Bourbon's El Alza del Pendón Real (a raising of the royal banner) ceremonies.[15]

The era of independence was signified by a reenactment of the entrance of the Trigarante Army, under the command of General Agustín de Iturbide, who received great applause when he marched past the crowds. Along with the general, other actors played historical figures of the independence movement, such as Vicente Guerrero, Manuel Mier y Terán, Guadalupe Victoria, and Anastasio Bustamante who also appear among the historical figures depicted on the *I Carta política*. Also in this section, floats from the various states displayed allegorical figures along with their local historical events or personages.

90 "El Emperador Moctecuhzoma," from Genaro García, *Cronica oficial de las fiestas del primer Centenario de la Independencia de México* (1911). Courtesy of the John Carter Brown Library at Brown University.

91 "El oidor decano [high judge], el virrey y el alférez real [royal officer]," from Genaro García, *Cronica oficial de las fiestas del primer Centenario de la Independencia de México* (1911). Courtesy of the John Carter Brown Library at Brown University.

Viewed as a whole, the Desfile de carros alegóricos del comercio, the Gran procesión cívica, and the Desfile histórica constitute a striking conundrum: their format was both fundamentally old and radically new. It was old not only because of the origins of the themes and imagery but also because of its format. Beginning with Cortés's conquest of Tenochtitlán and throughout subsequent centuries, great formal processions were a regular part of viceregal practices for asserting and enacting monarchical power in New Spain.[16] For example, the 1666 funeral procession marking King Philip IV's death took place in the Plaza Mayor in front of Mexico City's Cathedral. It began with the black and mulatto confraternities, followed by the Indian and Spanish confraternities and the Spanish elites. Next came the royal tribunals and, finally, the viceroy. The procession's order reflected the metaphor of the human body, constituted of disparate, independent yet intradependent members with a common head—the king, represented by the viceroy.[17]

Such pageants reaffirmed the sovereignty of the king and demonstrated that New Spain was a timeless, harmonious, and hierarchized manifestation of the corporate body of monarchy. These large-scale public spectacles imagined an idealized viceregal polity where everyone knew their place in the social hierarchy.[18] Through the control of both the message and the medium, viceregal authorities also used these opportunities to visualize political and social concepts that benefited their continued rule. Public processions placed the ideal government on display in order to impress onlookers and instill hope in the possibility of change and renewed prosperity.[19]

Likewise, the Porfiriato's Centenario parades demonstrated the power of the state to organize, orchestrate, and display; however, these processions were also radically new in their conception because this great panorama incorporated specific mid-nineteenth-century imagery and it was formed by "todos los elementos de la sociedad Mexicana." Here, sovereignty was visualized not as a hierarchical body made up of royal subjects but as a body of citizens that formed the Mexican state. Ostensibly, unlike a royal subject, a citizen's place in society was not located in corporate hierarchy but through his or her participation in and observation of the state and its projects. Thus, as we saw emerge in the mid-nineteenth century, the Centenario events—especially its parades—provided citizens en masse with ways to remember what could not be remembered and see themselves as members of the nation.[20] Further, this revised body of the state was set within historical epochs reenacted in the Desfile histórica; that is, sovereignty was located within the time of the nation.

The power of the state to produce this great centenary celebration marks its ability to exhibit and perform the nation as it had been fabricated through diverse mapping practices of the previous century. Mexico developed as a nation not only through political and economic independence but also through the emergence of a visual culture that engaged viewers, activating them to question and recognize their identity and destiny through intensified mapping practices. Here, Mexico was performed as a place—both real and imagined.

On 15 September each year, concluding Mexico's annual celebrations of independence, the current president repeats the chant of El Grito, "¡Viva la Libertad! ¡Viva la Independencia! ¡Vivan los héroes de la Patria! ¡Viva la República! ¡Viva el Pueblo Mexicano!" to an awaiting crowd in the Plaza de la Constitución. Future presidents will continue this tradition because at this annual reenacting moment, Mexico and Mexican citizens continue the journey of becoming Mexico as they hear and repeat answers to Mexico's enduring questions: "¿Quienes somos? ¿de donde venimos? ¿para donde caminamos?" Through this tradition, Mexican citizens remember and connect to the past from the present in order to assure the nation's future. Here, Mexico again is reenacted as a place both real and imagined. This is both the vision and legacy of the mapping practices of nineteenth-century Mexico.

NOTES

INTRODUCTION

Unless otherwise specified, translations of foreign-language quotations are my own. Further, the use of accents and other diacritical marks can be inconsistent in nineteenth-century publications. I have retained the diacritical marks of original texts when quoting or referring to them and I use contemporary markings in my narrative.

1 It should be noted that the history of nineteenth-century cartography throughout the Western world is not well studied.

2 The writings of José Rabasa, Walter Mignolo, and Mauricio Tenorio-Trillo allude to such a connection but do not develop it thoroughly.

3 See Svetlana Alpers, "The Mapping Impulse in Dutch Art."

4 Matthew Edney, *Mapping an Empire*, 23.

5 John Pickles, *A History of Spaces*, 46.

6 Edney, "Rewriting the Colonial Mapping of New England," 4–8.

7 Ibid., 4–8.

8 J. B. Harley, "Silences and Secrecy," 71.

9 Harley and David Woodward, *The History of Cartography*, 1: xv–xxi, xvi.

10 Examples include Thongchai Winichakul's *Siam Mapped* (1994), which traces the evolution of Thai nationhood through its discourses on geography. This influential work integrates the history of cartography into an evolving theory and literature of nation building. Matthew Edney's *Mapping an Empire* (1997) offers a detailed study of the British imperialism's mechanism of mapping to create the physical and ideological place called India as it reified the empire. The works of Winichakul, Edney, Ricardo Padrón, and Raymond Craib—just to mention a few researchers—reflect the implementation of Harley's call to

"interrogate maps as actions . . ." Harley, "Silences and Secrecy," 71. This work also comes out of Benedict Anderson's ideas about nation building.

11 Pickles, *A History of Spaces*, 4–5.

12 Ibid., 5.

13 Ibid., 7.

14 Ibid., 8.

15 Theresa Brennan and Martin Jay, *Vision in Context*, 67. For an important discussion of the distinction between northern and southern Renaissance ways of seeing, see Alpers, "The Mapping Impulse in Dutch Art." See also Ken Hillis, "The Power of Disembodied Imagination," 1–17.

16 Pickles, *A History of Spaces*, 80.

17 Ibid., 12.

18 Ibid., 87–88.

19 Ibid., 89. Anne Godlewska also discusses this in her writings on cartography and geography. See "Map, Text and Image."

20 Pickles introduces the notion of shimmering in his text on page 184, but does not extend the full impact of this visual metaphor. I believe that the concept of palimpsest, emphasized by Rabasa in *Inventing A-m-e-r-i-c-a*, is a better visual metaphor for map viewing.

21 "Al abrir el 'Libro de mis recuerdos' se levanta el velo de lo pasado y aparece en la escena una sociedad que, por sus costumbres, difiere esencialmente de la actual. En aquel brillaba más el elemento moral y en ésta resalta el elemento material: fúndanse en uno ambos caracteres y la nación será grande." García Cubas, *El libro de mis recuerdos*, 11.

22 For an excellent overview of the notion of staged space and its implication for the mapping of Mexico, see Craib's introduction in *Cartographic Mexico*.

23 See Gillies, "Theatres of the World," 70–98.

24 "Quiero que observes los cuadros que te ofrezco, con tu vista natural." Antonio García Cubas, *El libro de mis recuerdos*, 453.

25 Historians are now exploring the importance of visual order in Porfirian Mexico. For example, in his *Visions of the Emerald City*, Mark Overmeyer-Velázquez writes, "The contemporaneous industrial expansion, the amplification of the state's administrative power, and visual technologies made visual regimes central to the process of state formation during this period" (155). See also Tenorio-Trillo, *Mexico at the World's Fairs*.

26 In the introduction to *Visual Culture*, Margaret Dikovitskaya provides an excellent historical overview of the field of visual culture in the United States.

27 Vanessa Schwartz and Jeannene Przyblyski, *The Nineteenth-Century Visual Culture Reader*, 6–7.

28 Dikovitskaya, *Visual Culture*, 56, 57.

29 Schwartz and Przyblyski, *The Nineteenth-Century Visual Culture Reader*, 7.

30 Changes in the use of visuality in Hapsburg versus Bourbon reigns is well

documented and explained by Curcio-Nagy in *The Great Festivals of Colonial Mexico City* and Daniela Bleichmar's articles, "Visible Empire" and "A Visible and Useful Empire."

31 Jonathan Crary, *Techniques of the Observer*, 74.

32 Ibid., 14. In a zootrope, revolving figures placed on the inside of a cylinder are viewed through slits in its side, resulting in the illusion of viewing a single figure in motion.

33 Ibid., 24. Crary actually suggests that these classical modes "collapsed." Recent studies suggest that Renaissance visual modes and these evolving visual modes coexisted, as evidenced in France by the coincidence of academic and impressionist artists.

34 See Crary's "Modernity and the Problem of Attention," in *Suspensions of Perception.*

35 Crary, *Techniques of the Observer*, 10–11.

36 See ibid., 10 n. 8, and Crary, *Suspensions of Perception*, 22–28.

37 See Tony Bennett, *The Birth of the Museum.*

38 For diverse criticisms of Crary's ideas and arguments see, for example, Heidi Nast and Audrey Kobayashi, "Re-Corporealizing Vision"; and David Philip, "Modern Vision," 129–38.

39 Deborah Poole, *Vision, Race, and Modernity*, 9.

40 Ibid., 11.

41 Ibid., 8.

42 Dilip Parameshwar Gaonkar, *Alternative Modernities*, 17–18.

1 MAKING THE INVISIBLE VISIBLE

1 García Cubas, *Atlas Geográfico, estadístico é histórico de la República Mexicana formado por Antonio García y Cubas*, preface, n.p.

2 García Cubas does mention sixteenth-century and seventeenth-century sources in his 1869 *Curso elemental de geografia universal.*

3 For an extended analysis of the early modern exploration of the Americas, see Nicolás Wey Gómez, *The Tropics of Empire.*

4 Edney, *Mapping an Empire*, 21.

5 The scholarly literature on space and place is extensive. A comprehensive overview and extended bibliography is found in Mike Crang and Nigel Thrift's *Thinking Space.*

6 For an important discussion of the move away from Edmundo O'Gorman's discussion in *La invención de América* toward an understanding of invention of a semiotics of cultural processes, which allows for identification of counter-discourses, see José Rabasa, *Inventing A-m-e-r-i-c-a*, 3–6.

7 For historical development of cartography, see J. B. Harley and David Woodward, *The History of Cartography*, vol. 1; Hans Wolff, *America: Early Maps of the New World*; and John Wolter and Ronald Grim, *Images of the World.* For

discussion on the development of the geographic imagination, see O'Gorman, *La invención de América*; Walter Mignolo, *The Darker Side of the Renaissance*; Rabasa, *Inventing A-m-e-r-i-c-a*; and Denis Cosgrove, "Mapping the World."

8 Francesca Fiorani, *The Marvel of Maps*, 73. For an important discussion of the difference between curiosity and wonder, see Lorraine Daston and Katharine Park, *Wonders and the Order of Nature, 1150–1750*.

9 Fiorani, *The Marvel of Maps*, 74–78. While there are inventories of the Cosimo I's holdings, the precise contents of each cupboard in the Guardaroba have not been established.

10 Norman Thrower, *Maps and Civilization*, 58.

11 Ibid., 23–26.

12 For an overview of the development of the sixteenth-century mapping of America, see Uta Lingren's "Trial and Error in the Mapping of America during the Early Modern Period"; and Rüdiger Finsterwalder's "The Round Earth on a Flat Surface."

13 It was published in German as *Von der neuw gefunden Region die wol ain welt genent mag weerden* (Of the Recently Discovered Region Which Probably Can Be Called a World).

14 Hans Wolff, "Newly Discovered Islands, Regions and People," 106.

15 Wolter and Grim, *Images of the World*, ix–x, 81–82. Mercator's first name appears as Gerardus and Gerhard in various sixteenth-century editions of his work and has been popularized as Gerard.

16 Ibid.

17 It should be noted that the second edition of Ortelius's *Theatrum* includes a frontispiece that shows a columned space with four female figures representing the continents. An elaborately clothed Europe is seated and holds a scepter in her right hand while her left hand rests on an orb; in contrast, a completely nude America reclines, with a weapon in one hand and a severed human head in the other. The frontispiece imagery implies that the known world, its lands, people, and resources can be integrated into a single visual and textual narrative.

18 Rabasa, *Inventing A-m-e-r-i-c-a*, 191.

19 Ibid., 186.

20 Ibid., 181–192.

21 Wolter and Grim, *Images of the World*, 85–91.

22 For further discussion of this print, see Michael J. Schreffler's "Vespucci Rediscovers America," 295–310; and José Rabasa, *Inventing A-m-e-r-i-c-a*, 23–48.

23 Hugh Honour, *The European Vision of America*, 157.

24 For an enlightening analysis of the history of natural science from the standpoint of wonder, see Daston and Park, *Wonders and the Order of Nature, 1150–1750*.

25 Lisbet Koerner, "Carl Linnaeus in His Time and Place," 146–49.

26 Quoted in ibid., 151.

27 Daniel Bleichmar, *Visual Culture in Eighteenth-century Natural History*, 103–8.

28 Ibid., 126.

29 Ibid., 114.

30 Ibid., 129.

31 Ibid., 138–39.

32 Ibid., 206–8.

33 Jorge Cañizares-Esguerra, *How to Write the History of the New World*, 12.

34 These arguments are outlined well in Antonello Gerbi's classic work *The Dispute of the New World*; and further analyzed in Cañizares-Esguerra's more recent *How to Write the History of the New World*.

35 Cañizares-Esguerra, *How to Write the History of the New World*, 36.

36 Ibid., 41.

37 Ibid., 41.

38 Jacques Grasset de Saint-Sauveur, *Tableaux des Principaux Peuples de l'Europe, de l'Asie, de l'Afrique de l'Amerique. . . .* , n.p. See also Michelle Pastoureau's "French School Atlases," 109–34.

39 Grasset de Saint-Sauveur, *Tableaux des Principaux Peuples*, 9–14.

40 See Antonio Barrera-Osorio, "Empire and Knowledge."

41 Bleichmar, *Visual Culture in Eighteenth-Century Natural History*, 162, identifies this process as specific to natural history. I suggest that it was inherent in European attempts to narrate America in general and New Spain specifically.

2 LOCATING NEW SPAIN: SPANISH MAPPING

1 See Ricardo Padrón, *The Spacious Word*.

2 Bleichmar, *Visual Culture in Eighteenth-Century Natural History*, 24; Luisa Martín Merás, *Cartografía marítima hispana*, 76–77.

3 Thrower, *Maps and Civilization*, 67–68; Martín Merás, *Cartografía marítima hispana*, 78–80.

4 See Martín Merás, *Cartografía marítima hispana*; and Wolff, *America: Early Maps of the New World*, for excellent examples of the shifting image of the Americas in manuscript maps.

5 Padrón, *The Spacious Word*, 8. See also María Portuondo's *Secret Science*.

6 Mignolo discusses López de Velasco's work throughout his book *The Darker Side of the Renaissance*. See Juan López de Velasco, *Geografía y descripción universal de las Indias*.

7 Barbara Mundy, *The Mapping of New Spain*. See also Alessandra Russo, *El Realismo Circular*.

8 Padrón, *The Spacious Word*, 137–38.

9 For an excellent overview and assessment of this Spanish information-gathering process, see Bleichmar, *Visual Culture in Eighteenth-Century Natural History*; and Paula De Vos, "Research, Development, and Empire," and "Natu-

ral History and the Pursuit of Empire in Eighteenth-Century Spain." In his article "Empire and Knowledge," Antonio Barrera-Osorio argues that early modern empires and science were born together and that "empirical practices became quantitatively significant in the context of the Atlantic world. In the Spanish case, in particular, the establishment of the American kingdoms fostered the development of empirical practices for the purpose of studying, exploiting, and controlling the New World" (39).

10 The scholarly studies of the *relaciones geográficas* are extensive. Some of the more interesting recent works include Francisco de Solano, *Relaciones geográficas de Arzobispado de México, 1743*; de Solano and Pilar Ponce, *Cuestionarios para la formación de las relaciones geográficas de Indias siglos XVI–XIX*; Romero Navarrete and Echenique March, *Relaciones geográficas de 1792*; Áurea Commons and Atlántida Coll-Hurtado, *Geografía histórica de México en el siglo XVIII*; and Mundy, *The Mapping of New Spain*.

11 See de Solano and Ponce, *Cuestionarios para la formación de las relaciones geográficas de Indias siglos XVI–XIX*.

12 Mundy, *The Mapping of New Spain*, 12–16. See also Martín Merás, *Cartografía marítima hispana*, 101–111, for magnificent images of Santa Cruz's map productions. For a reading of Santa Cruz's work in a broader natural history context, see Barrera-Osorio, "Empire and Knowledge," 46–47.

13 Mundy, *The Mapping of New Spain*, 17.

14 These translations are by Barbara Mundy from the 1577 questionnaire. See *Relaciones geográficas del siglo XVI*, edited by René Acuña.

15 Mundy, *The Mapping of New Spain*, 113.

16 Such as Francisco Hernández's *Rerum medicarum Novae Hispaniae thesaurus, seu*. See Varey, Chabrán, and Weiner, *Searching for the Secret of Nature*.

17 Mundy, *The Mapping of New Spain*, 19–20.

18 See Barrera-Osorio, *Experiencing Nature*, chap. 2.

19 See also ibid., for an explanation of interest in the Americas during the seventeenth century.

20 Herrera y Tordesillas, *Historia general de los hechos de los castellanos en las islas y tierra firme del mar oceano*, 1: n.p.

21 For an excellent discussion of Herrera y Tordesillas's work as marking a new kind of historiography, see Tom Cummins, "De Bry and Herrera."

22 *Interrogatorio para todas las ciudades, villas y lugres de españoles, y pueblos de naturales de las Indias occidentales, islas, y tierra firme; al cual se ha de satisfacer, conforme a las preguntas siguientes habiendolas averiguado en cada pueblo con puntualidad y cuidado*. Reproduced in *de Solano and Ponce, Cuestionarios para la formación de las relaciones geográficas de Indias siglos XVI–XIX*, 99–101:

53 Cuántos indios tributarios hay en cada parcialidad.

54 Cuántos indios administra y tiene a su cada cargo cacique.

55 Cuántos casados, cuántos solteros, cuántos viejos reservados, cuántos menores que no tributan.
75 Cuál es la comida y bebida más ordinaria de los indios de este pueblo.
77 Qué ropa se labra en este pueblo y qué ganados se crían.
102 Cuántos son españoles; de los españoles, cuántos hombres, y cuántas mujeres, con la distinción de las edades y de los estados y cuántos son criollos, y cuántos nacidos en España, y de qué provincias de España.
106 Cuántos son mulatos y zambaygos, declarando el número de los hombres y el número de las mujeres, con la distinción de las edades y de los estados; y de estos mulatos, cuántos son libres y cuántos esclavos.

23 De Vos, "Research, Development, and Empire," 57.
24 Bleichmar, *Visual Culture in Eighteenth-Century Natural History*, 29–30. See also Gabriel Paquette, *Enlightenment, Governance, and Reform in Spain and Its Empire, 1759–1808.*
25 Bleichmar, *Visual Culture in Eighteenth-Century Natural History*, 41.
26 *Real cédula ordenando se envien completos informes sobre nucleos urbanos, demograficos, economicos y eclesiasticos de todos los territorios de Indias. Madrid, 19 julio 1741.* Reproduced in de Solano and Ponce, *Cuestionarios para la formación de las relaciones geográficas de Indias siglos XVI–XIX*, 141.
27 Linda Curio-Nagy's discussion of the structuring of Bourbon festivals in contrast to earlier Habsburg festivals is particularly instructive in understanding the underlying philosophy of Bourbon good government. See Curio-Nagy, *The Great Festivals of Colonial Mexico City*, as well as Román Gutiérrez, *Las reformas borbónicas y el nuevo orden colonial.*
28 Commons and Coll-Hurtado, *Geografía histórica de México en el siglo XVIII*, 12.
29 Ramón María Serrera, "Estudio preliminar," in José Antonio de Villaseñor y Sánchez, *Suplemento al Theatro Americano*, 9. This essay contains further details of Villaseñor y Sánchez's life.
30 The following list is reproduced in de Solano and Ponce, *Cuestionarios para la formación de las relaciones geográficas de Indias siglos XVI–XIX*, 143–44:

1. Lo primero, expresando la distancia de la Cabecera de esa jurisdicción a esta capital, y a qué rumbo está situada, y así mismo las de todos los Pueblos, Villas y Lugares sujetos a dicha cabecera y a todas las demás de su jurisdicción, con sus temperamentos, leguas y rumbos.
2 Qué familias se hallan en el Vecindario de cada Pueblo, por pequeño que sea, así de españoles como de indios, y demás Naciones que lo compongan.
3 Cuáles son los frutos, que en cada parte sirven de comercio a aquellas Repúblicas; si han tenido alguna decadencia de los tiempos pasados a éste, y en qué ha consistido, y qué remedios son los más proporcionados para sus mayores aumentos.

4 Qué minerales contiene dentro de su distrito, y de qué especies de metales y la naturaleza de cada recinto.
5 Por qué sujetos, y doctrineros están administrados en lo espiritual. Si hay falta de ellos. Las imágenes milagrosas y su origen.
6 Qué misión, o misiones se hallan en la propagación de nuestra Santa Fe Católica, las que son ya establecidas, y las que son nuevas reducciones.
7 Qué misioneros asisten en ellas a la dilatación del Santo Evangelio, sus idiomas, y su Estatuto.
8 Y por último, la distancia que tiene cada partido de la Jurisdicción de la Alcaldía Mayor, o Corregimiento, que es a cargo de V. y si hay necesidad por la incomodidad de grandes distancias de algunas poblaciones nuevas.

31 Reproduced in de Solano, *Relaciones geográficas de Arzobispado de México, 1743*.
32 *Mestizo*, *español*, and *castizo* were terms which identified an individual according to their blood ancestry: a *mestizo* refers to a person of one-half Spanish and one-half Indian blood; *español* refers to a person of pure Spanish blood; and a *castizo*, the offspring of a mestiza/o and español/a, thus refers to a person of three-quarters Spanish blood.
33 De Solano, *Relaciones geográficas de Arzobispado de México, 1743*, 75.
34 Villaseñor y Sánchez, *Theatro Americano descripcion general de los reynos y provincias de la Nueva España y sus juridicciones*, n.p.
35 Ibid., n.p.
36 "A la Dedicatoria ha añadido una pulida Lamina en que à mi juicio compendiza el designio todo de la Obra, y el intento de Soberano, que la mandò executar." Villaseñor y Sánchez, *Theatro Americano descripcion general de los reynos y provincias de la Nueva España y sus juridicciones*, n.p.
37 Villaseñor y Sánchez, *Theatro Americano descripcion general de los reynos y provincias de la Nueva España y sus juridicciones*, n.p. Michel Antochiw discusses this map in "La visión total de la Nueva España," 71–77.
38 "Assi como para conocer las partes de un cuerpo, es necessario aver discurrido primero, en el todo anathomico para distinguir sus miembros, es tambien necessario para tratar las partes de un Reyno hacerse primero cargo de la figura, y distancias del todo, y tamaño su contenido, para discernir sus lugares, y para que el entendimiento se satisfaga." Villaseñor y Sánchez, *Theatro Americano descripcion general de los reynos y provincias de la Nueva España y sus juridicciones*, 20.
39 Ibid., 148.
40 See *Real ordenanza para el establecimiento é instruccion de intendentes de exército y provincia en el reino de la Nueva-España*, repr. ed.
41 In *Geografía histórica de México en el siglo XVIII: Análisis del* Theatro Ameri-

cano, Commons and Coll-Hurtado analyze the data from Villaseñor's chapters from the points of view of ecclesiastic information, political and population divisions, and economic activities; they put this data into twenty-first-century graphic format. The study includes twenty maps that show the jurisdictions delineated by Villaseñor. This meticulous summary and graphic visualization of the information of the *Theatro* highlight the significant accomplishment of Villaseñor's compendium. As the authors write in the introduction to their study, the *Theatro Americano* can be considered as one of the primary regional geographies of Mexico, a point that establishes the contribution of each region made to the whole. Commons and Coll-Hurtado, *Geografía histórica de México en el siglo XVIII*, 13.

42 Ibid., 15.

43 Ibid., 13.

44 Villaseñor y Sánchez, *Theatro Americano descripcion general de los reynos y provincias de la Nueva España y sus juridicciones*, 1.

45 Ibid., 5.

46 Ibid., 13.

47 Ibid., 15.

48 D. A. Brading, *Mexican Phoenix*, 111.

49 See García Sáiz, *Las castas mexicanas*; Carrera, *Imagining Identity in New Spain*; Ilona Katzew, *Casta Painting*; Susan Deans-Smith, "Creating the Colonial Subject"; Carlos López Beltrán, "Hippocratic Bodies"; Renato Mazzonlini, "*Las castas*"; and María Elena Martínez, *Genealogical Fictions*.

50 For dating, see García Sáiz, *Las castas mexicanas*, 66; and Katzew, *Casta Painting*, 194.

51 De Española y Yndio, nase Mestiso
De Española y Mestiso, nase Castisa
De Castisa y Español, nase Española
De Negra y Español, nase Mulato

De Española y Mulato, nase Morisca
De Morisca y Español, nase Albino tornatrás
De Mestisa y Yndio, nase Lobo
De Yndio y Lobo, nase Yndio.

52 D. A. Brading, *Mexican Phoenix*, 69.

53 Bleichmar, "Painting as Exploration," 82. Further, while cuadros de casta are often identified as racial depictions, there is some carelessness in this use of the term *race*. Often, the present-day notion of race, which originated in the nineteenth century and is associated with an endogamous gene pool, is confused with notions about lineage and genealogy prevailing in the eighteenth century. See Ruth Hill, *Hierarchy, Commerce, and Fraud in Bourbon Spanish America*, chap. 5, "Before Race: Hierarchy in Bourbon Spanish American," for a thorough discussion of the differences between *casta*, *estado*, and

raza, as well as Hill, "Towards an Eighteenth-Century Transatlantic Critical Race Theory" for a discussion of the "prehistory" of race in the Hispanic world.

54 For a brief overview of the development of the study of geography in late eighteenth-century New Spain, see José Omar Moncada Maya's "La profesionalización de la Geografía Mexicana durante el siglo XIX," 63–67, and *Ingenieros militares en Nueva España*. An overview of the history of late eighteenth-century cartography is presented in María de Carmen León García, "Cartografía de los ingenieros militares en Nueva España."

55 See Alberto Saldino García, *Dos científicos de la ilustración hispanoamericana*; and Fiona Clark, "The *Gazeta de Literatura de México* (1788–1795)," 7–9.

56 For an interesting, although dated, discussion of the Sigüenza y Góngora map, see Sánchez Lamego, *El primer mapa general de México elaborado por un mexicano*. Mitchell Codding's "Perfecting the Geography of New Spain" provides a clear outline of the origins, development, and circulation of Alzate's cartographic work. Dated to 1769 by Antochiw.

57 "Estado de la geografía de la Nueva España, y modo de perfeccionarla." José Antonio de Alzate y Ramírez, *Asuntos varios sobre ciencias y artes* (Monday, 7 December 1772), núm. 7, n.p.

58 This map has an extended history of corrections and publications. See Antochiw, "La visión total de la Nueva España," 77–82.

59 See also Magali Carrera, "Creole Landscapes."

60 del Río, *Description of the ruins of an ancient city discovered near Palenque in the kingdom of Guatemala, in Spanish America*, 13–14. These ruins would fascinate John Lloyd Stephens and Frederick Catherwood in the late 1830s and appear in the title block of García Cubas's *Carta general de la República Mexicana* of his 1858 *Atlas geográfico*.

61 Ignacio Bernal, "La historia póstuma de Coatlicue," 31–34. An excellent overview of the early excavations in viceregal Mexico City is found in Elizabeth Hill Boone's "Templo Mayor Research 1521–1978," 5–69.

62 Antonio León y Gama, *Descripcion histórica y cronológica de las dos piedras*, 1792 ed.

63 Ibid., introduction to 1832 ed., 13–15.

64 Bleichmar, *Visual Culture in Eighteenth-Century Natural History*, 28–27.

3 TOURING MEXICO

1 There is a thread in scholarly literature that posits that Spanish authorities banned travel to and through New Spain. While it is the case that New Spain was not an open travel destination, recent research indicates that in the eighteenth-century, Spain sponsored various expeditions to their American kingdoms for the purposes of natural history in search of commercial resources. See Portuondo, *Secret Science*, chap. 7. Through the beginning of

the nineteenth century, Bourbon reform-minded administrators would continue to request data through comprehensive questionnaires such as those of 1777 formed by Antonio Ulloa and of 1791 developed by Alejandro Malaspina for his scientific expedition. Even with the demise of the Bourbon government under Napoleon, the Cortes de Cádiz continued the practice of these extensive questionnaires.

2 James Duncan and Derek Gregory, *Writes of Passage*, 3–4.

3 Ibid., 3–4.

4 Chard, *Pleasure and Guilt on the Grand Tour*, 7.

5 Ibid., 29, 86. See also Daston and Park, *Wonders and the Order of Nature, 1150–1750*; and Leask, *Curiosity and the Aesthetics of Travel Writing, 1770–1840*.

6 Chard, *Pleasure and Guilt on the Grand Tour*, 3.

7 Ibid., 48–61.

8 Michael Dettlebach, "Humboldtian Science," 289. See also Nicolaas Rupke's *Alexander von Humboldt*, and "A Geography of Enlightenment."

9 Godlewska, "Map, Text and Image," 7. Derek Gregory also offers an excellent analysis of the *Description* in "Emperors of the Gaze," 195–225.

10 Dettlebach, "Humboldtian Science," 288.

11 For example, Charles Darwin carried Humboldt's *Personal Narrative of Travels to the Equatorial Regions of the New Continent* with him on the HMS *Beagle* as he initiated his influential study. Dettelbach, "Humboldtian Science," 287.

12 Humboldt's popularity would flag by the beginning of the twentieth century and not grow again until the 1970s. As a result, the body of research on Humboldt and his work is extensive but not comprehensive, and is even disparate between popular biographies and scholarly analyses of his contributions. For example, a clear study on the history of the publication, translation, and republication of his works is not available. This makes it difficult to ascertain how his works and ideas would spread.

13 Humboldt, *Researches concerning the institutions & monuments of the ancient inhabitants of America*, 37.

14 Quoted in Dettlebach, "Humboldtian Science," 289.

15 Poole, *Vision, Race, and Modernity*, 67; and Dettlebach, "Humboldtian Science," 289.

16 Dettlebach, "Humboldtian Science," 289. On the importance of instrumentation in the early nineteenth century, see Mauricio Nieto Olarte, "Scientific Instruments, Creole Science, and Natural Order in the New Granada of the Early Nineteenth Century."

17 Humboldt, *Cosmos*, 2: 434. Cited in Leask, *Curiosity and the Aesthetics of Travel Writing 1770–1840*, 247.

18 Humboldt, *Atlas géographique et physique du Royaume de la Nouvelle-Espagne*,

lxxxiii–lxxiv. Quoted in Anne Godlewska, *Geography Unbound*, 257. Also see pages 242–65 for Godlewska's excellent analysis of Humboldt's use of graphs and charts.

19 Godlewska, *Geography Unbound*, 253.

20 Humboldt, *Researches concerning the institutions & monuments of the ancient inhabitants of America*, 1:37.

21 This is mentioned in the introduction to the abridged English version of the *Essai*. See also the introduction to Romero Navarrete and Echenique March, *Relaciones geográficas de 1792*.

22 Rupke, "A Geography of Enlightenment," 327–36. Analyzing reviews of Humboldt's publication in fifteen European periodicals between 1790 and 1865, Rupke finds that the *Essai politique* received the most attention. Rupke does not account for abridged versions of Humboldt's work.

23 Humboldt, *Political essay on the kingdom of New Spain*, ii.

24 Humboldt, *Essai politique sur le royaume de la Nouvelle-Espagne*, xviii–xliv.

25 Ibid., cv.

26 Three other graph charts included were *Production of the Mines of America from its discovery*; *Proportion in which gold and silver is produced in the different areas of the Americas*; and *Comparison of silver from, Americas, Asia and Africa*.

27 Godlewska, "From Enlightenment Vision to Modern Science?" 257–58.

28 See the discussion of the concept of population in Foucault's "Governmentality," 215–16.

29 German versions also appeared.

30 Humboldt, *Political essay on the kingdom of New Spain*, v.

31 Humboldt, *Minerva. Ensayo politico sobre el reyno de la Nueva-España*, 5. "Pero yo no propongo estractar todas estas obras, sino las de interés más general, contentándome con dar una ligera noticia de las otras; y en quanto á los mismos extractos procuraré reunir en ellos lo más substancial, curioso, é importante, presentando los resultados más generales, dexando los de menos interés, y omitiendo las discusiones y pormenores científicos, con lo que el comun de los lectores hallará en breve volúmen, y á poco precio, el espíritu ó análisis de unas obras que tanto conviene generalizar en España, y aun de las quales es como vergonoso el que no tengamos exâcta idea."

32 I thank Ray Craib for pointing out these changes to me.

33 See Mary Louise Pratt, *Travel Writing and Transculturation*; and Leask, *Curiosity and the Aesthetics of Travel Writing 1770–1840*.

34 See Godlewska, "From Enlightenment Vision to Modern Science?"

35 Ibid., 246.

36 See Dettlebach, "Humboldtian Science," 302, fig.17.3; and Leask, *Curiosity and the Aesthetics of Travel Writing 1770–1840*, 259.

37 Pratt, *Travel Writing and Transculturation*. In "Landscapes and Identities,"

Jorge Cañizares-Esguerra strongly disagrees with Pratt's analysis of Humboldt's work.

38 See Cañizares-Esguerra, "How Derivative Was Humboldt?" 148–65; and Antonio Barrera-Osorio, "Empire and Knowledge."

39 Rupke, "A Geography of Enlightenment," 334–36. In his *The Humboldt Current: Nineteenth-Century Exploration and the Roots of American Environmentalism*, Aaron Sachs offers a palliative for Pratt's and other's postcolonial reading of Humboldt's work and argues more positively for Humboldt's contribution to ecological perspectives.

40 This material is becoming a focus of recent scholarly research. See, for example, María Soledad Arbeláez A., *Mexico in the Nineteenth Century through British and North American Travel Accounts*; R. Tripp Evans, *Romancing the Maya*; Arturo Aguilar Ochoa, "La influencia de los artistas viajeros en la litografía mexicana (1837–1849)"; José Enrique Covarrubias, *Visión extranjera de México, 1840–1867*; and Robert D. Aguirre, *Informal Empire*. In addition, there are a number of excellent articles about travelers in Mexico in *Viajeros europeas del siglo XIX en México*. For a discussion in the context of the development of *costumbrismo*, see the work of María Esther Pérez Salas C, *Costumbrismo y litografía en México*; and Erica Segre, "The Development of *Costumbrista*."

41 Bullock, *Six months' residence and travels in Mexico*, v–vii.

42 Ibid., vi.

43 Ibid., 166.

44 Bullock, *A Description of the Unique Exhibition*, 2–3.

45 Stephan Oetterman, *The Panorama*, 5. In this book, Oetterman provides detailed information about the origins, development, and spread of the panorama in Europe.

46 Leask, *Curiosity and the Aesthetics of Travel Writing 1770–1840*, 303–4. See also Jonathan King, "William Bullock: Showman"; and Michael Costeloe, "William Bullock and the Mexican Connection."

47 Once the Mexico exhibition closed, this panorama was set up again as the *View of the City of Mexico and Surrounding Countryside* in the Leicester Square Panorama building from 1825 to 1827. See Scott Wilcox, "El Panorama de Leicester Square." It also appeared in New York City in 1827 in John Vanderlyn's New York rotunda. Oetterman, *The Panorama*, 315.

48 *Literary Gazette*, 14 August 1824, 521. Quoted in Leask, *Curiosity and the Aesthetics of Travel Writing 1770–1840*, 307.

49 Leask, *Curiosity and the Aesthetics of Travel Writing 1770–1840*, 314.

50 Mignel Mathes, "La litografía y los litógrafos en México, 1826–1900," 45–47.

51 Claudio Linati, *Costumes civils, militaires et réligieux du Mexique*. Parts of the following material on Linati and Nebel were published in Carrera, "Fabricating Specimen Citizens." I thank the Sussex Academic Press for reproduction rights.

52 "Tous le pays offrent quelques usages dont on ne sait pas se rendre raison, soit à cause de leurs incommodité, soit à cause de leur bizarrerie. Le porteur d'eau du Mexique est un des objets qui frappent le plus les yeux de l'étrange: on a peine à concevoir comment, pour porter 50 livres d'eau." Linati, *Costumes civils, militaires et réligieux du Mexique*, plate 7 commentary, n.p.

53 For examples of *casta* water carrier and tortilla maker, see García Sáiz, *Las castas mexicanas*, 108–9.

54 "Le froment n'était pas connu des anciens Mexicaines. Les régions situées sous les Tropiques ne sont pas favorables à sa culture; le défaut de gelées, les chaleurs excessives, les pluies périodiques, et d'autres causes, le font croître avec trop de luxe, et nuisent au développement et à la maturité des épis." Linati, *Costumes civils, militaires et réligieux du Mexique*, plate 5, Tortilleras commentary, n.p.

55 *Grisette* refers to a young French woman of the working class. "Sexe charmant, aimable moitié du genre humain, sous tous les climats de la terre, en dépit de l'ignorance et de la barbarie, n'importe sous quelles couleurs, et sous quel costume l'empire de tes grâces étend sa bienfaisante influence, et rend meilleurs les hommes, en imposant une trêve aux passions haineuses qui les agitent. Malgré son teint pâle e olivâtre, la jeune ouvrière mexicaine ne renonce pas au privilège de plaire, et sait, par sa vivacité naturelle, par ses mouvemens rapides et gracieux, faire oublier parfois la gentille grisette parisienne." Linati, *Costumes civils*, plate 1, Jeune ouvrière commentary, n.p.

56 Carl Nebel, *Voyage pittoresque et archéologique, dans la partie la plus intéressante du Mexique.*

57 "Nous voyons ici de femmes, dont l'une Indienne, l'autre Créole, occupées à faire la cuisine; la femme Créole écrase sur une pierre des grains de mais; il en résulte une pâte, dont l'autre forme un espèce de crêpe ou omelette, qu'elle jette sur une poèle en terre cuite pour la faire griller. Ce mets, qui n'a ni sel, ni beurre, sert de pain au peuple dans toute la république. On voit un pot-au-feu où l'on fait cuire un mauvais morceau de viande séchée au soleil. Ajoutez à cela quelques piments verts, qu'on appelle Chili, et une boisson nommée Pulqué, prise du jus de l'aloès, et vous avez le rapas habituel du bas people." Ibid., n.p.

58 "Ce sont des femmes de la classe ouvrière, quoique les dames de première classe adoptent souvent cette misse dans leur intérieur. La partie essentielle du costume est le *ruboso*, ou châle léger, qu'elles portent sur le tête et qu'elles ne quittent presque jamais, même pour fair la cuisine, sans que cela les gêne le moins de monde. Elles ne mettent pas de corset, et même les grandes dames ne s'en servent qu'aux heures de représentation. . . . toutes portant de garnitures d'or ou d'argent a la robe. . . . . Il est dans le caractère national de dépenser l'argent comme il vient e de ne songer qu'à ses plaisirs. Heureux le pays où le climat et la grande facilité d'acquérir permettent une pareille insouciance!" Ibid., n.p.

59 Ibid., preface, n.p.

60 Nebel also produced images of the Mexican-American War. See José Luis Juárez López, *Las litografías de Karl Nebel.*

61 Evans, *Romancing the Maya*, 36. See also Claude François Baudez, *Jean-Frédéric Waldeck, peitre le premier explorateur des ruines mayas.*

62 See Aguilar Ochoa, *El escenario urbano de Pedro Gualdi 1808–1857*; Esther Acevedo with Helia Bonilla, "La gráfica"; and Roberto Mayer, "Los dos álbumes de Pedro Gualdi."

63 Evans, *Romancing the Maya*, 49–50.

64 Ibid., 51. See also Victor Wolfgang Van Hagen, *Frederick Catherwood, Architect.*

65 Stephens and Catherwood also attempted to take daguerreotype images of the ruins, but the results did not render sufficient detail and were considered unusable.

66 Evans, *Romancing the Maya*, 45.

67 Ibid., 58–60.

68 William Prescott, *The Conquest of Mexico*, n.p.

69 These are found in Harvey Gardiner's *The Literary Memoranda of William Hickling Prescott.*

70 Ibid., 2: 8.

71 Ibid., 2: 22.

72 Prescott to John Lloyd Stephens, Boston, 2 August 1841, in Wolcott, *The Correspondence of William Hickling Prescott 1833–1847*, 240–42.

73 Gardiner, *The Literary Memoranda of William Hickling Prescott*, 2: 29.

74 Ibid., 2: 29.

75 Ibid., 2: 32.

76 Prescott, *The Conquest of Mexico*, 2: 478.

77 From Fanny Calderón de la Barca, Mexico 15 October, 1840, Wolcott, *The Correspondence of William Hickling Prescott 1833–1847*, 167–69.

78 To Fanny Calderón de la Barca, Boston, 5 December 1840, ibid., 187.

79 Prescott, *The Conquest of Mexico*, 2: 51–52.

80 Ibid., 2:51, n.11.

81 Aguirre, *Informal Empire*, talks about the visuality of Prescott in literary terms, too, but does not contextualize such visuality into the long history of visual description of the Americas nor the comprehensive context of nineteenth-century visual culture.

82 *The Conquest of Mexico* was carried by soldiers into the Mexican-American war, a kind of blueprint for implementing Manifest Destiny. In fact the term Manifest Destiny, alluding to the predestination of the United States to expand its territory from the Atlantic to Pacific Ocean, was coined right after the publication of Prescott's history. See Janet E. Buerger, "Ultima Thule," 89. Buerger further argues that Prescott's books highlighted the inheritance by the United States from the first Age of Discovery.

83 Corydon Donnavan, *Adventures in Mexico*, 128.
84 Ibid., 129.
85 Ibid., 132.
86 Ibid., 119. Quoted in Aguirre, *Informal Empire*, 56.
87 See Rebert, *La Gran Línea*, 3–9.
88 García Cubas, *The Republic of Mexico in 1876*, preface, n.p.
89 Carl Sartorius, *Mexico: Landscapes and popular sketches*, vi.
90 Ibid., 46–47. See also Sartorius, *México y los mexicanos.*
91 Sartorius, *Mexico: Landscapes and popular sketches*, 56.
92 Ibid., 57.
93 Ibid., 59.
94 Ibid., 61.

4 MEXICAN NATIONALIST IMAGERY

1 "En el estado actual de las cosas es todavía dificil formar una idea exacta del carácter mexicano que por estarse formando aún no es possible fijarlo; . . . . la sociedad Mexicana todavía en embrión no presenta hasta ahora sino una confusa mezcla de hábitos, usos y costumbres de la metropolis, Francia e Inglaterra dominando en ciertas líneas los de una nación." Mora, *Méjico y sus revoluciones*, 85, 128. Also quoted in Raymond Hernández-Durán, *Reframing Viceregal Painting in Nineteenth-Century Mexico*, 70.
2 Some of the material in the following section appears in Carrera, "Affections of the Heart."
3 *Descripción de las fiestas celebradas en la imperial corte de México con motivo de la solemne colocación de una estatua equestre de nuestro augusto soberano el Señor Don Carlos IV en la Plaza Mayor.* This is reproduced in facsimile in *Documentos varios para la historia de la Ciudad de México a fines de la época colonial (1769-1815)*. For further analyses of this statue, see Stacie Widdifield's "Manuel Tolsá's Equestrian Portrait of Charles IV," 61–83; and "Modernizando el pasado." See also Barbara Tenenbaum, "Streetwise History."
4 For a discussion of the imagery of the ancien régime, see de Antoine de Baecque, *The Body Politic.*
5 For further discussion of this painting, see Acevedo, "Entre la tradición alegórica y la narrativa factual." For an overview of women in nationalist imagery, see Joan Landes, *Visualizing the Nation: Gender, Representation, and Revolution in Eighteenth-Century France.*
6 For a comprehensive discussion of the Virgin of Guadalupe iconology, see D. A. Brading, *Mexican Phoenix*. See also Jaime Cuadriello, "Del escudo de armas al estandarte armado," and "Los umbrales de la nación y la modernidad de sus artes."
7 Timothy Anna, *Forging Mexico 1821-1835*, 53–54. The bibliography on this early

period is extensive; see Anna bibliography. Paul Garner also discusses the transition from the ancien régime. See his *Porfirio Díaz*, 19–24.

8 "La celebración del día 12 de diciembre de todos los pueblos dedicando a la patrona de nuestra libertad, María santísima de Guadalupe." José María Morelos, "Sentiment of the Nation," 189–90.

9 Anna, *Forging Mexico 1821–1835*, 81–82.

10 Ibid., 96–97.

11 Ibid., 95.

12 President Victoria's birth name was Félix Fernández. He took Guadalupe Victoria to identify with the Virgin of Guadalupe and the victory of independence.

13 See Michael Costeloe, *The Central Republic in Mexico, 1835–1846*.

14 Will Fowler, *Mexico in the Age of Proposals, 1821–1853*, 220.

15 See Guy Thomson, "Liberalism and Nation-Building in Mexico and Spain during the Nineteenth Century."

16 Brian Hamnett, *Juárez*, 13–27.

17 Fowler, *Mexico in the Age of Proposals, 1821–1853*, 266–67.

18 Such allegorical images were derived from viceregal period imagery of the Americas. See Acevedo, "Los comienzos de una historia laica en imagines."

19 Acevedo, "Entre la tradición alegórica y la narrativa factual."

20 Carlos María de Bustamante, *Mañanas de la Alameda de México. Publícalas para facilitar a las señoritas el estudio de la historia de su pais*, 1: 1.

21 Ibid., 1: 13, 19.

22 Hamnett, *Juárez*, 27.

23 Bustamante, *Mañanas de la Alameda de México*, 1: n.p.

24 James Scott, *Seeing like a State*, 79.

25 For the legislation augmenting the academy's, see Ida Rodríguez Prampolini, *La crítica de arte en México en el siglo XIX*, 1: 180–81. For a good overview of the academy's changes in the early nineteenth century, see Uribe, "Claves para leer la escultura mexicana."

26 On the notion of remembering as part of nation formation, see Shelley Garrigan's *Collecting the Nation*.

27 Luis Gerardo Morales Moreno, "El primer Museo Nacional de México," 47–50.

28 "Anuncio de mano de la primera publicación de Museo," in *Coleccion de las Antigüedades Mexicanas que existen en el Museo Nacional*, n.p.

29 Morales Moreno, "El primer Museo Nacional de México," 54–55.

30 The original intention was to publish four prints each month. Due to lack of subscribers to the publication, only four editions were published. María José Esparza Liberal, "La historia de México en el calendario de Ignacio Díaz Triujeque de 1851 y la obra de Prescott," 154; Mathes, "La litografía y los litógrafos en México, 1826–1900," 47.

31 In chapter 1 of *The Birth of the Museum*, Tony Bennett explores the importance of reordering and the public sphere, which are foundational concepts to the *Colección*. See also Scott's *Seeing like a State*.

32 Héctor Mendoza Vargas, "Las opciones geográficas al inicio de México independiente," 93.

33 Ibid., 96–98.

34 Raymond Craib provides an excellent overview of the history of the SMGE in his *Cartographic Mexico*, 20–29. See also Luz Fernanda Azuela Bernal, "La *Sociedad Mexicana de Geografía y Estadística*, la organización de la ciencia, la institucionalización de la geografía y la construcción del país en el siglo XIX."

35 Miguel Ángel Castro and Guadalupe Curiel, *Publicaciones periódicas mexicanas del siglo XIX: 1822–1855*, 484–85.

36 Mendoza Vargas, "Las opciones geográficas al inicio de México independiente," 101; Craib, *Cartographic Mexico: A History of State Fixations and Fugitive Landscapes*, 22–23.

37 Mendoza Vargas, "Las opciones geográficas al inicio de México independiente," 102.

38 Wolfgang Ernst, "Archi(vi)textures of Museology," 29. See also Craib, *Cartographic Mexico*, chap. 1 on the importance of history and geography.

39 It should be noted that new printing techniques and freedom of the press also led to a multiplication of readers.

40 The stone is moistened with water, which the stone accepts in areas not covered by the crayon. Oily ink is applied and adheres only to the drawing and is repelled by the wet parts of the stone.

41 Miguel Mathes, "La litografía y los litógrafos en México, 1826–1900," 45–47. See also Jiménez Codinach, "La litografía mexicana del siglo XIX"; and Aguilar, "La litografía en la ciudad de México." On earlier printing techniques, see Silvia Fernández Hernández, "La transición del diseño gráfico colonial al diseño moderno en México."

42 Mathes, "La litografía y los litógrafos en México, 1826–1900, 47.

43 Esther Acevedo, "La gráfica," 222.

44 Pérez Salas, "Los secretos de una empresa exitosa," 145–56.

45 Clark, "The *Gazeta de Literatura de México* (1788–1795)," 7–9.

46 For excellent scholarly studies of print culture in early nineteenth-century Mexico, see Laura Suárez de la Torre, *Constructores de un cambio cultural*; and *Empresa y cultura en tinta y papel (1800–1860)*. For a catalogue of print publications, which includes bibliographic data and summary of contents, see Castro and Curiel, *Publicaciones periódicas mexicanas del siglo XIX: 1822–1855*; and *Publicaciones periódicas mexicanas del siglo XIX: 1856–1876*.

47 For a comprehensive list of nineteenth-century lithographers in lithographic workshops see Apéndices (157–59) as well as essays in the Museo Nacional de

Arte's exhibition catalogue *Nación de imágenes.* See also Carlos Monsiváis's *Casmiro Castro y su taller.*

48 Tomás Pérez Vejo, "La invención de una nación." See also Amada Carolina Pérez Benavides, "Actores, escenarios y relaciones sociales en tres publicaciones periódicos mexicanas de mediados del siglo XIX."

49 Suárez de la Torre, "Una imprenta florencia en la calle de la Palma número 4," 137, fn.16.

50 Othón Nava Martínez, "Origen y desarrollo de un empresa editorial," 127–28. In fact, between 1822 and 1855, 342 publications appeared. For an excellent reference on periodical publications in the nineteenth century, see Castro and Curiel, *Publicaciones periódicas mexicanas del siglo XIX: 1856–1876*; Irma Lombardo García, *El siglo de Cumplido*; and Castro and Curiel, *Publicaciones periódicas mexicanas del siglo XIX: 1822–1855.*

51 Jeffrey Pilcher, *¡Que vivan los tamales!*, 45.

52 Laura Solares Robles, "Prosperidad y quiebra," 116; Nava Martínez, "Origen y desarrollo de un empresa editorial," 127.

53 Silvia Marina Arrom, *The Women of Mexico City, 1790–1857*, 14.

54 Jane Herricks, "Periodicals for Women in Mexico," 137–40.

55 *El Semanario*, III, 215. "Ya es tiempo de salir de esa situacion ignorante y limitada que no os permitia ser considerados compañeras inteligentes del hombre, y como una mitad de ser social: el mundo y nuestro pais mismo, se tranforman al rededor de vosotros y toda clama por otra organizacion en favor de los progresos sociales. . . . Feliz yo sí una sola señorita al leer estas lineas, se convence de la necesidad que tiene de cultivar su educacion y de perfeccionar su intelegencia para llegar á ser verdaderamente un muger de casa." Quoted in Herricks, "Periodicals for Women in Mexico," 140–41. (Misspellings occur in original text.)

56 Diego Álvarez, "Discurso sobre la influencia de la instrucción pública en la felicidad de las naciones," *Revista Mensual de la Sociedad Promovedora de Mejoras Materiales* 1 (1852): 376–77. Quoted in Arrom, *The Women of Mexico City, 1790–1857*, 23. See also Pérez Benavides's discussion of the identification of female archetypes in the *Álbum Mexicano* in "Actores, escenarios y relaciones sociales en tres publicaciones periódicos mexicanas de mediados del siglo XIX," 1191–92.

57 Sueann Caulfield, "The History of Gender in the Historiography of Latin America," 475. Caulfield defines gender as ways in which female and male subjects are socially constructed and positioned and how representation of femininity and masculinity structure institutional power.

58 Over the last twenty years, numerous scholars have researched the history and roles of women in Mexico in the late eighteenth century and the nineteenth. Examples of the range of academic studies include Arrom, *The Women of Mexico City, 1790–1857* (1985); Asunción Lavrin, *Sexuality and Marriage in*

Colonial Latin America, (1989); Carmen Ramos Escandón, *Presencia y transparencia: La mujer en la historia de México* (1987), and *Género e historia* (1992); Julia Tuñón, *El álbum de la mujer* (1991); Martha Eva Rocha, *El álbum de la mujer* (1991); Stacie Widdifield, *The Embodiment of the National in Late Nineteenth-Century Mexican Painting* (1996); and Montserrat Galí Boadella, *Historias del bello sexo* (2002).

59 Acevedo with Helia Bonilla, "La gráfica: Testigo de lo cotidiano" 227.

60 See Erica Segre's discussion of the introduction of the cultural press in her "The Development of *Costumbrista*"; and Pérez Salas, "El costumbrismo literario en México," in her *Costumbrismo y litografía en México*, 165–210.

61 Pérez Salas, "Los secretos de una empresa exitosa," 143–48. See also Pérez Benavides, "Actores, escenarios y relaciones sociales en tres publicaciones periódicos mexicanas de mediados del siglo XIX."

62 For a discussion of the development of the museo, see Edgar Mejía, *Políticas del espacio en México*, 91–98.

63 See Stephen Bann, *Parallel Lines*, 93–125, for a thoughtful analysis of the development of the daguerreotype in relation to lithography. See also John Tagg, *The Burden of Representation*, for a more theoretical overview of photography.

64 Initially, exposure times for daguerreotypes ran three to fifteen minutes; subsequently, modification in the process and improvement of the photographic lenses reduced this time to as little as a minute. Its presence in Mexico is well documented; see Deborah Dorotinsky, "Los tipos sociales desde la austeridad del estudio," 14–25. For a description of the early use of daguerreotype in Mexico, see Oliver Debroise, *Mexican Suite*, 19–21; Rosa Casanova, "Un nuevo modo de representar," 191–98; and José Antonio Rodríguez, "Fotógrafos viajeros camino abierto," 56–66.

65 Examples of these daguerreotypes are found in the George Eastman House's Photography Collection, in Rochester, New York.

66 Rodríguez, "Fotógrafos viajeros camino abierto," 59.

67 Ibid., 58–64.

68 "Venido a sorprender al hombre por su originalidad, y a proporcionarle un medio, el más sencillo y el más acertado, para retratar a la naturaleza con toda su propiedad con una exactitud que es prodigiosa." From the periodical *El Museo Mexicano*, 1843, 145. Quoted in Rosa Casanova, "Ingenioso descubrimiento," 9–10.

69 Debroise, *Mexican Suite*, 248n11.

70 *Diario de Gobierno*, dated Sunday, 1 August 1847 and 2 August 1847. Tomo 4: 141, 142. "Se venden un buen daguerrotipo, que acaba de llegaron de Francia y que hace los retratos de todos tamaños, muy baratos. En el calle Zulete, núm. 15. . . ."

71 See a portfolio of late 1850s photographic ads reproduced in *Alquimia*, May–August 1999, 26–30.

72 For discussion of *tarjetas de visita*, see Patricia Massé Zendejas, *Simulacro y elegancia en tarjetas de visita*, and *Cruces y campa*. For growth of photography in contemporary Europe, see Elizabeth McCauley, *A. A. E. Disdéri and the Carte de Visite Portrait Photography*, and *Industrial Madness*.

73 See Casanova, "Ingenioso descubrimiento," 7–14, 27–34.

74 Although the study of the early Mexican photographic history is limited, it is evident that the greatest expansion of this reproductive technology occurred in the late '60s (and will be discussed in the next chapter); however, its introduction and presence in the 1840s and 1850s would amplify the evolving visualization of the nation of Mexico and extend its underlying discourses on history and geography.

75 For an excellent discussion of the images included with these two Mexican publications, see Elena Estrada de Gelero, "La litografía y el Museo Nacional como armas del nacionalismo."

76 Esparza Liberal, "La historia de México en el calendario de Ignacio Díaz Triujeque de 1851 y la obra de Prescott," 154.

77 *Conquest of Mexico!*, 17.

78 Ibid., 18.

79 Ibid., 18.

80 Ibid., 22.

81 Fausto Ramírez, "Pintura e historia en México a mediados del siglo XIX," 84.

82 See ibid., 84.

83 Morales Moreno in "El primer Museo Nacional de México," 57, cites the date 1856 for publication.

84 *México y sus Alrededores*, 9. For a discussion on the circulation of European journals, see Thomas Gretton, "European Illustrated Weekly Magazines, c. 1850–1900."

Ramírez's reference to Egypt may reflect the popularization of Napoleon's expedition; see Derek Gregory, "Scripting Egypt"; and Anne Godlewska, "Map, Text and Image." Another context of Ramírez's antiquities interest may be the expansion of information about Mesopotamia initiating at this time. See Fredrick Bohrer, *Orientalism and Visual Culture*. Tomás Pérez Vejo also looks at orientalism's influence in the early nineteenth century in "La invención de una nación."

85 Craib, *Cartographic Mexico*, 28. See also Azuela Bernal, "La *Sociedad Mexicana de Geografía y Estadística*."

86 Craib, *Cartographic Mexico*, 26. This was not published.

87 Cited in Mendoza Vargas, "Las opciones geográficas al inicio de México independiente," 103.

88 Rodríguez Prampolini, *La crítica de arte en México en el siglo XIX*, 1: 197.

89 Ibid., 200–201; Hernández-Durán, *Reframing Viceregal Painting in Nineteenth-Century Mexico*, 48.

90 See Pablo Diener, "La pintura de paisajes entre los artistas viajeros," 137–57.
91 See Mayer, "Phillips, Rider y su album *Mexico Illustrated*."
92 María Elena Altamirano Piolle, *National Homage—José María Velasco (1840–1912)*, 1: 59–78. Altamirano provides detailed information about Landesio's lessons.
93 The bibliography on landscape history and theory is large. More interesting studies in the context of this discussion include Denis Cosgrove's *Social Formation and Symbolic Landscape*, and "Prospect, Perspective and the Evolution of the Landscape Idea," 45–62; and W. J. T. Mitchell, *Landscape and Power*.
94 Rodríguez Prampolini, *La crítica de arte en México en el siglo XIX*, 1: 320.
95 Widdifield, "Modernizando el pasado," 75–76.
96 Bennett, *The Birth of the Museum*, 62.
97 Ibid., 63.
98 Ibid., 67.
99 Mike Crang and Nigel Thrift, *Thinking Space*, 12.
100 For thought-provoking perspectives on travel writers in Mexico, see Arbeláez A., *Mexico in the Nineteenth Century through British and North American Travel Accounts*.
101 *Álbum Pintoresco de la República Mexicana*, 12.
102 For a full description of the calendar's images and their sources, see Esparza Liberal, "La historia de México en el calendario de Ignacio Díaz Triujeque de 1851 y la obra de Prescott."
103 Antonia Pi-Suñer Llorens, "Una gran empresa cultural de mediados del siglo XIX."
104 "Ranchero," *La Ilustración Mexicana* 1 (1851): 131. Quoted in Pérez Benavides, "Actores, escenarios y relaciones sociales en tres publicaciones periódicos mexicanas de mediados del siglo XIX," 1189–1190.
105 Pérez Salas, "Los secretos de una empresa exitosa,"151–53.
106 Similar publications were produced in Europe: *Los españoles pintados por sí mismos 1839*, and *Les Français peints par eux-mêmes* (1840–42). See W. S. Hendrix, "Notes on Collections of Types, a Form of Costumbrismo."
107 "Esa hija de México tan linda como su cielo azul; tan fresca como sus jardines floridos, y tan risueña y alegre como las mañanas deliciosas de esta tierra bendita de Dios y de sus santos." *Los mexicanos pintados por sí mismos* 90. For a discussion of the *china poblana*, see María del Carmen Vázquez Mantecón, "La *china* mexicana, mejor conocida como *china poblana*."
108 A thorough study of changes in *México y sus Alrededores* is found in Robert L. Mayer's "Nacimiento y desarrollo del álbum *México y sus Alrededores*."
109 A notice in *Siglo Diez y Nueve*, Monday, 19 June 1858, tomo 3565, announces its publication: "*México y sus Alrededores* con testo 16.0 pesos; sin testo $12.0 pesos."
110 "No somos aztecas, no somos españoles; raza bastarda de las dos, tenemos la

indolencia de la una, la arrogancia de la otra; pero aun no constituimos una raza propia, distinta de las demas, con cualidades peculiars, buenas ó malas. Pueblo de ayer, sin tradiciones, sin grandes recuerdos, nuestro historia de pocas años es la crónica de la inesperiencia, de la locura y de la discordia y falta á nuestros acontecimientos mas notables ese prestigio fascinador de la distancia que dan á los hombres y á las cosas los montones de siglos que se interponen entre las generaciones." *México y sus Alrededores*, 4.

111 Marcos Ernst, "Archi(vi)textures of Museology," 18.

112 Arróniz also wrote *Manual de biografía mejicana ó galería de hombres célebres de Méjico* (1857), and *Manual de historia y cronología de Méjico* (1858).

113 Arróniz, *Manual del Viajero en Méjico*, introduction, n.p.

114 Arróniz proudly comments that the equestrian statue took fifteen days and 15,000 pesos to move to its location in front of the Plaza de Torres in 1852; Arróniz, *Manual del Viajero en Méjico*, 112.

115 Ibid., 41–43.

116 Ibid., 129–76.

117 Ibid., 292.

118 See Carlos Córdova, *Arqueología de la imagen*.

119 See Nancy West, "Fantasy, Photography, and the Marketplace."

120 Córdova, *Arqueología de la imagen*, 25.

121 Ibid., 25.

122 Ibid., 25, 39, 142.

123 Casanova, "Un nuevo modo de representar," 215.

124 Cited in Casanova, "Ingenioso descubrimiento," 28.

125 See Juana Craib, *Cartographic Mexico*; and Gutiérrez Haces, "Etnografía y costumbrismo en las imágenes de los viajeros."

126 Joan Schwartz and James Ryan, *Picturing Place*, 5–8.

5 THE GARCÍA CUBAS PROJECTS

1 The *y* of García y Cubas appears and disappears across his early publication and was ultimately dropped.

2 García Cubas, *El libro de mis recuerdos*, 629.

3 Ibid., 426.

4 A thorough summary of García Cubas's life is provided in María del Carmen Collado's "Antonio García Cubas."

5 Richard Sinkin, *The Mexican Reform 1855–1876*, 9.

6 Ibid., 40, 45.

7 Hamnett, *Juárez*, 71–73.

8 García Cubas, *El libro de mis recuerdos*, 146.

9 Collado, "Antonio García Cubas," 427, 428.

10 García Cubas, *El libro de mis recuerdos*, 630. "Una niña, contrahecha y desme-

drada. . . . Curar á tan desgraciado ser era asunto que ofrecía serias dificultades, pues había que atender, al mismo tiempo, á su nutrición y al arreglo de todos sus miembros dislocados. A corregir los desperfectos de la niña y á curarla de su profunda anemia dirigí todos mis esfuerzos."

11 Ibid., 452.

12 Ibid. This story is not verified by other sources.

13 Craib, *Cartographic Mexico*, 27–29.

14 Ray Craib pointed out to me the precise relationship of the two meridians. See ibid., 33. In his memoir, García Cubas attests that his friend Francisco Díaz Covarrubias, an engineer, actually did the work to locate the capital. García Cubas, *El libro de mis recuerdos*, 630.

15 Matthew Edney points out, "As lines established by human agency to define longitude, . . . prime meridians were held to mediate between the concrete extent of the territorial state and the abstract extent of the cultural nations." "Cartographic Culture and Nationalism in the Early United States," 385.

16 Garcia Cubas, *Atlas geográfico, estadístico é histórico de la República Mexicana formado por Antonio Garcia y Cubas*, preface, n.p.

17 For a brief overview, see Mendoza Vargas, "Las opciones geográficas al inicio de México independiente," 105.

18 Craib, *Cartographic Mexico*, 31.

19 "El principal objeto con que se ha formado el presente Atlas, es el de dar á conocer este hermoso país tan rico por su producciones naturales. La falta de cartas y de noticias geográficas ha sido uno de los obstáculos para la realizacion de grandes proyectos. Bien conocido el país, las empresas de colonizacion, las de caminos, las de minas que poseemos ricas y abundantes, las de agricultura y otras muchas, darán el resultado de la prosperidad á que deben aspirar los votos de todos los mexicanos." García Cubas, *Atlas geográfico, estadístico é histórico de la República Mexicana formado por Antonio Garcia y Cubas*, introduction, n.p.

20 Because the atlas is made up of separate prints that were to be later collated, copies I reviewed at two research institutions and one held by a private collector are not ordered as I describe. This, of course, raises an interesting problem as to how and why this change occurred—one that cannot be addressed in this present study.

21 "Parte geográfica. Situacion, límites y estension, aspectos físicos, clima y producciones. Golfos, bahías, ensenadas, baras, esteros, penínsulas, islas, bancos, ó arrecifes, canales, cabos, puertos, rios, lagos, lagunas, montañas, volcanos. Parte estadística. Division territorial y poblacion. Industria minera, agrícola y fabril. Noticia sobre minerales y haciendas. Fábricas de algodon, lana seda, papel, vidrio, loza. etc. Comercio." García Cubas, *Atlas geográfico, estadístico é histórico*, introduction, n.p.

22 Copies of the *Atlas* at the Library of Congress and the Newberry Library place

these maps in alphabetical order by states and then territory. The Roman numerals on the state maps, however, indicate that this was not to be their original intended order.

23 I thank Ray Craib for this reference.

24 "Tercera. Carta general en mayor escala, en la se manifiestan las principales poblaciones de la República, sus rios, sus montañas, caminos, etc. etc." García Cubas, *Atlas geográfico, estadístico é histórico*, introduction, n.p.

25 The change in the northern border may reflect that García Cubas worked on the *Atlas geográfico* over a number of years to bring in updated information into his material.

26 "Cuarta. Cuadro con noticias sobre la historia antigua de México. Orígen, peregrinacion, establecimiento, época de mayor esplendor y destruccion de los primeros habitantes de este país. Fiestas y juegos públicos. Mitologia mexicana. Barbarie y civilizacion, manifestada la primera en su religion y la segunda en sus obras. Calendario civil, siglo, años y meses. Guerras mas famosas. Antigüedades mexicanas. Noticias cronológicas." García Cubas, *Atlas geográfico, estadístico é histórico*, introduction, n.p.

27 Craib also notes in his study of the *Carta general* that the landscape cartouche has a "theater setting" quality and the format of a diorama or panorama. See Craib, *Cartographic Mexico*, 38–40.

28 Antonio García Cubas, *El libro de mis recuerdos*, 452.

29 Ida Rodríguez Prampolini, *La crítica de arte en México en el siglo XIX*, tomo 2 (1810–58): 47–48.

30 Ibid., 78.

31 Hernández-Durán, *Reframing Viceregal Painting in Nineteenth-Century Mexico*, 38.

32 Ibid., 290, 143–45, 109, n. 63. Also see José Bernardo Couto, *Diálogo sobre la historia de la pintura en México*.

33 Hernández-Durán, *Reframing Viceregal Painting in Nineteenth-Century Mexico*, 244.

34 Ibid., 232–33, 232 n. 33.

35 Gary S. Dunbar, "'The Compass Follows the Flag,'" 229–37.

36 García Cubas, *El libro de mis recuerdos*, 452.

37 Widdifield, "Modernizando el pasado"; and Tenenbaum, "Streetwise History," 129–30.

38 Robert H. Duncan, "Embracing a Suitable Past," 256–60.

39 Ibid., 272.

40 Ibid., 268–70. See also Duncan's "Maximilian and the Construction of the Liberal State, 1863–1866."

41 Casanova, "El éxito del comercio fotográfico," 28. Arturo Aguilar Ochoa provides a list of photographers in *La fotografía durante el imperio de Maximiliano*, 154–55.

42 Aguilar Ochoa, *La fotografía durante el imperio de Maximiliano*, 57–78. See also Casanovo, "Las fotografías se vuelven historia."

43 Two useful comparative studies of the impact of photography in Mexico and Guatemala are Deborah Poole's "An Image of 'Our Indian'"; and Greg Grandin's "Can the Subaltern Be Seen?"

44 Cited in Patricia Massé Zendejas's *Simulacro y elegancia en tarjetas de visita*, which provides an excellent overview of the work of Cruces y Campa.

45 Ibid., 45.

46 "La circulación de imágenes durante el periodo del Imperio, significó un salto cualitativo en la manera en que los mexicanos se acostumbraron a pensar y usar la imagen fotográfica." Casanova, "El éxito del comercio fotográfico," 30.

47 "Panorama universal," in *Periódico Oficial del Departamento de Yucatán*, Mérida, núm. 122 (16 August 1865): 4. Quoted in Córdova, *Arqueología de la Imagen*, 28.

48 Dorotinsky, "Los tipos sociales desde la austeridad del estudio," 24.

49 Cited and discussed in Aguilar Ochoa, *La fotografía durante el imperio de Maximiliano*, 79–91. See also Mark Overmeyer-Velázquez's *Visions of the Emerald City*; and William E. French's "Prostitutes and Guardian Angels."

50 Sinkin, *The Mexican Reform, 1855–1876*, 166.

51 See Hamnett, *Juárez*, chap. 9.

52 These images are illustrated in Widdifield, *The Embodiment of the National in Late Nineteenth-Century Mexican Painting*, chap. 3.

53 In *National Homage* by Altamirano Piolle, the author provides a comprehensive illustrated overview of Velasco's work. Also, a provocative interpretation of Velasco's work as depicting the history of the nations is provided by Cañizares-Esguerra in "Landscapes and Identities," 156–64.

54 Widdifield, *The Embodiment of the National in Late Nineteenth-Century Mexican Painting*, 68. See also Cañizares-Esguerra, "Landscapes and Identities."

55 It should be noted that other encyclopedia efforts were under way in Europe. For example, the French geographer Elisée Reclus was working on his *Nouvelle Géographie universelle, la terre et les hommes*, 19 vols. (1875–94).

56 García Cubas, *Curso Elemental de Geografia Universal*, 163.

57 For a summary of the ethnocentric geography textbooks written in the United States, see Susan Schulten's *The Geographical Imagination in America, 1880–1950*, chap. 5.

58 "Conviccion, y el deseo de contribuir al progresso de la instruccion pública, base de nuestra futura felicidad, y el anhelo de hacerme útil á mis compatriotas." García Cubas, *Curso Elemental de Geografia Universal*, xiii.

59 Claudia Agostoni, *Monuments of Progress*, 90–91. For a discussion of the cult of heroes, see Rebecca Earle's, "*Sobre Héroes y Tumbas*," and "'Padres de la Patria' and the Ancestral Past."

60 *Rojo* may refer to a radical popular liberator who opposed abuse to the Constitution of 1857. See Garner, *Porfirio Díaz*, 32.

61 For a brief biography, see Judith de la Torre Rendón, "Manuel Rivera Cambas."

62 In the 1870s Garcia Cubas also produced short writings that were collected in *Escritos diversos de 1870 a 1874*.

63 The ethnographic information is taken in part from Manuel Orozco y Berra's, *Geografía de las lenguas y carta etnográfica de México*, as well as Fransciso Pimental's *Sobre las causas que han originado la situacion actual de la raza indígna de México y medios de remeidarla*. For a short analysis of this publication, see Collado, "Antonio García Cubas," 433–34; and Laura Pérez Rosales, "Manuel Orozco y Berra." Further, neither García Cubas himself nor archival documents elucidate the reason for the publication of this text in English or information about its translator. It should be noted, however, that *The Republic of Mexico in 1876* appeared at the same time as writers in the United States were publishing materials on the indigenous languages of the United States and was perhaps prepared for the 1876 Centennial International Exhibition of 1876 held in Philadelphia. See chap. 4 in Steven Conn, "Fade to Silence: Indians and the Study of Language," in *History's Shadow*, 79–115.

64 García Cubas, *The Republic of Mexico in 1876*, introduction, n.p. Parts of the following discussion come from Carrera, "Fabricating Specimen Citizens."

65 García Cubas, *The Republic of Mexico in 1876*, 15. Louis Figuier (1819–94) published *The Human Race* in 1872. The publication had 243 wood engravings and eight chromolithographs. See also Conn's chap. 6, *History's Shadow* for an overview of Figuier's association with history writing of North America.

66 García Cubas, *The Republic of Mexico in 1876*, 15.

67 Ibid., introduction, n.p.

68 Ibid., 14.

69 García Cubas attributes the idea of this map to Orozco y Berra. García Cubas, *El libro de mis recuerdos*, 248.

70 García Cubas, *The Republic of Mexico in 1876*, 14.

71 Ibid.

72 García Cubas's use of the designation of three "races" also ties this work to the so-called scientific literature, which was used to justify colonial expansion of European powers in the late nineteenth century.

73 García Cubas, *The Republic of Mexico in 1876*, 18.

74 Ibid., 61.

75 Ibid., 62.

76 Ibid. As an example of these degraded customs, he cites that "Indian women, even far advanced into their pregnancy, do not abstain from hard labor, and without any care for their offspring, continue grinding their corn: an occupation that cannot be otherwise than injurious to parturition" (62–63).

77 García Cubas, *The Republic of Mexico in 1876*, 61.

78 Ibid., 128.
79 Ann Laura Stoler, *Carnal Knowledge and Imperial Power*, 97.
80 Tenorio-Trillo, *Mexico at the World's Fairs*, 91.
81 García Cubas, *Álbum del ferrocarril Mexicano*, 1–2.
82 Ibid., 112.
83 Marking change is the fact that the railroad company promoted a statue of Columbus, as a metaphor for the new possibilities. Agostoni, *Monuments of Progress*.
84 Harley, "Silences and Secrecy," 59, 70.
85 Recall that García Cubas was producing teaching manuals by the 1870s. See also Arrom, *The Women of Mexico City, 1790–1857*, 16–22. She notes that lower-status girls were taught basic literacy and household skills. See also Mílada Bazant's *Historia de la educación durante el Porfiriato*.

6 FROM NEW SPAIN TO MEXICO

1 Garner, *Porfirio Díaz*, 20, 68. Garner argues that by 1876 Mexico still lacked the basic requirements of political stability: secure frontiers, financial stability, restructured rural society, and national identity. See also Leslie Bethell, *Mexico since Independence*.
2 See Garner, *Porfirio Díaz*; and William Beezley, *Judas at the Jockey Club and Other Episodes of Porfirian Mexico*.
3 Although dated, a useful article on the historiography of writing about the Porfiriato is Thomas Benjamin's and Marcial Ocasio-Meléndez's "Organizing the Memory of Modern Mexico."
4 Díaz's rule is equated with a positivist approach, defined as " the application of scientific method not only to the analysis of social, economic and political conditions but also in the formulation of policies, which would remedy their deficiencies and ensure material and social progress." Garner, *Porfirio Díaz*, 71. See also Eric Van Young, "Conclusion: The State as Vampire," 346.
5 Van Young, "Conclusion: The State as Vampire," 346.
6 See Duncan, "Maximilian and the Construction of the Liberal State, 1863–1866."
7 Tenorio-Trillo, *Mexico at the World's Fairs*.
8 For a discussion of consumption by Porfirian society, see William Beezley, "The Porfirian Smart Set Anticipates Thorstein Veblen in Guadalajara."
9 John Lear, "Mexico City; Space and Class in the Porfirian Capital," 473.
10 For an excellent discussion of the Porfiriato reorganization of urban space, see John Lear, "Mexico City."
11 Craib, *State Fixations, Fugitive Landscapes*, 130. Craib provides an excellent overview of the Comisión-Geográfico-Exploradora in chap. 4 of his book.
12 Paolo Riguzzi, "México próspero," 143, 152.

13 See ibid., 142–49, and the comprehensive study of Tenorio-Trillo, in *Mexico at the World's Fairs.*

14 Editorial, *El Siglo XIX* tomo 87, núm. 14158 (Monday 15 June 1885).

15 José Ortiz Monasterio, *México eternamente*, 196. Tenorio-Trillo, *Mexico at the World's Fairs*, points out other possible initiation points for the work, 287 n.15.

16 The archive is located at the University of Texas, Austin. Ortiz Monasterio, in *México eternamente*, points out that the document in not signed by Riva Palacio. The viewpoint and content however, he suggests, make it likely written by Riva Palacio.

17 "El movimiento intelectual de México y el adelanto notable, en ciencias y en literatura, que cada día toma mayor vuelo, hacen indispensable la publicación de una Historia completa que reúna todas las condiciones de las obras de este género, en época como la presente, en que ni el gusto ni el buen criterio se conforman ya con trabajos medianos o incompletos.

"Además, la necesidad de dar a luz la Historia general de México desde sus más remotos tiempos se halla ligada con la gran conveniencia de que se conozca en todo el mundo civilizado la marcha progresiva y fecunda en acontecimientos de un pueblo cuyo origen, desenvolvimiento y cultura, son desconocidos en los países más ilustrados de Europa y aun en el mismo continente americano." Riva Palacio (attributed), "Historia general de México. Prospecto," quoted by Ortiz Monasterio, *México eternamente*, 200–201.

18 Tenorio-Trillo, *Mexico at the World's Fairs*, 70.

19 "En esta nueva Historia los autores, separándose de aquel método, dan lugar no secundario a los medios que han servido para el progreso de la inteligencia y de la moral en México, siguiendo la fundación, desarrollo y utilidad de las Universidades fundadas, de los principales Colegios, de las escuelas de artes, de las Academias, museos, y sociedades literarias que han florecido o florecen actualmente en la República, de las casas de Beneficencia, hospitales, institutos de instrucción pública, de los literatos, poetas, hombres científicos, publicistas y artistas que han distinguídose y dádose a conocer en el mundo de las letras, de las ciencias y de las artes." Riva Palacio (attributed), "Historia general de México. Prospecto," quoted by Ortiz Monasterio, *México eternamente*, 204.

20 A very good discussion of Chavero and his work is found in Christina Maria Bueno's *Excavating Identity*.

21 In his article "Landscapes and Identities," Cañizares-Esguerra also looks at these frontispieces in the context of landscape and nationalism.

22 As pointed out to me by Ray Craib, these "pristine civilizations" emphasized sedentary groups in contrast to nomadic groups such as the Apache, Yaqui, and Maya, on whom the government was waging war at this time.

23 "Crónica Charlamentaria" from *El Monitor Republicano*, núm. 4162 (Mexico City; Sunday, 17 January 1869). In Rodríguez Prampolini, *La crítica de arte en México en el siglo XIX*, Vol. 2, 141.

24 For example, Chavero's tomo 1 images of Xochicalco and La Quemada are copied from Nebel. In the prospectus, illustrations are envisioned as "Cromos, grabados, planos, autógrafos, todo en abundancia y todo ejecutado por los mejores artistas y tomado de los mejores modelos; paisajes, vistas de ciudades, de edificios, de monumentos, retratos, representación de armas, de objetos de arte, numismática, antiguos geroglífos e inscripciones, todo cuanto sea necesario para la perfecta inteligencia del texto, todo aparecerá en la obra." Riva Palacio (attributed), "Historia general de México. Prospecto." (Chromolithographs, engravings, plans, autographs, all in abundance and all executed by the best artists and taken from the best models; landscapes, views of cities, of buildings, of monuments, portraits, representation of weapons, of objects of art, numismatics, old hieroglyphs and inscriptions, all that is necessary for the perfect intelligence of the text, all will appear in the work.) Quoted in Ortiz Monasterio, *México eternamente*, 202.

25 Collado, "Antonio García Cubas," 437. It should also be noted that while completing these translations, García Cubas was working on his five-volume opus, the *Diccionario geográfico, histórico, y biográfico de los Estados Unidos Mexicanos*.

26 The 1996 facsimile edition of the *Atlas pintoresco* includes the *Cuadro geográfico* text. Further, beginning with the tenth carta page, there are many discrepancy between the visual content of the *Atlas*'s pages and the associated *Cuadro geográfico* text.

27 García Cubas, *Cuadro geográfico, estadístico, descriptivo é histórico del los Estados Unidos Mexicanos*, 1. García Cubas also identifies Miguel R. Hernández, Vicente Calderón, Francisco Mendoza, and Santiago Hernández as artists who produced the images.

28 Ortiz Monasterio, *México eternamente*, 188–98.

29 The boundary line with Guatemala is annotated "Limites no determinados," indicating that this line was in dispute at the time. The text associated with the Carta I, however, summarizes the boundaries agreed upon in an 1883 treaty with the Republic of Guatemala. García Cubas was part of the team that determined this line.

30 In 1885 the Aztec Sun Stone was removed from the wall of the Mexico City Cathedral and placed in the National Museum.

31 See Tenorio-Trillo's *Mexico at the World's Fairs*.

32 See Poole, *Vision, Race and Modernity*, chaps. 5–7; and Eleanor Hight and Gary D. Sampson, *Colonialist Photography*, introduction.

33 A similar coat of arms is depicted in *Alegoría de las autoridades españoles e indígenas* (1809), Patricio Suárez de Peredo of figure 39.

34 The theme may also relate to the fact that in 1869, concordant with the incorporation of New Spanish history into Mexican history, Ignacio Altamirano pronounced churches to be like viceregal art, treasures of Mexico's national

glory that not only honor the nation but are sites that travelers might find interesting. Raymond Hernández-Durán, "Entre el espacio y el texto," 59.

35 See Overmeyer-Velázquez, *Visions of the Emerald City*, chap. 3.

36 Here, the city with its grand Paseo de la Reforma and passing public monuments becomes a monument to the Porfiriato, too. See Agostoni, *Monuments of Progress*, 45–46, 52–61, 157–58.

37 Ibid., 30.

38 Overmeyer-Velázquez, *Visions of the Emerald City*, 63. These words were stated by Díaz when he was governor of Oaxaca during his four-year hiatus from the Mexican presidency; nevertheless, they are indicative of his view of the nation as proceeding on a journey.

39 The Porfiriato's recapitulating of the past and previewing of the future are discussed by Tenorio-Trillo, *Mexico at the World's Fairs*, 6.

40 Ibid., 66.

41 See Tenenbaum, "Streetwise History," 141–44.

42 Gaonkar, *Alternative Modernities*, 23.

43 Collado, "Antonio García Cubas," 430.

44 José María Vigil, "Educación de la mujer," in *El Corréo de las Señoras*, 1884, 601–2. Reprinted in Rocha, *El álbum de la mujer*, tomo 4: *El Porfiriato y la Revolución*, 137–38.

45 "¿Quiere que ejerza la medicina, que desempeñe la abogacía, que sea represente del pueblo y que asuma el importante papel de gobernante? . . . cada seccion de la humanidad tiene su objeto bien definido. El papel de la mujer es tan fundamentalmente importante en el santuario del hogar, que cualquiera otra actividad á que quisiera consagrársele (por importante que se suponga) sería pequeña y miserable en comparación de sus grandiosos deberes domésticos. . . ." Luís E. Ruiz, "Artículo del señor Luis E. Ruiz en respuesta al que sobre 'La educación de la mujer' publica el señor J. M. Vigil," in *El Corréo de las Señoras*, 1888, 630–31. Reprinted in Rocha, *El álbum de la mujer*, Vol. 4:140.

46 Gimeno de Flaquer's novels had moralizing titles such as *Victorina; ó Heroismo del Corazon, novel original* (1879) and *Mujer juzgada por una mujer* (1882). Cited in Lisa Carolyn Glowacki, *Writing Gender and Nation*.

47 In the second edition, Gimeno de Flaquer was the director but no longer owner.

48 *El Álbum de la Mujer* 2, núm. 22 (1 June 1881): 318.

49 Ibid.: 319. "La familia es la sociedad, concentrada, reducida á su más delicada esencia, y que el hombre ha nacido para vivir en sociedad no es cuestionable."

50 García Cubas, *El libro de mis recuerdos*, 632.

7 PERFORMING THE NATION

1 Although there is significant archival material in Mexico and the United States, the Centenario de la Independencia has not been studied comprehensively; nor can it be thoroughly analyzed in the space of this conclusion.

Tenorio-Trillo's analysis "1910 Mexico City" locates how, using the city as a stage, "cultural axioms . . . were envisioned at a specific historical moment" through Centenario (75). While identifying the context of the world's fairs on the planning of the Centenario, he does not trace the origin of the imagery to the earlier part of the century and the establishment Mexico's visual culture. See also Thomas Reese and Carol Reese, "Revolutionary Urban Legacies." Michael J. Gonzalez's "Imagining Mexico in 1910" looks at the centennial impact in the post-Porfiriato context as well as a global phenomenon associated with celebration of revolutionary epochs.

2 Genaro García, *Cronica oficial de las fiestas del primer Centenario de la Independencia de México*, 110–23.

3 Ibid., 198–205.

4 Ibid., 248–69.

5 Ibid., 193–94.

6 Ibid., 182–87.

7 Ibid., 123.

8 Ibid., 129.

9 Ibid., 129–31.

10 Ibid., 130–31.

11 Ibid., 134.

12 "Los calles estaban henchidas de gente; los edificios lucían adornos del mejor gusto, y en los balcones y aparadores de las casas comerciales se apiñaba un multitud ávida de presenciar el desfile cívico." Ibid., 135.

13 Ibid., 138.

14 Ibid., 140.

15 See Curio-Nagy, *The Great Festivals of Colonial Mexico City*, 32, 78.

16 See Seed, *Ceremonies of Possession in Europe's Conquest of the New World 1492–1640*.

17 Cañeque, *The King's Living Image*, 324–26.

18 Curio-Nagy, *The Great Festivals of Colonial Mexico City*, 7.

19 Ibid., 146.

20 Bennett, *The Birth of the Museum*, 63. See chap. 4.

BIBLIOGRAPHY

PRIMARY SOURCES

Ackerman, R. *Cartas sobre la educación del bello sexo*. London: R. Ackerman, 1824.

Alamán, Lucas. *Disertaciones sobre la historia de la República Megicana, desde la epoca de la conquista que los españoles hicieron, a fines del siglo XV y principios del siglo XVI, de las islas y continente americano hasta la independencia*. 3 tomos. Mexico City: J. M. Lara, 1844–49.

———. *Historia de Méjico desde los primeros movimientos que prepararon su independencia en el año de 1808, hasta la época presente. Por don Lúcas Alamán*. 5 tomos. Mexico City: Impr. de J. M. Lara, 1849–52.

Álbum Pintoresco de la República Mexicana. Mexico City: Estamperia de Julio Michaud y Thomas, ca. 1849–52. Facsimile. Mexico City: Grupo Condumex, 2000.

Álvarez, Diego. "Discurso sobre la influencia de la instrucción publica en la felicidad de las naciones." *Revista Mensual de la Sociedad Promovedora de Mejoras Materiales* 1 (1852): 370–80.

Anderson, William M. *An American in Maximillian's Mexico 1865-1861: The Diaries of William Marshall Anderson* (1865). Edited by Ramón Eduardo Ruiz. San Marino: Huntington Library, 1959.

Arróniz, Marcos. *Manual de biografía mejicana ó galería de hombres célebres de Méjico*. Paris: Librería de Rosa y Bouret, 1857.

———. *Manual de historia y cronología de Méjico*. Paris: Librería de Rosa y Bouret, 1858.

———. *Manual del Viajero en Méjico, ó compendio de la historia de la ciudad de Méjico con La descripcion é historia de sus Templos, Conventos, Edificios públicos, las Costumbres de sus habitantes, etc., y con el plan de dicha ciudad.* Paris: Librería de Rosa y Bouret, 1858.

———. *Atlas de las antigüedades mexicanas halladas en el curso de los tres viajes de la Real Expedición de Antigüedades de la Nueva España, emprendidos en 1805, 1806 y 1807: contiene la reproducción facsimilar de las litograficas ejecutadas a partir de los dibujos de José Luciano Castañeda e impresas en Paris, en 1834, por Jules Didot, así como la relación de dichos viajes por el capitán Guillermo Dupaix, jefe de la Real Expedición.* Facsimile. Mexico City: San Angel Ediciones, 1978.

Becher, Henry. *A trip to Mexico: being notes of a journey from Lake Erie to Lake Tezcuco and back, with an appendix containing and being a paper about the ancient nations and races who inhabited Mexico before and at the time of the Spanish conquest, and the ancient stone and other structures and ruins of ancient cities found there.* Toronto: Willing and Williamson, 1880.

Beltrami, Giacomo. *Le Mexique.* 2 vols. Paris: Crevot, 1830.

Blaeu, Willem Janszoon. *Theatrum Orbis Terrarum, sive, Atlas Novus.* Amsterdam: Guiljelmum e Iohannem Blaeu, 1638.

Buffon, Georges-Louis Leclerc, compte de. *De l'homme.* 1747–77. Edited by Michèle Duchet. Paris: Maspero, 1971.

Bullock, William, F. L. S. *Catalogue of the exhibition called Modern Mexico; containing a panoramic view of the city, with specimens of the natural history of New Spain, and models of the vegetables produce, costume, &c., &c.: and now open for public inspection at the Egyptian Hall, Piccadilly.* London: Printed for proprietor, 1824.

———. *A description of the unique exhibition called Ancient Mexico: collected on the spot in 1823 by the assistance of the Mexican government, and now open for public inspection at the Egyptian Hall, Piccadilly.* London: Printed for proprietor, 1824.

———. *Six months' residence and travels in Mexico; containing remarks on the present state New Spain, its natural products, state of society, manufacturers, trade, agriculture, and antiquities, &c.: with plates and maps.* London: John Murray, Albemarle Street, 1824.

Cabrera, Miguel. *Maravilla americana y conjunto de raras maravillas observadas con la direccion de las Reglas de el Arte de la Pintura en la prodigiosa imagen de nuestra Sra. de Guadalupe de Mexico.* Mexico City: Impr. del Colegio de San Ildefonso, 1756.

Carey, H. C. *A Complete historical, chronological, and geographic American atlas being a guide to the history of North and South America, and the West Indies: exhibiting an accurate account of the discovery, settlement and the progress of their various kingdoms, states, provinces, &c together with the wars, celebrated battles, and remarkable events, to the year 1821. According to the plan of Le Sage's Atlas, and intended as an companion to Lavoisine's improvement of that celebrated work.* Philadelphia: H.C. Carey and I. Lea, 1822.

Castiglioni, Luigi. *Storia delle piante forestiere le più importanti nell'uso medico, od economico*. Milan: Nella Stamperia di Giuseppe Marelli, 1791–1794.

Castro, Casimiro. *Álbum del ferrocarril mexicano: Colección de vistas pintadas del natural*. Description by Antonio García Cubas. 1877. Facsimile. Mexico City: Cartón y Papel de México, S.A., 1977.

Cátalogos de la exposiciones de la antigua Ácademia de San Carlos de México (1850–1898). Edición de Manuel Romero de Torreros. Mexico City: Imprenta Universitaria, 1963.

Catherwood, Fredrick. *Views of ancient monuments in Central America, Chiapas and Yucatan*. London: Published by F. Catherwood, 1844.

Charnay, Désiré. *The Ancient Cities of the New World; being travels and explorations in Mexico and Central America from 1857–1882*. Translated by J. Gonino and Helen S. Conant. London: Chapman and Hall, 1887.

Clavijero, Francisco Xavier. *Historia antigua de México y de su conquista, sacada de los mejores historiadores españoles*. 2. vols. Mexico City: Impr. de Lara, 1844.

Colección de las antigüedades mexicanas que existían en el Museo nacional y dieron a luz el Pbro. y dr. d. Isidro Ignacio y el br. d. Isidro Rafael Gondra en 1827. Facsimile. Mexico City: Talleres gráficos de Museo Nacional de Arqueología, Historia y Etnografía, 1927.

Conquest of Mexico! An Appeal to the citizens of the United States, on the justice and expediency of the conquest of Mexico; with historical and descriptive information respecting the country. Boston: Jordan and Wiley, 1846, 1–32.

Consideraciones sobre la situación política y social de la república mexicana, en el año 1847. [*Considerations on the Political and Social Situation of the Mexican Republic 1847*.] Author/s unknown. Translated and edited by Dennis E. Berge. *Southwestern Studies*, Monograph 45. El Paso: Texas Western Press. 1975.

Cortés, Hernán. *Cartas del famoso conquistador Hernan Cortes al emperador Carlos Quinto*. Mexico City: Imprenta de I. Escalante y Cia, 1870.

———. *Historia de Nueva-España, escrita por su esclarecido conquistador Hernán Cortés; aumentada con otros documentos, y notas, por el ilustrissimo señor don Francisco Antonio Lorenzana*. Mexico City: en la Imprenta de Superior Gobierno, 1770.

Couto, José Bernardo. *Diálogo sobre la historia de la pintura en México*. (1860?). Mexico City: Fondo de Cultura Económica, 1947.

de Alzate y Ramírez, José Antonio. *Asuntos varios sobre ciencias y artes: Obra periódica dedicado al Rey N sr*. Mexico City: Josef de Jauregui, 1772.

de Bustamante, Carlos María. *Mañanas de la Alameda de México. Publícalas para facilitar a las señoritas el estudio de la historia de su pais*. 2 tomos. Mexico City: Imprenta de la Testamentaria de Vallés, 1835–36.

de Charnay, Désiré. *Álbum fotográfico mexicano*. México: Julio Michaud, 1860.

de Pauw, Corneilus. *Recherches philosophiques sur les américains, o Mémoires inté-*

ressants pour servir à l'histoire de l'espèce humaine. 1768–69. New ed. in 3 vols. that includes *Dissertation critique par Dom Pernety & de la Défense de l'auteur des Recherches contre cette dissertation.* Berlin: G. J. Decker, 1770.
de Vaugondy, Robert. *Atlas universel.* Paris: Chez les auteurs, 1757.
del Río, Antonio. *Description of the ruins of an ancient city discovered near Palenque in the kingdom of Guatemala, in Spanish America; translated from the original manuscript report of Don Antonio del Rio; followed by the Teatro critico Americano, or a critical investigation and research into the history of the Americans,* by Doctor Felix Cabrera, London: Henry Berthoud, 1822.
Descripción de las fiestas celebradas en la imperial corte de México con motivo de la solemne colocación de una estatua equestre de nuestro augusto soberano el Señor Don Carlos IV en la Plaza Mayor. Archivo General de la Nación. *Bandos.* 1796. Tomo 18, exp. 108, fojas 456–63v.
Description de l'Égypte, ou, Recueil de observations et des recherches qui ont été faites en Égypte pendant l'éxpédition de l'armée française, publié par les ordres de Sa Majesté l'empereur Napoléon le Grand. 21 vols. Paris: Imprimerie impériale, 1809–28.
Díaz, Agustín. *Memoria de la Comisión Geográfico-Exploradora presentada al official mayor, encargada de la Secretaría de Fomento, sobre los trabajos ejecutados durante el año fiscal 1878 a 1879.* Mexico City: Imprente de Francisco Díaz de León, 1880.
Diccionario universal de historia y de geografia, obra dada a luz en España por una sociedad de literatos distinguidos, y refundida y aumentada considerablemente para su publicacion en Mexico con noticias históricas, geográficas, estadísticas y biográficas sobre las Americas en general y especialmente sobre la Republica Mexicana. 10. vols. México: Tipografia de Rafael, 1853–56.
Documentos varios para la historia de la Ciudad de México a fines de la época colonial (1769–1815). Facsimile. Mexico City: Rolston-Bain, 1983, XIV.
Domenech, Emmanuel. *Le mexique tel qu'il est; la vérité sur son climat, ses habitants et son gouvernement.* Paris: E. Dentue, 1867.
Donnavan, Corydon. *Adventures in Mexico; experienced during a captivity of seven months.* 12th ed. Boston: G. R. Holbrook and Co., 1848.
D'Orbigny, Alcides Dessaignes. *Viaje Pintoresco á las dos Americas.* Barcelona: Juan Oliveres, 1842.
Dunn, Archibald Joseph. *Mexico and her resources.* London: A. Boot & Son, 1890.
Dupaix, Guillermo. *Antiquités mexicaines: relation des trois expéditions du colonel Dupaix, ordonnées en 1805, 1806, et 1807, par le roi Charles IV, pour la recherche des antiquités du pays, notamment celles de Mitla et de Palenque: avec les dessins de Castañeda . . . et une carte du pays explorés. . . .* 2 vols. Paris: Au Bureau des antiquités mexicaines . . . , Imprimerie de Firmin Didot, Frères, 1844.
———. *Expediciones acerca de los antiguos monumentos de la Nueva España, 1805–1808.* Madrid: Ediciones J. Porrúa, 1969.

El Álbum de la mujer; illustración hispano-mexicana. Mexico City: Imprinta de F. Díaz de León, 1885.
El Mosaico mexicano, ó Coleccion de amenidades curiosas é instructivas. 7 tomos. Mexico City: L. Cumplido, 1840–42.
El Museo mexicano, ó Miscelánea pintoresca de amenidades curiosas é instructivas. 4 vols. Mexico City: I. Cumplido, 1843–44.
Fernández de Oviedo y Valdés, Gonzalo. *La historia general de las Indias*. Seville: en la empré[n]ta de Iuam Cromberger, 1535.
———. *Sumario de la natural historia de las Indias*. [1526] José Miranda ed. Mexico City: Fondo de Cultura Económica, 1950.
Figuier, Louis. *The Human Race*. London: Chapman and Hall, 1872.
———. *Primitive Man: illustrated with thirty scenes of primitive life, and two hundred and thirty-three figures of objects belonging to prehistoric ages*. London: London, Chapman and Hall, 1870.
Folsom, Charles. *Mexico in 1842: A Description of the Country, its Natural and Political Features; with a Sketch of its History brought Down to the Present Year. To which is added, An account of Texas and Yucatan; and of the Sante Fé Expedition*. New York: Charles J. Folsom, 1842.
Fossey, Mathieu de, comte. *Le Mexique*. Paris: Henri Plon, 1857.
García, Genaro. *Apuntes sobre la condición de la mujer*. Mexico City: Compañia Limit. de Tipografos, 1891.
———. *Cronica oficial de las fiestas del primer Centenario de la Independencia de México*. Mexico City: Talleres del Museo Nacional, 1911. Facsimile. Mexico City: Centro de Estudios de Historia de México Condumex, 1991.
García Cubas, Antonio. *Álbum del ferrocarril Mexicano*. Colección de vistas pintadas del natural por Casimiro Castro. Mexico City: Victor Debray y Compania, 1877.
———. *Atlas geográfico, estadístico é histórico de la República Mexicana formado por Antonio Garcia y Cubas*. Mexico City: Imprenta de José Mariano Fernandez de Lara, 1858.
———. *Atlas geográfico, estadístico é histórico de los Estados Unidos Mexicanos*. Mexico City: Debray Sucesores, 1887.
———. *Atlas metodico para la enseñanza de la geografía de la república mexicana*, formado y dedicado á la Sociedad mexicana de geografía y estadistica por el ingeniero Antonio García Cubas. Mexico City: Sandoval y Vazquez, Impresores, 1874.
———. *Atlas mexicano*. Mexico City: Debray Suc., ca. 1884–86.
———. *Atlas pintoresco é historico de los Estados Unidos Mexicanos por Antonio García Cubas*. Mexico City: Debray, 1885.
———. *Carta general de la República Mexicana*. Mexico City: 1863.
———. *Cuadro geográfico, estadístico, descriptivo é histórico del los Estados Unidos Mexicanos*. Mexico City: Oficina Tip. de la Secretaría de Fomento, 1884.

———. *Cuadro geográfico, estadístico, descriptivo é histórico del los Estados Unidos Mexicanos, obra que sirve de texto al Atlas pintoresco de Antonio Garcia Cubas,* Mexico City: Oficina Tip. de la Secretaría de Fomento, 1885.

———. *Curso de dibujo topografica y geografico.* Mexico City: Imprenta del Gobierno, 1868.

———. *Curso elemental de geografía dispuesto con arreglo á un Nuevo método de facilite su enseñanza en los establecimientos de instrucción de la República.* 2nd ed. Mexico City: Imprenta de la Viuda é Hijos de Murguía, 1876.

———. *Curso elemental de geografía universal dispuesto con arreglo á un nuevo método que facilite su enseñanza en los establecimientos de instrucción de la República.* Mexico City: Imprenta del Gobierno, en Palacio, 1869.

———. *Diccionario geográfico, histórico, y biográfico de los Estados Unidos Mexicanos.* 5 vols. Mexico City: Antigua Imprenta de Murguía, 1888.

———. *El libro de mis recuerdos.* 1904. Reproduction. Mexico City: Porrúa, 1986.

———. "Ensayo de un estudio comparativo entre las pirámides egipcias y mexicanas." In *Escritos diversos de 1870 a 1874,* 269–77. Mexico City: Imprenta de Ignacio Escalante, 1874.

———. *Escritos diversos de 1870 a 1874.* Mexico City: Imprenta de Ignacio Escalante, 1874.

———. *Étude géographique, statistique, descriptive et historique des États Unis Mexicains.* Translated by William Thompson. Mexico City: Oficina Tip. dé la Secretaría de fomento, 1889.

———. *Memoria para servir á la Carta General de la República Mexicana.* Mexico City: Andrade y Escalante, 1861.

———. *Memoria para servir á la Carta General del Imperio Mexicano y demás naciones descubiertas y conquistadas por los españoles durante el siglo XVI perteneciente hoy á la República Mexicana.* Mexico City: Secretaría de Fomento, 1892.

———. *Mexico: its trade, industries and resources.* Mexico City: Oficina Tip. dé la Secretaría de fomento, 1893.

———. *The Republic of Mexico in 1876. A political and ethnographical division of the population, character, habits, costumes and vocations of its inhabitants.* Translated by George F. Henderson. Mexico City: La Enseñanza, 1876.

Gemelli Careri, Giovanni Francesco. *Voyage du tour du monde, traduit de l'italien de Gemelli Careri, par L.M.N.* 6 vols. Paris: Étienne Ganeau, 1719.

Grasset de Saint-Sauveur, Jacques. *Tableaux des Principaux Peuples de l'Europe, de l'Asie, de l'Afrique de l'Amerique. . . .* Paris: chez l'auteur Bordeaux, 1798.

Gualdi, Pedro. *Monumentos de Méjico / tomados del natural y litografiados por Pedro Gualdi.* Mexico City: Masse y Decaen, 1841.

Hernández, Francisco. *Rerum medicarum Novae Hispaniae thesaurus.* Rome: Ex Typographeio Vitalis Mascardi, 1651.

Herrera y Tordesillas, Antonio de. *Descripción de las Indias Ocidentales*. Madrid: N. R. Franco, 1601.

———. *Historia general de los hechos de los castellanos en las islas y tierra firme del mar oceano*: 1492–1531. 4 vols. Madrid: Emprenta Real, 1601–15.

Humboldt, Alexander von. *Atlas géographique et physique du nouveau continent*. Paris: F. Schoell [et al.], 1814 [1820].

———. *Atlas géographique et physique du Royaume de la Nouvelle-Espagne*. Paris: F. Schoell [et al.], 1811.

———. *Cosmos: A Sketch of a Physical Description of the Universe*. 5 vols. Translated by E. C. Otte. London: Bohn, 1849–52.

———. *Ensayo politico sobre el reino de la Nueva-España*. Translated into Spanish by Vicente Gonzalez Arnao. Paris: Rosa, 1822.

———. *Essai politique sur le royaume de la Nouvelle-Espagne; / par Alexandre de Humboldt. Avec un atlas physique et géographique, fondé sur des observations astronomiques, des mesures trigonométriques et des nivellemens barométriques*. Tome 1. Paris: Chez F. Schoell, 1811.

———. *Essai sur la géographie des plantes: accompagné d'un tableau physique des régions équinoxiales, fondé sur des mesures exécutées, depuis le dixième degré de latitude boréale, jusqu'au dixième degrée de latitude australe, pendant les années 1799, 1800, 1801, 1802 et 1803. / Par Al. de Humboldt et A. Bonpland; rédigée par Al. de Humboldt. Avec une planche*. Paris: Chez Levrault, Schoell et Compagnie, 1807.

———. *Minerva. Ensayo político sobre el reyno de la Nueva-España*. Written in French by Alexandro de Humboldt, and translated into Spanish by Don Pedro Maria de Olive. 2 vols. Madrid: Imbara, Impresor de Cámara de S. M. con Privilego Real, 1818.

———. *Political essay on the kingdom of New Spain. With physical sections and maps founded on astronomical observations and trigometrical and barometrical measurements*. 2 vols. Translated from the Original French by John Black. New York: I. Riley, 1811.

———. 4 vols. London: Printed for Longman, Hurst, Rees, Orme and Brown, 1811.

———. *Researches concerning the institutions & monuments of the ancient inhabitants of America: with descriptions & views of some of the most striking scenes in the Cordilleras!* Written in French by Alexander de Humboldt, and translated into English by Helen Maria Williams. 2 vols. London: Longman, 1814.

———. *The travels and researches of Alexander von Humboldt: being a condensed narrative of his journeys in the equinoctial regions of America, and in Asiatic Russia: together with analysis of his more important investigations / by W. Macgillivray*. Harper's Family Library, no. 54. New York: J. and J. Harper, 1833.

———. *Vistas de las Cordilleras y Monumentos de los pueblos indígenas de América*. (1822). 2 vols. Mexico City: Siglo XXI Editores, S.A. de C.V., 2000.

———. *Voyage de Humboldt et Bonpland. Partie 1, Voyage aux régions équinoxiales du Nouveau Continent, fait en 1799, 1800, 1801, 1802, 1803 et 1804*. Paris: F. Schoell [et al.], 1814.

———. *Vues des Cordillères et monumens des peuples indigènes de l'Amérique*. Paris: F. Schoell, 1810 [1813?].

———. 2 vols. Paris: Librairie grecque-latine-allemande, 1816.

Leclercq, Jules Joseph. *Voyage au Mexique, de New York à Vera-Cruz, en suivant les routes de terre*. Paris: Hachette et cie, 1885.

León y Gama, Antonio. *Descripcion histórica y cronológica de las dos piedras: que con ocasión del nuevo empedrado que se está formando en la Plaza Principal de México, se hallaron en ella el año de 1790. Explícase el sistema de los calendarios de los indios, el método que tenian de dividir el tiempo, y la correcion que hacian de él para igualar el año civil, de que usaban, con el año solar trópico. Noticia muy necesaria para la perfecta inteligencia de la segunda piedra: á que se añaden otras curiosas é instructivas sobre la mitología de los mexicanos, sobre su astronomía, y sobre los ritos y ceremonias que acostumbraban en tiempo de su gentilidad. / Por don Antonio de Leon y Gama*. Mexico City: En la Imprenta de Don Felipe de Zúñiga y Ontiveros, 1792.

———. *Descripción histórica y cronológica de las dos piedras, que con ocasión del nuevo empedrado que se está formando en la plaza principal de México, se hallaron en ella el año de 1790.* . . . Reproduction. Edited by Carlos Maria de Bustamante. Mexico City: Impr. del Ciudadano A. Valdés, 1832.

Linati, Claudio. *Costumes civils, militaires et réligieux du Mexique; dessinés d'après nature par C. Linati*. Brussels: C. Sattanino; Imprimés à la Lithographie Royale de Jobard [1828].

Linne, Carl von. *Systema naturae per regna tria naturae, secundum classes, ordines, genera, species, cum characteribus, differentiis, synonymis, locis*. Stockholm: Impensis L. Salvii, 1758–59.

The Literary Memoranda of William Hickling Prescott. Edited by C. Harvey Gardiner. 2 vols. Norman: University of Oklahoma Press, 1961.

López de Gómara, Francisco. *Hispania victrix. Primera y segvnda parte de la historia general de las Indias con todo el descubrimiento, y cosas notables que han acaescido desde que se ganaron hasta el año de 1551. Con la conquista de Mexico, y de la Nueua España*. 2 vols. Medina del Campo, [Spain]: Guillermo de Millis, 1553.

López de Vargas Machuca, Tomás. *Atlas geographico de la America Septentrional y meridional: dedicado a la Catholica Sacra Real Magestad de el Rey Nuestro Senor Don Fernando VI*. Madrid: A. Sanz, 1758.

López de Velasco, Juan. *Descripción y demarcación de las Indias Ocidentales* [1574]. Edited by Marcos Jiménez de la Espada. Madrid: Ediciones Atlas, 1971.

———. *Geografía y descripción universal de las Indias, recopilada por el cosmógrafo-cronista Juan López de Velasco, desde el año de 1571 al de 1574, publicada por primera vez en el Boletín de la Sociedad geográfica de Madrid, con adiciones é ilustraciones, por don Justo Zaragoza*. Madrid: Fortanet, 1894.

Los mexicanos pintados por sí mismos. Tipos y costumbres nacionales por varios Authores. Mexico City: Imprente de M. Murguía y Comp, 1854. Facsimile. Mexico City: Biblioteca Nacional, 1935.

Mayer, Brantz. *Mexico as it was and as it is: with numerous illustrations on Wood. Engraved by Butler from Drwgs by the Author*. New York: J. Winchester, New World Press, 1844.

Mercator, Gerhard. *Atlas sive Cosmographicae meditationes de fabrica mvndi et fabricati figvra*. Dvisbvrgi Clivorvm [1595].

———. Amsterdam: Excusum in Aedibus, 1607.

México y sus alrededores. Coleccion de monumentos, trajes y paisajes. Drawn in their natural setting and lithographed by the Mexican artists C. Castro, J. Campillo, L. Auda, y G. Rodriguez. Mexico City: Decaen, 1855–56. Facsimile. Mexico City: Editorial del Valle de México, 1974.

Montanus, Arnoldus. *America: being an accurate description of the New World; Containing the original of the inhabitants; the remarkable voyages thither: the conquest of the vast empires of Mexico and Peru; . . . by John Ogilby*. Translated by John Ogilby. London: Printed by Tho. Johnson for the author, 1671.

Mora, José María Luis. *Méjico y sus revoluciones, obra por José María Luis Mora*. 3 vols. Paris: Librería de Rosa, 1836.

Morelos, José María. "Sentiment of the Nation." In *The Mexico reader: history, culture, politics*, edited by Gilbert M. Joseph and Timothy J. Henderson, 189–91. Durham: Duke University Press, 2002.

Nebel, Carl. *Voyage pittoresque et archéologique, dans la partie la plus intéressante du Mexique,par C. Nebel; 50 planches lithographiées avec texte explicatif*. Paris: Chez M. Moench, 1836.

O'Crouley, Pedro Alonso. *Idea Compendiosa del Reyno de Nueva España (1774), A Description of the Kingdom of New Spain*. Translated and edited by Seán Galvin. Dublin: Allen Figgs, 1972.

Orozco y Berra, Manuel. *Apuntes para la Historia de la Geografía en México*. Mexico City: Imprente de Francisco Dias de Leon, 1881.

———. *Geografía de las lenguas y carta etnográfica de México: precedidas de un ensayo de clasificación de las mismas lenguas y de apuntes para las inmigraciones de las tribus*. Mexico City: Imprente de J. M. Andrade y Escalante, 1864.

Ortelius, Abraham. *Theatrvm orbis terrarvm*. Antwerp: Apud Aegid. Coppenium Diesth, 1570.

Pimental, Franciso. *Sobre las causas que han originado la situacion actual de la raza indígena de México y medios de remediarla*. Mexico City: Imprenta de Andrade y Escalante, 1864.

Prescott, William. *The Conquest of Mexico, with a preliminary view of the ancient Mexican civilization, and the life of the conqueror, Hernando Cortés*. 3 vols. New York: Harper and Brothers, 1843.

———. *Historia de la conquista de México, con bosquejo preliminar de la civiliza-*

ción de los antiguos mejicanos y la vida de su conquistador Hernán Cortés, escrita en inglés por William H. Prescott, autor de la Historia de Fernando e Isabel. Translated by José María González de la Vega. 3 vols. Mexico City: Vicente García Torres, 1844–46.

———. *Historia de la conquista de México, con una ojeada sobre la Antigua civilizacion de los Mexicanos, y con la vida de su conquistador, Fernando Cortes.* Translated by Joaquín Navarro. 3 vols. Mexico City: Ignacio Cumplido, 1844–46.

Primer almanaque histórico, artístico y monumental de la República Mexicana 1883–1884. Manuel Caballero, ed. Mexico City and New York: Charles M. Green, 1883.

Ptolemy, Claudius. *Geographia.* Rome: Petrus de Turre, 1490.

Ramírez, José Fernando. *Descripción de algunos objetos del museo nacional de antiguedades de Mexico.* Mexico City: J. M. Andrade y F. Escalante, 1857.

Ramírez, Santiago. *Noticia histórica de la riqueza minera de México y de su estado de explotación.* Mexico City: Secretaría de Fomento, 1884.

Ramusio, Giovanni Battista. *Delle navigationi et viaggi.* 3 vols. Venice: Heredi di L. Givnti, 1550–65.

Raynal, Guillaume-Thomas-François. *Histoire philosophique et politique, des éstablissemens et du commerce de Européens dans les deux Indes.* 7 vols. Amsterdam: n.p., 1772–74.

Real ordenanza para el establecimiento é instruccion de intendentes de exército y provincia en el reino de la Nueva-España. Madrid: [n.p.], Año de 1786.

———. Mexico City: Universidad Nacional Autónoma de México, Instituto de Investigaciones Históricas; repr. ed., 1984.

Reclus, Elisée. *Nouvelle géographie universelle; la terre et les hommes, par Elisée Reclus.* 19 vols. Paris: Hachette, 1875–94.

Relaciones geográficas de Arzobispado de Mexico. 1743. Transcriptions by Catalina Romero, Belen Bañas, Manuel Lucena, Giraldo, Eduardo Moyano, and Francisco de Solano. 2 vols. Madrid: Consejo Superior de Investigaciones Cientificas, 1988.

Relaciones geográficas del siglo XVI. René Acuña, ed. Mexico City: Universidad Nacional Autónoma de México, Instituto de Investigaciones Antropológicas, 1982–87.

Riva Palacio, Vicente, ed. *México a través de los siglos: Historia general y completa del desenvolvimiento social, politico, religioso, military, artistic, científico y literario de México desde la antigüedad más remota hasta la época actual.* 5 vols. Barcelona: Espasa, 1883–90.

Riva Palacio, Vicente, and Manuel Payno. *El Libro Rojo 1520–1867.* Mexico City: Diaz de Leon y White, 1870.

Rivera Cambas, Manuel. *Atlas y catecismo de geografía y estadistíca de la República Mexicana.* 1st ed. Mexico City: Flores and Monsalvo, 1874.

———. *Los gobernantes de México. Galeria de biografias y retratos de los vireyes,*

emperadores, presidentes y otros gobernantes que ha tenido México, desde don Hernando Cortes hasta el C. Benito Juarez. Mexico City: Imprenta de J. M. Aguilar Ortiz, 1872–73.
———. *México pintoresco, artistico y monumental. Vistas, descripción, anecdotas, y episodios de los lugares mas notables de la capital y de los estados, aun de las poblaciones cortas, pero de importancia geográfico ó histórico. Obra ilustrada con gran número de hermosas litografías, representando las iglesias, ... todo cuanto puede señalar el grado de nuestro adelanto y el aspecto fisico, moral é intellectual de la República. Las descripciones contienen datos científicos, históricos y estadísticos.* 3 tomos. Mexico City: Imprenta de la Reforma, 1880–83.
Robertson, William. *The History of America.* 2 vols. London: W. Strahan, 1777.
Rodríguez de Campomanes, Pedro. *Reflexiones sobre el comercio español a Indias* [1762].Vicente Llombart Rosa, ed. Madrid: Instituto de Estudios Fiscales, 1988.
Ruiz, Ramón Eduardo. *An American in Maximilian's Mexico 1865–1866. The Diaries of William Marshall Anderson.* Edited by Ramón Eduardo Ruiz. San Marino, Calif.: Huntington Library, 1959.
Sanson, Nicolas. *L'Amerique Septentrionale divisée en ses principales parties*, 1674.
Sartorius, Carl. *Mexico: Landscapes and popular sketches.* London: Trübner and Company, 1859.
Sartorius, Carlos. *México y los mexicanos.* 1859. With 18 illustrations by M. Rugendas. Version, selection, and notes by Marita Martínez del Rio de Redo. Mexico City: San Ángel Ediciones, S.A., 1995.
Semanario de las Señoritas Mejicanas. Educacion científica, moral y literaria del bello sexo. Mexico City: Imp. de Vicente G. Torres, 1840–1842.
Sevin, Charles. "Journey in Mexico." *Proceedings of the Royal Geographical Society of London* 3 (1858–59): 108–11.
Stephens, C. A. *The Knockabout Club in the Tropics. The adventures of a party of young men in New Mexico, Mexico and Central America.* Boston: Estes and Lauriat, 1884.
Stephens, John Lloyd. *Incidents of travel in Egypt, Arabia Petraea, and the Holy Land.* 2 vols. New York: Harper and Brothers, 1838.
———. *Incidents of travel in Greece, Turkey, Russia, and Poland. By the author of "Incidents of travel in Egypt, Arabia, Petraea, and the Holy Land."* 2 vols. New York: Harper and Brothers, 1838.
———. *Incidents of Travel in Central America, Chiapas and Yucatan.* 2 vols. New York: Harper and Brothers, 1841.
———. *Incidents of Travel in Yucatan*, 2 vols. New York: Harper and Brothers, 1843.
Temsky, Gustav Ferdinand von. *Mitla. A narrative and personal adventure on a journey in Mexico, Guatemala, and Salvador in the years 1853 to 1855.* Edited by J. S. Bell. London: Longman, Brown, Green, Longmans, and Roberts, 1858.
Ticknor, George. *Papers discussing the comparative merits of Prescott's and Wilson's*

histories, pro and con.: as laid out before the Massachusetts Historical Society. Boston: [n.p.], 1861.

Valle, Juan. *El viajero en México.* Mexico City: Tipografía de M. Castro, 1859.

Vespucci, Amerigo. *Von der neu gefunden Region die wol ein Welt genent mag werden, durch den cristenlichen Künig von Portigal, wunderbarlich erfunden.* [Known as *Mundus Novus*] Nuremberg: W. Huber, 1506.

Villaseñor y Sánchez, José Antonio de. *Suplemento al Theatro Americano (La Ciudad de México en 1755).* Edited by Ramón María Serrera. Mexico City: Universidad Nacional Autónoma de México, 1980.

———. *Theatro Americano: descripcion general de los reynos y provincias de la Nueva España y sus juridicciones. Dedicala al Rey nuestro señor El Señor D. Phelipe Quinto monarcha de las Españas. por Joseph Antonio Villaseñor y Sanchez, Contador General de la Real Contraduria de Azoguez y Cosmographo de este Reyno quien la escribó de orden del escelentissimo Señor Conde de Fuen-Clara, Virrey Gobernador y Capitan General de esta Nueva España y Presidente de su Real Audiencia, &c.* 2 vols. Madrid: 1746–48.

Waldeck, Jean-Frédéric, Comte de. *Voyage pittoresque et archéologique dans la province d'Yucatan pendant les années 1834–1836.* Paris: [n.p.], 1838.

Wineburgh, Michael. *Where to spend the winter months. A birdseye view of a trip to Mexico, via Havana.* New York: M. Wineburgh, 1880.

SECONDARY SOURCES

Abbey, Kristen L. *Recipe for Narrative: Representations of Culture in Culinary Literature.* Ph.D. diss., Rutgers, 2006.

Acevedo, Esther. "Entre la tradición alegórica y la narrativa factual." In *Los pinceles de la historia. De la patria criolla a la nación mexicana 1750–1860,* edited by Esther Acevedo, Jaime Cuadriello, and Fausto Ramírez, 115–31. Mexico City: Instituto Nacional de Bellas Artes, 2000.

———, ed. *Hacia otra historia del arte en México. De la estructuración colonial a la exigencia nacional (1780–1860).* Tomo 1. Mexico City: Consejo Nacional para la Cultura y las Artes, 2001.

———. "Los comienzos de una historia laica en imagines." In *Los pinceles de la historia: La fabricación del estado, 1864–1910,* edited by Esther Acevedo, Jaime Cuadriello, and Fausto Ramírez, 35–53. Mexico City: Instituto Nacional de Bellas Artes, 2003.

Acevedo, Esther, in collaboration with Helia Bonilla. "La gráfica: Testigo de lo cotidiano." In *Hacia otra historia del arte en México. De la estructuración colonial a la exigencia nacional (1780–1860),* 218–40. Tomo 1. Mexico City: Consejo Nacional para la Cultura y las Artes, 2001.

Acevedo, Esther, Jaime Cuadriello, and Fausto Ramírez. *Los Pinceles de la historia. De la patria criolla a la nación mexicana 1750–1860.* Mexico City: Instituto Nacional de Bellas Artes, 2000.

Acevedo, Esther, and Fausto Ramírez. *Los Pinceles de la historia: La fabricación del estado 1864–1910*. Mexico City: Instituto Nacional de Bellas Artes, 2003.

Agostoni, Claudia. *Monuments of Progress: Modernization and Public Health in Mexico City, 1876–1910*. Calgary: University of Calgary Press, 2003.

Aguilar Ochoa, Arturo. *La fotografía durante el imperio de Maximiliano*. Mexico City: Instituto de Investigaciones Estéticas, 1996.

———. "Los inicios de la litografía en México: El periodo oscuro (1827–37)." *Anales de Instituto de Investigaciones Estéticas*, núm. 60 (2007): 66–100.

———. "La influencia de los artistas viajeros en la litografía mexicana (1837–1849)." *Anales del Instituto de Investigaciones Estéticas*, núm. 76 (2000): 113–41.

———. *La litografía en la ciudad de México, los años decisivos: 1827–1847*. Ph.D. diss., Universidad Autónoma de México, Facultad de Filosofía y Letras, 2001.

———. "Pedro Gualdi, Pintor de perspectiva en México." In *El escenario urbano de Pedro Gualdi 1808–1857*, 33–67. Catalogue. Museo Nacional de Arte Abril-Julio 1997. Mexico City: Instituto Nacional de Bellas Artes, 1997.

———. "Preguntas a un fotógrafo: François Aubert en México." *Alquimia* (May–August 2004): 7–14. Mexico City: INAH Sistema Nacional de Fototecas.

Aguirre, Robert D. *Informal Empire: Mexico and Central America in Victorian Culture*. Minneapolis: University of Minnesota Press, 2005.

Akerman, James, and Robert W. Karrow. *Maps: Finding Our Place in the World*. Chicago: University of Chicago Press, 2007.

Alba, Jacobo Stuart Fitz-James y Falcó, ed. *Mapas españoles de América, siglos XV–XVII*. Madrid: 1951.

Alpers, Svetlana. "The Mapping Impulse in Dutch Art." In *Art and Cartography: Six Historical Essays*, edited by David Woodword, 51–96. Chicago: University of Chicago Press, 1987.

Altamirano Piolle, María Elena. *National Homage—José María Velasco (1840–1912)*. 2 vols. Mexico City: Museo Nacional de Arte, 1993.

Altick, Richard. *The Shows of London*. Cambridge: Belknap Press of Harvard University Press, 1978.

Alvarado, María de Lourdes. "Mujeres y educación superior en el México del siglo XIX." *Diccionario de historia de la educación en México*. Visited 1 July 2007. Universidad Nacional Autónoma de México, 2002. Web pages of Publicaciones Digitales, DGSCA UNAM, http://biblioweb.dgsca.unam.mx. Printouts on file with author.

Ambler, Louise Todd, and Melissa Banta. *The Invention of Photography and Its Impact on Learning: Photographs from Harvard University and Radcliffe College and from the Collection of Harrison D. Horblit*. Cambridge: Harvard University Library, 1989.

Anderson, Benedict. *Imagined Communities: Reflections on the Origin and Spread of Nationalism*. Rev. ed. London: Verso, 1991.

Andrews, Malcolm. *The Search for the Picturesque: Landscape Aesthetics and Tourism in Britain, 1760–1800*. Stanford: Stanford University Press, 1989.

Anna, Timothy E. *Forging Mexico 1821–1835*. Lincoln: University of Nebraska Press, 1998.

Antochiw, Michel. "La visión total de la Nueva España: Las mapas generales del signo XVIII." In *México a través de los mapas*, edited by Héctor Mendoza Vargas, 71–88. Mexico City: Instituto de Geografía, Universidad Nacional Autónoma de México, 2000.

Appelbaum, Nancy P., Anne S. Macpherson, and Karin A. Rosemblatt, eds. *Race and Nation in Modern Latin America*. Chapel Hill: University of North Carolina Press, 2003.

Arbeláez A., María Soledad. *Mexico in the Nineteenth Century through British and North American Travel Accounts*. Ph.D. diss, University of Miami, 1995.

Arias, Santa, and Mariselle Meléndez. *Mapping Colonial Spanish America: Places and Commonplaces of Identity, Culture and Experience*. London: Associated University Presses, 2002.

Arrom, Silvia Marina. *The Women of Mexico City, 1790–1857*. Stanford: Stanford University Press, 1985.

Atlas cartográfico histórica. Mexico City: Instituto Nacional de Estadística Geografía e Informática, ca. 1982.

Ayala Alonso, Enrique. "Cómo la casa se convirtió en hogar: Vivienda y ciudad en el México decimónico." *Scripta Nova* (Revista Electrónica de Geografía y Ciencias Sociales, Universidad de Barcelona) 7, núm. 146 (107): 1 August 2003. Web pages of *Scripta Nova*, www.ub.es, visited April 2007. Printouts on file with author.

Azuela Bernal, Luz Fernanda. "La *Sociedad Mexicana de Geografía y Estadística*, la organización de la ciencia, la institucionalización de la geografía y la construcción del país en el siglo XIX." *Investigaciones Geográficas* [Boletín del Instituto de Geografía, UNAM], núm. 52 (2003):153–66.

Baecque, Antoine de. *The Body Politic: Corporeal Metaphor in Revolutionary France, 1770–1800*. Translated by Charlotte Mandell. Stanford: Stanford University Press, 1997.

Báez Macías, Eduardo. *Guía del archivo de la Antigua academia de San Carlos*. 3 vols. Mexico City: Universidad Nacional Autónoma de México, Instituto de Investigaciones, 1968–76.

Balakrishnan, Gopal, ed. *Mapping the Nation*. New York: Verso, 1996.

Bann, Stephen. "The Odd Man Out: Historical Narrative and the Cinematic Image." *History and Theory* 26, no. 4, Beiheft 26: The Representation of Historical Events (December, 1987): 47–67.

———. *Parallel Lines: Printmakers, Painters and Photographers in Nineteenth-Century France*. New Haven: Yale University Press, 2001.

Barrera-Osorio, Antonio. "Empire and Knowledge: Reporting from the New World." *Colonial Latin American Review* 15, no. 1 (June 2006): 39–54.

———. *Experiencing Nature: The Spanish American Empire and the Early Scientific Revolution*. Austin: University of Texas Press, 2006.

Baudez, Claude François. *Jean-Frédéric Waldeck, peitre le premier explorateur des ruines mayas*. Italy: Editiones Hazan, 1993.

Bauer, Ralph. *The Cultural Geography of Colonial American Literature: Empire, Travel and Modernity*. Cambridge: Cambridge University Press, 2003.

Bazant, Mílada. *Historia de la educación durante el Porfiriato*. Mexico City: El Colegio de México, 1993.

Beezley, William. *Judas at the Jockey Club and Other Episodes of Porfirian Mexico*. 2nd ed. Lincoln: University of Nebraska, 2004.

———. "The Porfirian Smart Set Anticipates Thorstein Veblen in Guadalajara." In *Rituals of Rule, Rituals of Resistance: Public Celebrations and the Popular Culture in Mexico*, edited by William H. Beezley, Cheryl English Martin, and William E. French, 173–90. Wilmington, Del.: Scholarly Resources, 1994.

Beezley, William, Cheryl English Martin, and William French, eds. *Rituals of Rule, Rituals of Resistance: Public Celebrations and Popular Culture of Mexico*. Wilmington, Del: Scholarly Resources, 1994.

Benjamin, Thomas, and Marcial Ocasio-Meléndez. "Organizing the Memory of Modern Mexico: Porfirian Historiography in Perspective, 1880s–1980s." *Hispanic American Historical Review* 64, no. 2 (May 1984): 323–64.

Bennett, Tony. *The Birth of the Museum*. New York: Routledge, 1995.

Bernal, Ignacio. "La historia póstuma de Coatlicue." In *Del arte: Homenaje a Justino Fernández*, 31–34. Mexico City: Universidad Autónoma de México, Instituto de Investigaciones Estéticas, 1977.

Bernstein, Marvin D. *The Mexican Mining Industry, 1890–1950: A Study of the Interactions of Politics, Economics, and Technology*. Albany: State University of New York, 1964.

Bethell, Leslie. *Mexico since Independence*. New York: Cambridge University Press, 1991.

Black, Jeremy. *Maps and History: Constructing Images of the Past*. New Haven: Yale University Press, 1997.

Bleichmar, Daniela. "Painting as Exploration: Visualizing Nature in Eighteenth-Century Colonial Science." In *Colonial Latin American Review* 15, no. 1 (June 2006): 81–104.

———. "Sixteenth-Century Transatlantic Encounters with New World *Materia Medica*." In *Colonial Botany: Science, Commerce, and Politics in the Early Modern World*, edited by Londa Schiebinder and Claudia Swan, 83–99. Philadelphia: University of Pennsylvania Press, 2005.

———. "A Visible and Useful Empire: Visual Culture and Colonial Natural History in the Eighteenth-Century Spanish World." In *Science in the Spanish and Portuguese Empires (1500–1800)*, edited by Daniela Bleichmar, Paula De Vos, Kristin Huffine, and Kevin Sheehan, 290–310. Stanford: Stanford University Press, 2009.

———. "Visible Empire." *Postcolonial Studies*, Vol. 12, No. 4, (December 2009), 441–66.

———. *Visual Culture in Eighteenth-Century Natural History: Botanical Illustrations and Expeditions in the Spanish Atlantic*. Ph.D. diss., Princeton University, 2005.

Bleichmar, Daniela, Paula De Vos, Kristin Huffine, and Kevin Sheehan, eds. *Science in the Spanish and Portuguese Empires*. Stanford: Stanford University Press, 2009.

Boelhower, William. "Inventing America: A Model of Cartographic Semiosis." *Word and Image* 2, no. 4 (1988): 475–97.

Bohrer, Fredrick N. *Orientalism and Visual Culture: Imagining Mesopotamia in Nineteenth-Century Europe*. New York: Cambridge University Press, 2003.

Boone, Elizabeth Hill. "Templo Mayor Research 1521–1978." In *The Aztec Templo Mayor*, edited by Elizabeth Hill Boone, 5–69. Washington, D.C.: Dumbarton Oaks, 1987.

Borges, Jorge Luis. *The Aleph and Other Stories*. Translated by Andrew Hurley. New York: Penguin Books, 1998.

Brading, D. A. *The First America: The Spanish Monarchy, Creole Patriots and the Liberal State 1492–1867*. Cambridge: Cambridge University Press, 1991.

———. *Mexican Phoenix: Our Lady of Guadalupe: Image and Tradition across Five Centuries*. Cambridge: Cambridge University Press, 2001.

Brennan, Theresa, and Martin Jay, eds. *Vision in Context: Historical and Contemporary Perspectives on Sight*. New York: Routledge, 1996.

Bueno, Christina Maria. *Excavating Identity: Archaeology and Nation in Mexico, 1876–1911*. Ph.D. diss., University of California Davis, 2004.

Buerger, Janet E. "Ultima Thule: American Myth, Frontier, and the Artist-Priest in Early American Photography." *American Art* 6. no. 1 (winter 1992): 82–103.

Buissert, David, ed. *Monarchs, Ministers and Maps: The Emergence of Cartography as the Tool of Government in Early Modern Europe*. Chicago: University of Chicago Press, 1992.

———. "Monarchs, Ministers and Maps in France before the Accession of Louis XIV." In *Monarchs, Ministers and Maps: The Emergence of Cartography as the Tool of Government in Early Modern Europe*, edited by David Buissert, 99–123. Chicago: University of Chicago Press, 1992.

Burkholder, Mark, and D. S. Chandler. "Creole Participation, Spanish Reaction." In *Latin American Revolutions 1808–1826*, edited by John Lynch, 50–57. Norman: University of Oklahoma Press, 1994.

Calatayud, Arinero. *Catálogo de las expediciones de viajes científicos españoles a América y Filipinas (siglos XVIII y XIX)*. Madrid: CSIS, 1984.

Cañeque, Alejandro *The King's Living Image: The Culture and Politics of Viceregal Power in Seventeenth-Century New Spain*. Ph.D. diss., New York University, 1999.

Cañizares-Esguerra, Jorge. "How Derivative Was Humboldt?" In *Colonial Botany: Science, Commerce, and Politics in the Early Modern World*, edited by Londa

Schiebinder and Claudia Swan, 148–65. Philadelphia: University of Pennsylvania Press, 2005.

———. *How to Write the History of the New World: Histories, Epistemologies, and Identities in the Eighteenth-Century Atlantic World*. Stanford: Stanford University Press, 2001.

———. "Landscapes and Identities: Mexico 1850–1900." In *Nature, Empire, and Nation: Explorations of the History of Science in the Iberian World*, by Jorge Cañizares-Esguerra, 129–68. Stanford: Stanford University Press, 2006.

———. "New World, New Stars: Patriotic Astrology and the Invention of Indian and Creole Bodies in Colonial Spanish America, 1600–1650." *American Historical Review* 104, no. 1 (February 1999): 33–68.

———. "Spanish America in Eighteenth-Century European Travel Compilations: A New 'Art of Reading' and the Transition to Modernity." *Journal of Early Modern History* 2 (1998): 329–49.

Caplan, Karen Deborah. *Local Liberalisms: Mexico's Indigenous Villagers and the State 1812–1857*. Ph.D. diss., Princeton University, 2001.

Carr, Gerald L. "Frederic Edwin Church and Mexico." In XVII Coloquio Internacional de Historia del Arte: *Arte, historia e identidad en América: Visiones Comparativas* edited by Gustavo Curiel, Renato González Mello, and Juana Gutiérrez Haces. Tomo 1: 253–366. Mexico City: Universidad Nacional Autónoma de México, Instituto de Investigaciones Estéticas, 1994.

Carrera, Magali. "Affections of the Heart: Female Imagery and the Notion of Nation in Mexico." In *Women and Art in Early Modern Latin America*, edited by Kellen McIntyre and Richard E. Phillips, 46–72. London: Brill Press, 2006.

———. "Creole Landscapes." In *Mapping Latin America: Space and Society, 1492–2000*, edited by Jordana Dym and Karl Offen. Chicago: University of Chicago Press, 2011.

———. "El Nuevo [Mundo] no se parece á el Viejo": Racial Categories and the Practice of Seeing." *Journal of Spanish Cultural Studies* 10, no 1 (March 2009): 59–73.

———. "Fabricating Specimen Citizens: Nation-building in Nineteenth-Century Mexico." In *The Politics of Dress in Asia and the Americas*, edited by Mina Roces and Louise Edwards, 215–35. Sussex: Sussex Academic Press, 2007.

———. *Imagining Identity in New Spain: Race, Lineage, and the Colonial Body in Portraiture and Casta Paintings*. Austin: University of Texas Press, 2003.

Carter, Paul. *The Road to Botany Bay: An Exploration of Landscape and History*. New York: Alfred A. Knopf, 1988.

Casanovo, Rosa. "El éxito del comercio fotográfico." *Alquimia* (May–August 2004): 27–34.

———. "Ingenioso descubrimiento: Apuntes sobre los primeros años de la fotografía en México." *Alquimia* (May–August 1999): 7–14.

———. "Las fotografías se vuelven historia: Algunos usos entre 1865 y 1910." In

Los pinceles de la historia: La fabricación del estado 1864–1910, edited by Esther Acevedo, Jaime Cuadriello, and Fausto Ramírez, 214–41. Mexico City: Instituto Nacional de Bellas Artes, 2003.

———. "Un nuevo modo de representar: Fotografía en México 1839–1861." In *Hacia otra historia del arte en México: De la estructuración colonial a la exigencia nacional (1780–1860)*, edited by Esther Acevedo. Tomo 1: 191–217. Mexico City: Consejo Nacional para la Cultura y las Artes, 2001.

Casid, Jill H. *Sowing Empire: Landscape and Colonization*. Minneapolis: University of Minnesota Press, 2005.

Castro, Miguel Ángel and Guadalupe Curiel, eds. *Publicaciones periódicas mexicanas del siglo XIX: 1822–1855*. Mexico City: Universidad Nacional Autónoma de México, 2000.

———. *Publicaciones periódicas mexicanas del siglo XIX: 1856–1876*. Pt. 1. Mexico City: Universidad Nacional Autónoma de México, 2003.

Catálogo de la Mapoteca Antonio García Cubas. 2 vols. Mexico City: Sociedad Mexicana de Geografía y Estadístico, 1994–95.

Caulfield, Sueann. "The History of Gender in the Historiography of Latin America." *Hispanic American Historical Review* 81, nos. 3–4 (August–November, 2001): 449–90.

Chard, Chloe. *Pleasure and Guilt on the Grand Tour: Travel Writing and Imaginative Geography 1600–1830*. Manchester: Manchester University Press, 1999.

Clark, Fiona. "The *Gazeta de Literatura de México* (1788–1795): The Formation of a Literary-Scientific Periodical in Late-Viceregal Mexico." *Dieciocho* 28, no. 1 (spring 2005): 7–30.

Codding, Mitchell A. "Perfecting the Geography of New Spain: Alzate and the Cartographic Legacy of Sigüenza y Góngora." *Colonial Latin American Review* 3, nos. 1–2 (1994): 185–219.

Collado, María del Carmen. "Antonio García Cubas." In *En busca de un discurso integrador de la nación, 1848–1884*, edited by Antonia Pi-Suñer Llorens, 425–48. Mexico City: Universidad Nacional Autónoma de México, 2001.

Comment, Bernard. *The Painted Panorama*. Translated by Anne-Marie Glasheen. New York: Harry N. Abrams, 2000.

Commons, Áurea, and Atlántida Coll-Hurtado. *Geografía histórica de México en el siglo XVIII: Análisis del* "Teatro Americano." Serie Libros, núm. 4. Mexico City: Instituto de Geografía, Universidad Nacional Autónoma de México, 2002.

Conn, Steven. *History's Shadow: Native Americans and Historical Consciousness in the Nineteenth Century*. Chicago: University of Chicago Press, 2004.

Córdova, Carlos A. *Arqueología de la imagen: México en las vistas estereoscópicas*. Mexico City: Museo de Historia de México, 2000.

Cosgrove, Denis E. "Mapping the World." In *Maps: Finding our Place in the World*, edited by James Akerman and Robert W. Karrow, 65–115. Chicago: University of Chicago Press, 2007.

———. "Prospect, Perspective and the Evolution of the Landscape Idea." *Transactions of the Institute of British Geographers*, n.s., 10, no. 1. (1985): 45–62.

———. *Social Formation and Symbolic Landscape*. Totowa, N.J.: Barnes and Noble Books, 1985.

Costeloe, Michael P. *The Central Republic in Mexico, 1835–1846*: Hombres de Bien *in the Age of Santa Anna*. Cambridge: Cambridge University Press, 1993.

———. "William Bullock and the Mexican Connection." *Mexican Studies / Estudios Mexicans*. Volume 22, issue 2 (summer 2006): 275–309.

Covarrubias, José Enrique. "De Fossey y Sartorius en la tierra de la nostalgia." *Artes de México* núm. 31 (1996): 48–55.

———. *Visión extranjera de México, 1840–1867*. Tomo 1: *El estudio de las costumbres y de la situación social*. Mexico City: Universidad Nacional Autónoma de México, 1998.

Craib, Raymond B. *Cartographic Mexico: A History of State Fixations and Fugitive Landscapes*. Durham: Duke University Press, 2004.

———. "Cartography and Power in the Conquest and Creation of New Spain." *Latin American Research Review* 35, no. 1 (2000): 7–36.

———. "El discurso cartográfica en el México de Porfiriato." In *México a través de los mapas*, edited by Héctor Mendoza Vargas, 131–50. Mexico City: Instituto de Geografía, Universidad Nacional Autónoma de México, 2000.

———. "The Nationalist Metaphysics: State Fixations, National Maps, and the Geo-Historical Imagination in Nineteenth-Century Mexico." *Hispanic American Historical Review* 82, no. 1 (2002): 33–68.

———. "Relocating Cartography." In *Postcolonial Studies*, Vol. 12, no. 4, (December 2009), 481–90.

———. *State Fixations, Fugitive Landscapes: Mapping, Surveying and the Spatial Creation of Modern Mexico, 1850–1930*. Ph.D. diss., Yale University, 2001.

Crane, Susan A., ed. *Museums and Memory*. Stanford: Stanford University Press, 2000.

Crang, Mike, and Nigel Thrift. *Thinking Space*. New York: Routledge, 2000.

Crary, Jonathan. *Suspensions of Perception: Attention, Spectacle and Modern Culture*. Cambridge: MIT, 1999.

———. *Techniques of the Observer: On Vision and Modernity in the Nineteenth Century*. Cambridge: MIT, 1990.

Cuadriello, Jaime. "Del escudo de armas al estandarte armado." In *Los pinceles de la historia de la patria criolla a la nación mexicana 1750–1860*, edited by Esther Acevedo and Jaime Cuadriello Fausto Ramírez, 33–49. Mexico City: Instituto Nacional de Bellas Artes, 2000.

———. "Los jeroglíficos de la Nueva España." In *Juegos de ingenio y agudeza: La pintura emblemática de la Nueva España*, edited by Jaime Cuadriello, 84–111. Mexico City: Consejo Nacional para la Cultura y las Artes, 1994.

———. "Los umbrales de la nación y la modernidad de sus artes: Criollismo,

Ilustración y academia." In *Hacia otra historia del arte en México: De la estructuración colonial a la exigencia nacional (1780–1860)*, edited by Esther Acevedo. Tomo 1:17–35. Mexico City: Consejo Nacional para la Cultura y las Artes, 2001.

Cummins, Tom. "De Bry and Herrera: 'Aguas Negras' or the Hundred Years of War over the Image of America." In XVII Coloquio Internacional de Historia del Arte: *Arte, historia e identidad en América: Visiones Comparativas*. Tomo 1:17–31. Mexico City: Universidad Nacional Autónoma de México, 1994.

Curcio-Nagy, Linda. *The Great Festivals of Colonial Mexico City: Performing Power and Identity*. Albuquerque: University of New Mexico Press, 2004.

Curiel, Gustavo, Fausto Ramírez, Antonio Rubial, Angélica Valázquez, eds. *Pintura y vida cotidiana en México 1650–1950*. Mexico City: Fomento Cultural Banamex, CONACULTA, 1999.

Darrah, William Culp. *Cartes de Visite in Nineteenth-Century Photography*. Gettysburg: W. C. Darrah, 1981.

———. *Stereo Views: A History of Sterographs in American and Their Collection*. Gettysburg[?]: Times and News Publishing, 1964.

———. *The World of Sterographs*. Gettysburg: W. C. Darrah, 1977.

Daston, Lorraine. "The Image of Objectivity." *Representations*, no. 40, special issue: "Seeing Science" (autumn 1992): 81–128.

Daston, Lorraine, and Katharine Park. *Wonders and the Order of Nature, 1150–1750*. New York: Zone Books, 2001.

Davis, Keith. *Désiré Charnay, Expeditionary Photographer*. Albuquerque: University of New Mexico Press, 1981.

Dean, Carolyn, and Dana Leibsohn. "Hybridity and Its Discontents: Considering Visual Cultures in Colonial Spanish America." *Colonial Latin American Review* 12, no. 1 (June 2003): 5–35.

Deans-Smith, Susan. "Creating the Colonial Subject: Casta Paintings, Collectors, and Critics in Eighteenth-Century Mexico and Spain." *Colonial Latin American Review* 14, no. 2 (2005): 169–204.

———. "Nature and Scientific Knowledge in the Spanish Empire: Introduction." *Colonial Latin American Review* 15, no. 1 (June 2006): 29–38.

Debroise, Oliver. *Mexican Suite: A History of Photography in Mexico*. Translated in collaboration with author by Stella de Sá Rego. Austin: University of Texas Press, 2001 [original ed. 1994].

de Certeau, Michel. *The Practice of Everyday Life*. Translated by Steven Rendell. Berkeley: University of California Press, 1984.

de la Torre Rendón, Judith. "Manuel Rivera Cambas." In *En busca de un discurso integrador de la nación, 1848–1884*, edited by Antonia Pi-Suñer Llorens, 295–309. Mexico City: Universidad Nacional Autónoma de México, 2001.

Dettlebach, Michael. "Humboldtian Science." In *Cultures of Natural History*, edited by Nicholas Jardin, J. A. Secord, and Emma C. Spary, 287–304. Cambridge: Cambridge University Press, 1996.

De Vos, Paula. "Natural History and the Pursuit of Empire in Eighteenth-Century Spain." *Eighteenth-Century Studies* 40, no. 2 (2007): 209–39.

———. "Research, Development, and Empire: State Support of Science in the Later Spanish Empire." *Colonial Latin American Review* 14, no. 1 (June 2006): 55–79.

Diener, Pablo. "La pintura de paisajes entre los artistas viajeros." In *Viajeros europeas del siglo XIX en México*, 137–57. Mexico City: Fomento Cultural Banamex, 1996.

Dikovitskaya, Margaret. *Visual Culture: The Study of the Visual after the Cultural Turn*. Cambridge: MIT Press, 2005.

Documentos gráficos para la historia de México. 2 vols. Mexico City: Editorial del Sureste, 1985.

Domenella, Ana Rosa. *Las voces olvidadas: Antología crítica de narradoras mexicanas nacidas en el siglo XIX*. Mexico City: El Colegio de México, 1997.

Dorotinsky, Deborah. "Los tipos sociales desde la austeridad del estudio." *Alquimia* (May–August 2004): 14–25.

Dunbar, Gary S. "'The Compass Follows the Flag': The French Scientific Mission to Mexico, 1864–1867." *Annals of the Association of Geographers* 78, no. 2 (June 1988): 229–40.

Duncan, James, and Derek Gregory. *Writes of Passage: Reading Travel Writing*. New York: Routledge, 1999.

Duncan, Robert H. "Embracing a Suitable Past: Independence Celebrations under Mexico's Second Empire, 1864–6." *Journal of Latin American Studies* 30, no. 2 (May 1998): 249–77.

———. *For the Good of the Country: State and Nation Building during Maximilian's Mexican Empire, 1854–1867*. Ph.D. diss., University of California, Irvine, 2001.

———. "Maximilian and the Construction of the Liberal State, 1863–1866." In *The Divine Charter: Constitutionalism and Liberalism in Nineteenth-Century Mexico*, edited by Jaime E. Rodríguez O., 133–67. Lanham, Md.: Rowman and Littlefield Publishers, 2005.

Dym, Jordana. "'Our Pueblos, Fractions with No Central Unity': Municipal Sovereignty in Central America, 1808–1821." *Hispanic American Historical Review* 86, no. 3 (2006): 431–66.

Earle, Edward. *Points of View: The Stereograph in America—A Cultural History*. New York: Visual Studies Workshop Press, 1979.

Earle, Rebecca. "'Padres de la Patria' and the Ancestral Past: Commemorations of Independence in Nineteenth-Century Spanish America." *Journal of Latin American Studies* 34, pt. 4 (November 2002): 775–805.

———. "*Sobre Héroes y Tumbas*: National Symbols in Nineteenth-Century Spanish America." *Hispanic American Historical Review* 85, no. 3 (August 2005): 375–416.

Edney, Matthew H. "Cartographic Culture and Nationalism in the Early United States: Benjamin Vaughan and the Choice for a Prime Meridian, 1811." *Journal of Historical Geography* 20, no. 4 (1994): 384–95.

———. *Mapping an Empire: The Geographical Construction of British India, 1765–1843*. Chicago: University of Chicago Press, 1997.

———. "Rewriting the Colonial Mapping of New England." In *Cartographies of Colonial New England: Geographical Practices, Public Discourse, Regional Construction*. Unpublished manuscript, 1 November 2003.

El escenario urbano de Pedro Gualdi 1808–1857. Museo Nacional de Arte. Exhibition catalogue. Mexico City: Instituto Nacional de Bellas Artes, 1997.

Elsner, Jaś, and Joan-Pau Rubiés, eds. *Voyages and Visions: Towards a Cultural History of Travel*. London: Reaktion Books, 1999.

Ernst, Wolfgang. "Archi(vi)textures of Museology." In *Museums and Memory*, edited by Susan Crane, 17–34. Stanford: Stanford University Press, 2000.

Esparza Liberal, María José. "La historia de México en el calendario de Ignacio Díaz Triujeque de 1851 y la obra de Prescott." *Anales del Instituto de Investigaciones Estéticas* 24, núm. 80 (2002): 149–67.

Estrada de Gelero, Elena Isabel. "El tema anticuario en lost pintores viajeros." In *Viajeros europeas del siglo XIX en México*. Mexico City: Fomento Cultural Banamex, 1996, 183–201.

———. "La litografía y el Museo Nacional como armas del nacionalismo." In *Los pinceles de la historia de la patria criolla a la nación mexicana 1750–1860*, edited by Esther Acevedo and Jaime Cuadriello Fausto Ramírez, 153–69. Mexico City: Instituto Nacional de Bellas Artes, 2000.

Evans, R. Tripp. *Romancing the Maya: Mexican Antiquity in the American Imagination, 1820–1915*. Austin: University of Texas Press, 2004.

Evans, William McKee. "From the Land of Canaan to the Land of Guinea: The Strange Odyssey of the 'Sons of Ham.'" *American Historical Review* 85, no. 1 (February 1980): 15–43.

Falcón, Romana, and Raymond Buve. *Don Porfirio presidente . . . , nunca omnipotente: Hallazgos, reflexiones y debates. 1876–1911*. Mexico City: Universidad Iberoamericana, 1998.

Fernández-Armesto, Felipe. Book review of Ricardo Padrón's *The Spacious World: Cartography, Literature and Empire in Early Modern Spain*. *Hispanic American Historical Review* 86, no. 2 (May 2006): 353–54.

Fernández Hernández, Silvia. "La transición del diseño gráfico colonial al diseño moderno en México (1777–1850)." In *Empresa y cultura en tinta y papel (1800–1860)*, edited by Laura Suárez de la Torre, 15–26. Mexico City: Instituto Mora, 2001.

Finsterwalder, Rüdiger. "The Round Earth on a Flat Surface: World Map Projections before 1550." In *America: Early Maps of the New World*, edited by Hans Wolff, 161–73. Munich: Prestel, 1992.

Fiorani, Francesca. *The Marvel of Maps: Art, Cartography and Politics in Renaissance Italy*. New Haven: Yale University Press, 2005.

Flores Olea, Aurora. "José Fernando Ramírez." In *En busca de un discurso integrador de la nación, 1848–1884*, edited by Antonia Pi-Suñer Llorens, 313–38. Mexico City: Universidad Nacional Autónoma de México, 2001.

Foucault, Michel. "Governmentality." In *Power: Essential Works of Foucault 1954–1984*, edited by James D. Faubion, vol. 3, 201–22. New York: New Press, 1994.

Fowler, Will. *Mexico in the Age of Proposals, 1821–1853*. Westport, Conn.: Greenwood Press, 1998.

Fowles, Jib. "Stereography and the Standardization of Vision." *Journal of American Cultures* 17, no. 2 (1994): 89–93.

French, William E. "Imagining and the Cultural History of Nineteenth-Century Mexico." *Hispanic American Historical Review* 79, no. 2 (May 1999): 249–67.

———. "Prostitutes and Guardian Angels: Women, Work, and the Family in Porfirian Mexico." *Hispanic American Historical Review* 72, no. 4 (November 1992): 529–53.

Fullerton, John, and Elaine King. "Local Views, Distant Scenes: Registering Affect in Surviving Mexican Actuality Films of the 1920s." *Film History* 17, no. 1 (2005): 66–87.

Galí Boadella, Montserrat. *Historias del bello sexo: La introducción de romanticismo en México*. Mexico City: Universidad Nacional Autónoma de México, Instituto de Investigaciones Estéticas, 2002.

Gaonkar, Dilip Parameshwar, ed. *Alternative Modernities*. Durham: Duke University Press, 2001.

García Luna, Margarita, and José N. Iturriaga. *Viajeros extranjeros en el estado de México*. Mexico City: Universidad Autónoma del Estado de México, 1999.

García Martínez, Bernardo. "La comisión Geográfico-Exploradora." *Historia Mexicana* 96 (April–June 1975): 485–555.

García Sáiz, María Concepción. "Imágenes palencanas." In *Viajeros europeas del siglo XIX en México*, 202–9. Mexico City: Fomento Cultural Banamex, 1996.

———. *Las castas mexicanas: Un género pictórico americano*. Milan: Olivetti, 1989.

Gardiner, C. Harvey, ed. *The Literary Memoranda of William Hickling Prescott*. 2 vols. Norman: University of Oklahoma Press, 1961.

Garner, Paul. *Porfirio Díaz*. London: Pearson Education Limited, 2001.

Garrigan, Shelley. *Collecting the Nation: From Objects to Meaning*. Ph.D. diss., New York University, 2003.

Gerbi, Antonello. *The Dispute of the New World: History of a Polemic*, translated by J. Moyle. Pittsburgh: University of Pittsburgh Press, 1973.

Gillies, John. "Theatres of the World." In *Shakespeare and the Geography of Difference*, 70–98. Cambridge: Cambridge University Press, 1994.

Glowacki, Lisa Carolyn. *Writing Gender and Nation*: El Album de la Mujer, *1883–1884*. M.A. thesis, University of British Columbia, 1996.

Godlewska, Anne. "From Enlightenment Vision to Modern Science? Humboldt's Visual Thinking." In *Geography and Enlightenment*, edited by David N. Livingstone and Charles W. J. Withers, 236–75. Chicago: University of Chicago Press, 1999.

———. *Geography Unbound: French Geographic Science from Cassini to Humboldt*. Chicago: University of Chicago Press, 1999.

———. "Map, Text and Image. The Mentality of Enlightenment Conquerors: A New Look at the *Description del'Egypte*." *Transactions of the Institute of British Geographers*, n.s., vol. 20, no. 1 (1995): 5–28.

Gold, Susanna. *Imaging Memory: Re-Presentations of the Civil War at the 1876 Centennial Exhibition*. Ph.D. diss., University of Pennsylvania, 2004.

Gonzalez, Michael J. "Imagining Mexico in 1910: Visions of the *Patria* in the Centennial Celebrations in Mexico City." *Journal of Latin American Studies* 29, no. 3 (August 2007): 495–534.

González Echevarría, Roberto. *Myth and Archive: A Theory of Latin American Narrative*. Durham: Duke University Press, 1998.

Granados, Luis Fernando. *Cosmopolitan Indians and Mesoamerican Barrios in Bourbon Mexico City: Tribute, Community, Family and Work in 1800*, Ph.D. diss., Georgetown University, 2008.

Grandin, Greg. "Can the Subaltern Be Seen? Photography and the Affects of Nationalism." *Hispanic American Historical Review* 84, no. 1 (February 2004): 83–111.

Greenblatt, Stephen. *Marvelous Possessions: The Wonder of the New World*. Chicago: University of Chicago Press, 1991.

Gregory, Derek. "Emperors of the Gaze: Photographic Practices and Productions of Space in Egypt, 1839–1914." In *Picturing Place: Photography and the Geographical Imagination*, edited by Joan M. Schwartz and James R. Ryan, 195–225. London: I. B. Tauris, 2003.

———. *Geographical Imaginations*. Cambridge, Mass.: Blackwell, 1994.

———. "Scripting Egypt: Orientalism and the Cultures of Travel." In *Writes of Passage: Reading Travel Writing*, edited by James Duncan and Derek Gregory, 151–64. New York: Routledge, 1999.

Gretton, Thomas. "European Illustrated Weekly Magazines, c. 1850–1900: A Model and a Counter-model for the Work of José Guadalupe Posada." *Anales del Instituto de Investigaciones Estéticas* 19, núm. 70 (1997): 99–125.

Grosz Elizabeth. *Space, Time and Perversions: Essays on the Politics of Bodies*. London: Routledge, 1995.

Gutiérrez Haces, Juana. "Etnografía y costumbrismo en las imágenes de los viajeros." In *Viajeros europeas del siglo XIX en México*, 159–81. Mexico City: Fomento Cultural Banamex, 1996.

Gwin, Minrose C. *The Woman in the Red Dress: Gender, Space and Reading*. Chicago: University of Illinois Press, 2002.

Hall, Stuart. "The West and the Rest: Discourse and Power." In *Modernity: An Introduction to Modern Societies*, edited by Stuart Hall, David Held, Don Hubert, and Kenneth Thompson, 184–228. Cambridge: Open University, 1995.

Hall, Stuart, David Held, Don Hubert, and Kenneth Thompson, eds. *Modernity: An Introduction to Modern Societies*. Cambridge: The Open University, 1995.

Hamnett, Brian. *Juárez*. New York: Longman, 1994.

Harley, J. B. "Deconstructing the Map." *Cartographica* 26, no. 2 (summer 1989): 1–20.

———. [posthumously]. *The New Nature of Maps: Essays in the History of Cartography*, edited by Paul Laxton. Baltimore: Johns Hopkins University Press, 2001.

———. "Power and Legitimation in English Geographic Atlases of the Eighteenth Century." In *Images of the World: The Atlas through History*, edited by John A. Wolter and Ronald E. Grim, 161–204. Washington: Library of Congress, 1997.

———. "Silences and Secrecy: The Hidden Agenda of Cartography in Early Modern Europe." *Imago Mundi* 40 (1988): 57–76.

Harley, J. B., and David Woodward, eds. *The History of Cartography*. Chicago: University of Chicago Press, 1987–.

Harvey, David. *Paris, Capital of Modernity*. New York: Routledge, 2003.

———. "Between Space and Time: Reflections on the Geographical Imagination." *Annals of the American Geographers* 80, no. 3 (September 1990): 418–34.

Hendrix, W. S. "Notes on Collections of Types, a Form of Costumbrismo." *Hispanic Review* 1, no. 3. (July 1933): 208–21.

Hernández-Durán, Raymond. "Entre el espacio y el texto: Ubicación del art virreinal en el México poscolonial." In *Hacia otra historia del arte en México. Tomo 2: Amplitud del modernismo y la modernidad (1861–1920)*, edited by Stacy Widdifield. 2: 33–68. Mexico City: Consejo Nacional para la Cultura y las Artes, 2001.

———. *Reframing Viceregal Painting in Nineteenth-Century Mexico: Politics, the Academy of San Carlos, and Colonial Art History*. Ph.D. diss., University of Chicago, 2005.

Herrejón Peredo, Carlos, ed. *Morelos: Antología documental*. Mexico City: Secretaría Educación Publica, 1985.

Herricks, Jane. "Periodicals for Women in Mexico." *Americas* 14, no. 3 (October 1957): 136–41.

Herzog, Tamar. *Defining Nations: Immigrants and Citizens in Early Modern Spain and Spanish America*. New Haven: Yale University Press, 2003.

Hight, Eleanor, and Gary D. Sampson, eds. *Colonialist Photography: Imag(in)ing Race and Place*. New York: Routledge, 2002.

Hill, Ruth. *Hierarchy, Commerce, and Fraud in Bourbon Spanish America: A Postal Inspector's Exposé*. Nashville: Vanderbilt University Press, 2005.

———. "Towards an Eighteenth-Century Transatlantic Critical Race Theory." In *Literature Compass* 3, no. 3 (2006): 53–64.

Hillis, Ken. "The Power of Disembodied Imagination: Perspective's Role in Cartography." *Cartographica* 31, no. 3 (1994): 1–17.

Holmes, Burton. *A Trip around the World through the Telebinocular*. Meadville, Pa.: Keystone View Company, ca. 1930.

Honour, Hugh. *The European Vision of America*. Cleveland: Cleveland Museum of Art, 1975.

Hudson, Nicholas. "From 'Nation' to 'Race': The Origin of Racial Classification in Eighteenth-Century Thought." *Eighteenth-Century Studies* 29, no. 3 (1996): 247–64.

Huerta, David. *Claudio Linatí: Acurelas y litografías*. Mexico City: Inversora Bursatil, S.A., 1993.

Iturriaga de la Fuente, José. *Anecdotario de viajeros extranjeros en México siglos XVI–XX*. 3 vols. Mexico City: Instituto Nacional de Bellas Artes, 1989.

Jardin, Nicholas, J. A. Secord, and Emma C. Spary, eds. *Cultures of Natural History*. Cambridge: Cambridge University Press, 1996.

Jiménez Codinach, Guadalupe. "La litografía mexicana del siglo XIX: piedra de toque de una época y de una pueblo." In *Nación de imágenes: La litografía mexicana del siglo XIX*, 138–49. Exhibition catalogue. Mexico City: Consejo Nacional para la Cultura y las Artes, 1994.

———. *México: The Projects of a Nation 1821–1888*. Mexico City: Fomento Cultural Banamex, 2001.

Juárez López, José Luis. *Las litografías de Karl Nebel: Versión estética de la invasión norteamericana 1846–1848*. Mexico City: Porrúa, 2004.

Kagan, Richard. "Prescott's Paradigm: American Historical Scholarship and the Decline of Spain." *American Historical Review* 101, no. 2 (April 1996): 423–46.

———. *Urban Images of the Hispanic World 1493–1793*. New Haven: Yale University Press, 2000.

Kaplan, Amy. "Manifest Domesticity." In *American Literature* 70, no. 3, special issue: "No More Separate Spheres!" (September 1998): 581–606.

Katzew, Ilona. *Casta Painting: Images of Race in Eighteenth-Century Mexico*. New Haven: Yale University Press, 2004.

———. *Una visión del México del siglo de las luces: La codificación de Joaquín Antonio de Basarás*. Mexico City: Landucci, 2006.

Kaufmann, Thomas DaCosta. *Toward a Geography of Art*. Chicago: University of Chicago Press, 2004.

Keen, Paul. "The 'Balloonomania': Science and Spectacle in 1780s England." *Eighteenth-Century Studies* 39, no. 4 (2006): 507–35.

Keith, Michael, and Steve Piles. *Place and the Politics of Identity*. New York: Routledge, 1993.

King, Jonathan. "William Bullock: Showman." In *Viajeros europeas del siglo XIX en México*, 117–25. Mexico City: Fomento Cultural Banamex, 1996.

Klein, Kerwin Lee. “On the Emergence of Memory in Historical Discourse.” *Representations*, no. 69, special issue: “Grounds for Remembering” (winter 2000): 127–50.

Koeman, Cornelius. *The History of Abraham Ortelius and His Theatrum Orbis Terrarum.* Lausanne: Sequoia, 1964.

Koerner, Lisbet. “Carl Linnaeus in His Time and Place.” In *Cultures of Natural History*, edited by Nicholas Jardin, J.A. Secord, and Emma C. Spary, 145–62. Cambridge: Cambridge University Press, 1996.

König, Hans-Joachim, and Marianne Wiesebron, eds. *Nation Building in Nineteenth Century Latin America: Dilemmas and Conflicts.* Leiden: Research School, Leiden University, 1998.

Kraus, Rosalind. “Photography’s Discursive Spaces: Landscape/View.” *Art Journal* 42, no. 4 (winter 1982): 311–19.

Lafuente, Antonio. “Enlightenment in an Imperial Context: Local Science in the Late-Eighteenth-Century Hispanic World.” *Osiris*, 2nd ser., 5 (2000): 155–73.

Lafuente, Antonio, and Nuria Valverde. “Linnaean Botany and Spanish Imperial Biopolitics.” In *Colonial Botany: Science, Commerce, and Politics in the Early Modern World*, edited by Londa Schiebinder, and Claudia Swan, 133–47. Philadelphia: University of Pennsylvania Press, 2005.

Landes, Joan. *Visualizing the Nation: Gender, Representation, and Revolution in Eighteenth-Century France.* Ithaca: Cornell University Press, 2001.

Lavrin, Asunción. *Sexuality and Marriage in Colonial Latin America.* Lincoln: University of Nebraska Press, 1989.

Lear, John. “Mexico City: Space and Class in the Porfirian Capital.” *Journal of Urban History* 22, no. 4 (May 1996): 454–92.

Leask, Nigel. *Curiosity and the Aesthetics of Travel Writing 1770–1840.* Oxford: Oxford University Press, 2002.

———. “‘The Ghost in Chapultepec’: Fanny Calderón de la Barca, William Prescott and Nineteenth-Century Travel Accounts.” In *Voyages and Visions: Towards a Cultural History of Travel*, edited by Jaś Elsner and Joan-Pau Rubiés, 184–209. London: Reaktion Books, 1999.

León García, María de Carmen. “Cartografía de los ingenieros militares en Nueva España, segunda mitad del siglo XIII.” In *Historias de la Cartografía de Iberomérica: Nuevos caminos, viejos problemas*, edited by Héctor Mendoza Varga and Carla Lois, 441–66. Mexico City: Instituto de Geografía, Universidad Nacional Autónoma de México, 2009.

Lestringant, Frank. *Mapping the Renaissance World.* Translated by David Fausett. Berkeley: University of California Press, 1994.

Lewis, Cynthia Neri. *Lotería and “La Alma Nacional”: Art and Identity in Nineteenth and Twentieth-Century Mexico.* M.A. thesis, California State University, Fullerton, 2004.

Lingren, Uta. “Trial and Error in the Mapping of America during the Early Modern Period." In *America: Early Maps of the New World*, edited by Hans Wolff, 145–60. Munich: Prestel, 1992.

Livingston, David. *Putting Science in Its Place: Geographies of Scientific Knowledge*. Chicago: University of Chicago Press, 2003.

Lombardo García, Irma. *El siglo de Cumplido: La emergencia del periodismo mexicano de opinión (1832–1857)*. Mexico City: Universidad Nacional Autónoma de México, 2002.

Lomnitz-Adler, Claudio. *Deep Mexico, Silent Mexico: An Anthropology of Nationalism*. Minneapolis: University of Minnesota Press, 2001.

López Beltrán, Carlos. "Hippocratic Bodies: Temperament and Castas in Spanish America (1570–1820). *Journal of Spanish Cultural Studies* 8, no. 2 (July 2007) 253–89.

López, Oresta. "Leer para vivir en este mundo: Lecturas modernas par alas mujeres morelianas durante el Porfiriato." *Diccionario de historia de la educación en México* (Universidad Nacional Autónoma de México, 2002). Web pages of Publicaciones Digitales, DGSCA, UNAM, biblioweb.dgsca.unam.mx, visited 24 June 2007. Printouts on file with author.

Lynch, John, ed. *Latin American Revolutions 1808–1826*. Norman: University of Oklahoma Press, 1994.

Mallon, Florencia. *Peasant Nation: The Making of Postcolonial Mexico and Peru*. Berkeley: University of California Press, 1995.

———. "The Promise and Dilemma of Subaltern Studies: Perspectives from Latin American History." *American Historical Review* 99, no. 5 (December 1991): 1491–1515.

Manthrone, Katherine. "A Transamerican Reading of 'The Machine in the Garden': Nature vs. Technology in 19th Century Landscape Art." In XVII Coloquio Internacional de Historia del Arte: *Arte, historia e identidad en América. Visiones Comparativas*. Tomo 3: 243–51. Mexico City: Universidad Nacional Autónoma de México, 1994.

Márquez Rodiles, Ignacio. "Don Antonio García Cubas, pintor de México." *Boletín de la Sociedad Geografía y Estadística* 123 (January–June 1976): 25–47.

Martín Merás, Luisa. *Cartografía marítima hispana: La imagen de América*. Madrid: Lunwerg Editores, 1993.

Martínez, María Elena. *Genealogical Fictions: Limpieza de Sangre, Religion, and Gender in Colonial Mexico*. Stanford: Stanford University Press, 2008.

Massé Zendejas, Patricia. *Cruces y Campa: Una experiencia mexicana del retrato tarjeta de visita*. Mexico City: Consejo Nacional para la Cultura y las Artes, Dirección General de Publicaciones, 2000.

———. *Simulacro y elegancia en tarjetas de visita: Fotografías de Cruces y Campa*. Mexico City: Instituto Nacional de Antropología y Historia, 1998.

Mathes, Miguel. "La litografía y los litógrafos en México, 1826–1900: Un resumen histórico." In *Nación de imágenes: La litografía mexicana del siglo XIX*, 42–55.

Exhibition catalogue. Mexico City: Consejo Nacional para la Cultura y las Artes, 1994.
Maybie, Roy W. *The Stereoscope and Stereograph*. New York: Mabies Stereographic Galleries, 1942.
Mayer Celis, Leticia. *Entre el infierno de una realidad y el cielo de un imaginario: Estadística y comunidad científica en el México de la primera mitad del siglo XIX*. Mexico City: El Colegio de México, 1999.
Mayer, Roberto L. "Los dos álbumes de Pedro Gualdi." *Anales del Instituto de Investigaciones Estéticas* 69 (1996): 81–102.
———. "Nacimiento y desarrollo del Álbum *México y sus Alrededores*." In *Casmiro Castro y su taller*, edited by Carlos Monsiváis, 135–57. Mexico City: Fomento Cultural Banamex, 1996.
———. "Phillips, Rider y su álbum *Mexico Illustrated*: ¿Quiénes fueron los autores de los dibujos originales?" *Anales del Instituto de Investigaciones Estéticas* XXII, núm. 76 (2000): 291–305.
Mazzonlini, Renato. "*Las Castas*: Interracial Crossing and Social Structure, 1770–1835." In *Heredity Produced: At the Crossroads of Biology, Politics, and Culture, 1500–1870*, edited by Staffan Muller-Wille and Hans-Jorg Rheinberger, 349–73. Cambridge: MIT Press, 2007.
McCauley, Elizabeth. *A. A. E. Disdéri and the Carte de Visite Portrait Photography*. New Haven: Yale University Press, 1985,
———. *Industrial Madness: Commercial Photography in Paris, 1848–1871*. New Haven: Yale University Press, 1994.
McDowell, Linda. *Gender, Identity and Place: Understanding Feminist Geographies*. Minneapolis: University of Minnesota Press, 1999.
Mejía, Edgar. *Políticas del espacio en México: Crónica de viaje y cartografía en el siglo XIX*. Ph.D. diss., Boston University, 2006.
Mendoza Vargas, Héctor. "Las opciones geográficas al inicio de México independiente." In *México a través los mapas*, edited by Héctor Mendoza Vargas, 89–110. Mexico City: Instituto de Geografía, Universidad Nacional Autónoma de México, 2000.
———, coordinador. *México a través de los mapas*. Mexico City: Instituto de Geografía, Universidad Nacional Autónoma de México, 2000.
Mendoza Vargas, Héctor, and Carla Lois, eds. *Historias de la Cartografía de Iberomérica: Nuevos caminos, viejos problemas*. Mexico City: Instituto de Geografía, Universidad Nacional Autónoma de México, 2009.
Mignolo, Walter. *The Darker Side of the Renaissance: Literacy, Territoriality and Colonization*. Ann Arbor: University of Michigan Press, 1995.
Mitchell, W. J. T. *Landscape and Power*. 2nd ed. Chicago: University of Chicago Press, 2002.
———. "Showing Seeing: A Critique of Visual Culture." In *Art History, Aesthetics, Visual Studies*, edited by Michael Ann Holly and Keith Moxey, 231–50. Williamstown, Mass.: Sterling and Francine Clark Art Institute, 2002.

Mompradé, Electra L., and Tonatiúh Gutiérrez. *Imagen de México: Mapas, grabados y litografías*. Mexico City: Salvat, 1976.

Moncada Maya, José Omar. "Construyendo el territorio: El desarrollo de la cartografía en Nueva España." In *Historias de la Cartografía de Iberomérica. Nuevos caminos, viejos problemas*, edited by Héctor In Mendoza Vargas, and Carla Lois, 161–82. Mexico City: Instituto de Geografía, Universidad Nacional Autónoma de México, 2009.

———. *El ingeniero Miguel Constanzó: Un militar ilustrado en la Nueva España del siglo XVIII*. Mexico City: Instituto de Geografía, Universidad Nacional Autónoma de México, 1994.

———. *Ingenieros militares en Nueva España: Inventario de su labor científico y espacial, siglos XVI a XVIII*. Mexico City: Instituto de Geografía, Universidad Nacional Autónoma de México, 1993.

———. "La profesionalización de la Geografía Mexicana durante el siglo XIX." *Ería* 48 (1999): 63–74.

Monsiváis, Carlos, ed. *Casmiro Castro y su taller*. Mexico City: Fomento Cultural Banamex, 1996.

Montellano, Francisco. *Antonio le Cosmes de Cossío: Un precursor del fotoreportaje*. Mexico City: Consejo Nacional para la Cultura y las Artes, 2001.

Morales Moreno, Luis Gerardo. "El primer Museo Nacional de México (1825–1847)." In *Hacia otra historia del arte en México: De la estructuración colonial a la exigencia nacional (1780–1860)*, edited by Esther Acevedo. Tomo 1: 36–60. Mexico City: Consejo Nacional para la Cultura y las Artes, 2001.

Mundy, Barbara. *The Mapping of New Spain: Indigenous Cartography and the Maps of the Relaciones Geográficas*. Chicago: University of Chicago Press, 1996.

Nación de imágenes: La litografía mexicana del siglo XIX. Exhibition catalogue, April–June 1994. Mexico City: Consejo Nacional para la Cultura y las Artes, 1994.

Nast, Heidi, and Audrey Kobayashi. "Re-Corporealizing Vision." In *Bodyspace: Destabilizing Geographies of Gender and Sexuality*, edited by Nancy Duncan, 75–98. New York: Routledge, 1996.

Nava Martínez, Othón. "Origen y desarrollo de una empresa editorial: Vicente García Torres, 1838–1841." In *Empresa y cultura en tinta y papel (1800–1860)*, edited by Laura Suárez de la Torre, 123–30. Mexico City: Instituto Mora, 2001.

Newhall, Beaumont. "Photography and the Development of Kinetic Visualization." *Journal of the Warburg and Courtauld Institutes* 7 (1944): 40–45.

Nieto Olarte, Mauricio. "Scientific Instruments, Creole Science, and Natural Order in the New Granada of the Early Nineteenth Century." *Journal of Spanish Cultural Studies*. 8, no. 2 (July 2007): 235–52.

Oetterman, Stephan. *The Panorama: History of a Mass Medium*. Translated by Deborah Lucas Schneider. New York: Zone Books, 1997.

O'Gorman, Edmundo. *La invención de América: Investigación acerca de la estructura histórica del Nuevo Mundo y del sentido de su porvenir*. 2nd ed. Mexico City: Fondo de Cultura Económico, 1977.

Ortiz Macedo, Luis. *Edouard Pingret: Un pintor romántico francés que retrató el México del mediar del siglo XIX*. Mexico City: Fomento Cultural Banamex, 1989.

Ortiz Monasterio, José. *México eternamente: Vicente Riva Palacio ante la escritura de la historia*. Mexico City: Instituto Mora, 2004.

Osorio, Alejandra. "Postcards in the Porfirian Imaginary." *Social Justice* 34, no. 1 (2007): 141–54.

Overmeyer-Velázquez, Mark. *Visions of the Emerald City: Modernity, Tradition, and the Formation of Profirian Oaxaca, Mexico*. Durham: Duke University Press, 2006.

Padrón, Ricardo. "A Sea of Denial: The Early Modern Spanish Invention of the Pacific Rim." In *Hispanic Review* (winter 2009): 1–27.

———. "Mapping Plus Ultra: Cartography, Space and Hispanic Modernity." In *Representations* 79 (summer 2002): 28–60.

———. *The Spacious Word: Cartography, Literature and Empire in Early Modern Spain*. Chicago: University of Chicago Press, 2004.

Paquette, Gabriel B. *Enlightenment, Governance, and Reform in Spain and Its Empire, 1759–1808*. New York: Palgrave Macmillan, 2008.

Parker, Geoffrey. "Maps and Ministers: The Spanish Hapsburgs." In *Monarchs, Ministers and Maps: The Emergence of Cartography as the Tool of Government in Early Modern Europe*, edited by David Buissert, 153–67. Chicago: University of Chicago Press, 1992.

Pastoureau, Mireille. "French School Atlases: Sixteenth to Eighteenth Centuries." In John A. Wolter and Ronald E. Grim, eds. *Images of the World: The Atlas through History*, 109–34. Washington: Library of Congress, 1997.

Pedley, Mary Sponberg. *Bel et Utile: The Work of the Robert de Vaugondy Family of Mapmakers*. Tring: Map Collector Publications, 1992.

Pérez Benavides, Amada Carolina. "Actores, escenarios y relaciones sociales en tres publicaciones periódicos mexicanas de mediados del siglo XIX." In *Historia Mexicana* 56, no. 4 (2007): 1163–99.

Pérez Escamilla, Ricardo. "Arriba el Telón. Los litógrafos mexicanos, vanguardia artística y política del siglo XIX." In *Nación de imágenes: La litografía mexicana del siglo XIX*, 19–41. Exhibition catalogue. Mexico City: Consejo Nacional para la Cultura y las Artes, 1994.

Pérez Rosales, Laura. "Manuel Orozco y Berra." In *En busca de un discurso integrador de la nación, 1848–1884*, edited by Antonia Pi-Suñer Llorens, 359–86. Mexico City: Universidad Nacional Autónoma de México, 2001.

Pérez Salas C, María Esther. *Costumbrismo y litografía en México: Un nuevo mundo de ver*. Mexico City: Universidad Nacional Autónoma de México, 2005.

———. "Ignacio Cumplido: Un empresario a cabalidad." In *Empresa y cultura en tinta y papel (1800–1860)*, edited by Laura Suárez de la Torre, 145–56. Mexico City: Instituto Mora, 2001.

———. "Los secretos de una empresa exitosa: La imprenta de Ignacio Cumplido." In *Constructores de un cambio cultural: Impresores-editores y libreros en la ciudad de México 1830–1855*, edited by Laura Suárez de la Torre, 145–56. Mexico City: Instituto Mora, 2003.

Pérez Vejo, Tomás. "La invención de una nación: La imagen de México en la prensa ilustrada de la primera mitad del siglo XIX (1830–1855)." In *Empresa y cultura en tinta y papel (1800–1860)*, edited by Laura Suárez de la Torre, 395–408. Mexico City: Instituto Mora, 2001.

Phelan, John Leddy. "Bourbon Innovations, American Responses." In *Latin American Revolutions 1808–1826*, edited by John Lynch, 41–49. Norman: University of Oklahoma Press, 1994.

Philip, David. "Modern Vision." *Oxford Art Journal* 16, no. 1 (1993): 129–38.

Pichardo Hernández, Hugo, and José Omar Moncada Maya. "La labor geográfica de Antonio García Cubas en el Ministerio de Hacienda, 1868–1876." *Estudios de Historia Moderna y Contemporánea de México*, núm. 31 (January–June) 2006: 83–107.

Pickles, John. *A History of Spaces: Cartographic Reason, Mapping and Geo-coded World*. New York: Routledge, 2004.

Pilcher, Jeffery M. *¡Que vivan los tamales! Food and the Making of Mexican Identity*. Albuquerque: University of New Mexico Press, 1998.

———. "Recipes for *Patria*: Cuisine, Gender, and Nation in Nineteenth-Century Mexico." In *Recipes for Reading: Community Cookbooks and Their Stories*, edited by Anne L. Bower, 200–215. Amherst: University of Massachusetts Press, 1997.

Pimental, Juan. "The Iberian Vision: Science and Empire in the Frameworks of a Universal Monarchy 1500–1800." *Osiris* 15 (2000): 17–30.

Pi-Suñer Llorens, Antonia, ed. *En busca de un discurso integrador de la nación, 1848–1884*. Mexico City: Universidad Nacional Autónoma de México, 2001.

———. "Una gran empresa cultural de mediados del siglo XIX: El *diccionario universal de historia y de geografía*." In *Empresa y cultura en tinta y papel (1800–1860)*, edited by Laura Suárez de la Torre, 409–18. Mexico City: Instituto Mora, 2001.

Podgorny, Irina. "The Reliability of Ruins. *Journal of Spanish Cultural Studies* 8, no. 2 (July 2007): 213–33.

Poole, Deborah. "An Image of 'Our Indian': Type Photographs and Racial Sentiments in Oaxaca, 1920–1940." *Hispanic American Historical Review* 84, no. 1 (February 2004): 37–82.

———. *Vision, Race, and Modernity: A Visual Economy of the Andean Image World*. Princeton: Princeton University Press, 1997.

Poole, Stafford. *Our Lady of Guadalupe: The Origins and Sources of a Mexican National Symbol 1531–1797*. Tucson: University of Arizona Press, 1995.

Portuondo, María M. *Secret Science: Spanish Cosmography and the New World.* Chicago: University of Chicago Press, 2009.

Powers, Karen Vieira. "Conquering Discourses of 'Sexual Conquest': Of Women Language and *Mestizaje*." *Colonial Latin American Review* 11, no. 1 (2002): 7–32.

Pratt, Mary Louise. *Travel Writing and Transculturation*. New York: Routledge, 1992.

Rabasa, José. *Inventing A-m-e-r-i-c-a: Spanish Historiography and the Formation of Eurocentrism*. Norman: University of Oklahoma Press, 1993.

Radcliffe, Sarah, and Sallie Westwood. *Remaking Nation: Place, Identity and Politics in Latin America*. New York: Routledge, 1996.

Ramensnyder, Amy G. "Our Lady of Colonization." Lecture, John Carter Brown Library, 3 December 2003.

Ramírez, Fausto. "México a través de los siglos (1881–1910): La pintura de historia durante el Porfiriato." In *Los pinceles de la historia: La fabricación del estado 1864–1910*, edited by Esther Acevedo and Fausto Ramírez, 110–49. Mexico City: Instituto Nacional de Bellas Artes, 2003.

———. "La construcción de la patria y el desarrollo del paisaje en el México decimonónico." In *Hacia otra historia del arte en México*. Tomo 2: *Amplitud del modernismo y la modernidad (1861–1920)*, edited by Stacy Widdifield. 2:269–91. Mexico City: Consejo Nacional para la Cultura y las Artes, 2001.

———. "Pintura e historia en México a mediados del siglo XIX: El programa artístico de los conservadores." In *Hacia otra historia del arte en México: De la estructuración colonial a la exigencia nacional (1780–1860)*, edited by Esther Acevedo. Tomo 1:82–104. Mexico City: Consejo Nacional para la Cultura y las Artes, 2001.

Ramos Escandón, Carmen, ed. *Género e historia: La historiografía sobre la mujer.* Mexico City: Universidad Autónoma Metropolitana, 1992.

———, ed. *Presencia y transparencia: La mujer en la historia de México*. Mexico City: El Colegio de México, 1987.

Rebert, Paula. *La Gran Línea: Mapping the United States–Mexico Boundary, 1849–1857*. Austin: University of Texas Press, 2001.

———. "Los ingenieros mexicanos en la frontera: Cartografía de los límites entre México y Estados Unidos, 1849–1857." In *México a través de los mapas*, edited by Héctor Mendoza Vargas, 111–29. Mexico City: Instituto de Geografía, Universidad Nacional Autónoma de México, 2000.

Rees, Peter. "Origins of Colonial Transportation in Mexico." *Geographical Review* 65, no. 3 (July 1975): 323–34.

Rees, Ronald. "Historical Links between Cartography and Art." *Geographical Review* 70, no. 1 (January, 1980): 60–78.

Reese, Thomas F., and Carol McMichael Reese. "Revolutionary Urban Legacies:

Porfirio Díaz's Celebrations of the Centennial of Mexican Independence in 1910." In XVII Coloquio Internacional de Historia del Arte: *Arte, historia e identidad en América. Visiones Comparativas.* Tomo 3: 361–72. Mexico City: Universidad Nacional Autónoma de México, 1994.

Reyna, María del Carmen. "Impresores y libreros extranjeros en la ciudad de México, 1821–1853." In *Empresa y cultura en tinta y papel (1800–1860)*, edited by Laura Suárez de la Torre, 259–71. Mexico City: Instituto Mora, 2001.

Riguzzi, Paolo. "México próspero: Las dimensiones de la imagen nacional en el Porfiriato." Translated by Francisco Pérez Arce. *Historias* 20 (1988): 137–57.

Ristow, Walter. "Lithography and Maps, 1796–1850." In *Five Centuries of Map Printing*, edited by David Woodward, 77–89. Chicago: University of Chicago Press, 1975.

Roa-de-la-Carrera, Cristían Andrés. *Histories of Infamy: Francisco López de Gómara and the Ethics of Spanish Imperialism.* Boulder: University Press of Colorado, 2005.

Robinson, Author H. *Early Thematic Mapping in the History of Cartography.* Chicago: University of Chicago Press, 1982.

Robinson, Cecil. *The View from Chapultepec: Mexican Writers on the Mexican-American War.* Tucson: University of Arizona Press, 1989.

Rocha, Martha Eva. *El álbum de la mujer: Antología ilustrada de las mexicanas.* Tomo 4: *El Porfiriato y la Revolución.* Mexico City: Instituto Nacional de Antropología e Historia, 1991.

Rodríguez, José Antonio. "Fotógrafos viajeros camino abierto." *Artes de México*, núm. 31 (1996): 56–66.

Rodríguez Benítez, Leonel. "La geografía en México independiente 1824–1835." In *Mundialización de la ciencia y cultura nacional: Actas del Congreso Internacional "Ciencia, Descubrimiento y Mundo Colonial,"* edited by A. Elena Lafuente and M. L. Ortega, 429–38. Madrid: Doce Calles, 1993.

Rodríguez O., Jaime. *The Divine Charter: Constitutionalism and Liberalism in Nineteenth-Century Mexico.* Lanham, Md.: Rowman and Littlefield Publishers, 2005.

———. *Down from Colonialism: Mexico's Nineteenth Century Crisis.* Los Angeles: University of California Chicano Studies Research Center Publications, 1983.

———, eds. *The Origins of Mexican National Politics 1808–1847.* Wilmington, Del.: Scholarly Resources, 1997.

Rodríguez Prampolini, Ida. *La crítica de arte en México en el siglo XIX.* 3 vols. 2nd ed. Mexico City: Universidad Nacional Autónoma de México, 1997.

Rojas, Rafael. *La escritura de la Independencia: El surgimiento de la opinión pública en México.* Mexico City: Taurus, Centro de Investigación y Docencia Económicas, 2003.

Román Gutiérrez, José Francisco, ed. *Las reformas borbónicas y el nuevo orden colonial.* Mexico City: Instituto Nacional de Antropología e Historia, 1998.

Romero Navarrete, Lourdes, and Filipe I. Echenique March. *Relaciones geográficas de 1792*. Mexico City: Instituto Nacional de Antropología e Historia, 1994.
Ross, Dorothy. “Historical Consciousness in Nineteenth-Century America.” *American Historical Review* 89, no. 4 (October 1984): 909–28.
Rubiés, Joan-Pau. “Futility in the New World: Narratives of Travel in Sixteenth-Century America.” In *Voyages and Visions: Towards a Cultural History of Travel*, edited by Jaś Elsner and Joan-Pau Rubiés, 74–100. London: Reaktion Books, 1999.
Ruiz Naufal, Víctor Manuel. “La faz del terruño: Planos locales y regionales siglos XVI–XVIII.” In *México a través de los mapas*, edited by Héctor Mendoza, 33–69. Mexico City: Instituto de Geografía, Universidad Nacional Autónoma de México, 2000.
Rupke, Nicolaas. *Alexander von Humboldt: A Metabiography*. New York: Peter Lang, 2005.
———. “A Geography of Enlightenment: The Critical Reception of Alexander von Humboldt’s Mexico Work.” In *Geography and Enlightenment*, edited by David N. Livingstone and Charles W. J. Withers, 319–39. Chicago: University of Chicago Press, 1999.
Russo, Alessandra. *El realismo circular: Tierras, espacios y paisajes de la cartografía novohispana, siglos XVI y XVII*. Mexico City: Universidad Autónoma de México, 2005.
Ryan, James R. *Picturing Empire: Photography and the Visualization of the British Empire*. Chicago: University of Chicago Press, 1997.
Saborit, Antonio. “Tipos y costumbres artes y guerras del callejero amor.” In *Nación de imágenes: La litografía mexicana del siglo XIX*, 56–69. Exhibition catalogue. Mexico City: Consejo Nacional para la Cultura y las Artes, 1994.
Sachs, Aaron. *The Humboldt Current: Nineteenth-Century Exploration and the Roots of American Environmentalism*. New York: Viking, 2006.
———. “The Ultimate ‘Other’: Post-Colonialism and Alexander von Humboldt’s Ecological Relationship with Nature.” *History and Theory*, theme issue 42 (December 2003): 111–35.
———. *Ciencia y prensa durante la ilustración latinoamericana*. Mexico City: Universidad Autónoma del Estado de México, 1996.
Saldino García, Alberto. *Dos científicos de la ilustración hispanoamericana: J. A. Alzate y F. J. De Caldas*. Mexico City: Universidad Autónoma del Estado de México, 1990.
Sánchez Lamego, Miguel A. *El primer mapa general de México elaborado por un mexicano*. Publicación núm. 10 de la Comisión de Cartografía. Mexico City: Instituto Panamericano de Geografía é Historia, publicación 175, 1955.
Schávelzon, Daniel, ed. *La polémica del arte nacional en México, 1850–1910*. Mexico City: Fondo de Cultura Económica, 1988.
Schiebinder, Londa. *Plants and Empire: Colonial Bioprospecting in the Atlantic World*. Cambridge: Harvard University Press, 2004.

Schiebinder, Londa, and Claudia Swan, eds. *Colonial Botany: Science, Commerce, and Politics in the Early Modern World*. Philadelphia: University of Pennsylvania Press, 2005.

Schilder, Günter. "Willem Jansz. Blaeu's Wall Map of the World, on Mercator's Projection 1606–07 and Its Influence." *Imago Mundi* 31 (1979): 36–54.

Schmidt-Nowara, Christopher, and John M. Nieto-Philips. *Interpreting Spanish Colonialism: Empires, Nations, and Legends*. Albuquerque: University of New Mexico Press, 2005.

Schreffler, Michael J. "Vespucci Rediscovers America: The Pictorial Rhetoric of Cannibalism in Early Modern Culture." *Art History* 28, no. 3 (June 2005): 295–310.

Schulten, Susan. "Emma Willard and the Graphic Foundation of American History." *Journal of Historical Geography* 33 (2007): 542–64.

———. *The Geographical Imagination in America, 1880–1950*. Chicago: University of Chicago Press, 2001.

Schwartz, Joan M. "*The Geography Lesson*: Photographs and the Construction of Imaginative Geographies." *Journal of Historical Geography* 22, no. 1 (1996): 16–45.

Schwartz, Joan M., and James R. Ryan. *Picturing Place: Photography and the Geographical Imagination*. London: I. B. Tauris, 2003.

Schwartz, Seymour I. *The Mapping of America*. New York: Harry N. Abrams, 1980.

Schwartz, Vanessa R., and Jeannene M. Przyblyski, eds. *The Nineteenth-Century Visual Culture Reader*. New York: Routledge, 2004.

Scott, James C. *Seeing like a State: How Certain Schemes to Improve the Human Conditions Have Failed*. New Haven: Yale University Press, 1998.

Seed, Patricia. *Ceremonies of Possession in Europe's Conquest of the New World 1492–1640*. Cambridge: Cambridge University Press, 1995.

Segre, Erica. "The Development of *Costumbrista*: Iconography and Nation-building Strategies in Literary Periodicals of the Mid-Nineteenth Century." In *Intersected Identities: Strategies of Visualization in Nineteenth and Twentieth Century Mexican Cultures*, by Erica Segre, 5–58. New York: Berghahn Books, 2007.

Sharp, Joanne. "Gendering Nationhood: A Feminist Engagement with National Identity." In *Bodyspace: Destabilizing Geographies of Gender and Sexuality*, edited by Nancy Duncan, 98–107. New York, Routledge, 1996.

Sinkin, Richard. N. *The Mexican Reform, 1855–1876: A Study in Liberal Nation-Building*. Austin: Institute of Latin American Studies, 1979.

Smith, Woodruff. *Consumption and the Making of Respectability 1600–1800*. New York: Routledge Press, 2002.

Solano, Francisco de, ed. *Relaciones geográficas de Arzobispado de México, 1743*. 2 vols. Madrid: Consejo Superior de Investigaciones Científicas, 1988.

Solano, Francisco de, and Pilar Ponce, eds. *Cuestionarios para la formación de las relaciones geográficas de Indias siglos XVI–XIX*. Madrid: Consejo Superior de Investigaciones Científicas, 1988.

Solares Robles, Laura. "Prosperidad y quiebra: Una vivencia constante en la vida de Mariano Galván Rivera." In *Empresa y cultura en tinta y papel (1800–1860)*, edited by Laura Suárez de la Torre, 110–21. Mexico City: Instituto Mora, 2001.

Sommer, Doris. *Foundational Fictions: The National Romances of Latin America*. Berkeley: University of California Press, 1991.

Stafford, Barbara. *Voyage into Substance: Art, Science, Nature and the Illustrated Travel Account 1760–1840*. Cambridge: MIT Press, 1984.

Stern, Steve J. *The Secret History of Gender: Women, Men, and Power in Late Colonial Mexico*. Chapel Hill: University of North Carolina Press, 1995.

Stoler, Ann Laura. *Carnal Knowledge and Imperial Power: Race and the Intimate in Colonial Rule*. Los Angeles: University of California Press, 2002.

Suárez de la Torre, Laura, ed. *Constructores de un cambio cultural: Impresores-editores y libreros en la ciudad de México 1830–1855*. Mexico City: Instituto Mora, 2003.

———, ed. *Empresa y cultura en tinta y papel (1800–1860)*. Mexico City: Instituto Mora, 2001.

———, "Una imprenta florencia en la calle de la Palma número 4." In *Empresa y cultura en tinta y papel (1800–1860)*, 136–44. Mexico City: Instituto Mora, 2001.

Tagg, John. *The Burden of Representation: Essays on Photographies and Histories*. Minneapolis: University of Minnesota Press, 1993.

Tanck de Estrada, Dorothy. *Atlas ilustrado de los pueblos de indios Nueva España, 1800*. Mexico City: Colegio de México, 2005.

Tenenbaum, Barbara. "Streetwise History: The Paseo de la Reforma and the Porfirian State, 1876–1910." In *Rituals of Rule, Rituals of Resistance: Public Celebrations and Popular Culture in Mexico*, edited by William H. Beezley, Cheryl English Martin, and William E. French, 127–50. Wilmington, Del.: Scholarly Resources, 1994.

Tenorio-Trillo, Mauricio. *Mexico at the World's Fairs: Crafting a Modern Nation*. Berkeley: University of California, 1996.

———. "1910 Mexico City: Space and Nation in the City of the Centenario." *Journal of Latin American Studies* 28, no. 1 (February 1996): 75–104.

Terán, María Isabel. "'La heroína mexicana': Una novel inédita novohispana de siglo XVII." *Anales del Instituto de Investigaciones Estéticas*, 21, núm. 74 (1999): 291–309.

Thomson, Guy. "Bulwarks of Patriotic Liberalism: The National Guard, Philharmonic Corps and Patriotic Juntas in Mexico, 1847–1888." *Journal of Latin American Studies* 22, no. 1 (February 1990): 31–68.

———. "Liberalism and Nation-Building in Mexico and Spain during the Nineteenth Century." In *Studies in the Formation of the Nation-State of Latin*

America, edited by James Dunkerly, 189–211. London: Institute of Latin American Studies, 2002.

Thrower, Norman J. W. *Maps and Civilization: Cartography in Culture and Society.* 2nd ed. Chicago: University of Chicago Press, 1999.

Todd Ambler, Louise, and Melissa Banta, eds. *The Invention of Photography and Its Impact on Learning.* Cambridge: Harvard University Library, 1989.

Trabulse, Elías, ed. *Cartografía mexicana: Tesoros de la nación siglos XVI a XIX.* Mexico City: Archivo General de la Nación, 1983.

———. "Introducción." In *Viajeros europeas del siglo XIX en México*, 23–38. Mexico City: Fomento Cultural Banamex, 1996.

Tuñón, Julia. *El álbum de la mujer: Antología ilustrada de las mexicanas.* Tomo 3: *El siglo XIX (1821–1880).* Mexico City: Instituto Nacional de Antropología e Historia, 1991.

Uribe, Eloisa. "Claves para leer la escultura mexicana: Periodo 1781–1861." In *Hacia otra historia del arte en México: De la estructuración colonial a la exigencia nacional (1780–1860)*, edited by Esther Acevedo. Tomo 1:165–90. Mexico City: Consejo Nacional para la Cultura y las Artes, 2001.

Valázquez Guadarrama, Angélica. "La historia patria en el Paseo de la Reforma: La propuesto de Francisco Sosa y la consolidación del estado en el Porfiriato." In XVII Coloquio Internacional de Historia del Arte: *Arte, historia e identidad en América. Visiones Comparativas.* Tomo 3: 333–44. Mexico City: Universidad Nacional Autónoma de México, 1994.

———. "La representación de la domesticidad burguesa: El caso de las her-manas Sanromán." In *Hacia otra historia del arte en México. De la estructura-ción colonial a la exigencia nacional (1780–1860)*, edited by Esther Acevedo. Tomo 1:122–48. Mexico City: Consejo Nacional para la Cultura y las Artes, 2001.

Valverde, Nuria, and Antonio Lafuente. "Space Production and Spanish Imperial Geopolitics." In *Science in the Spanish and Portuguese Empires (1500–1800)* 198–215. Stanford: Stanford University Press, 2009.

Van Hagen, Victor Wolfgang. *Frederick Catherwood, Architect.* New York: Oxford University Press, 1950.

Van Young, Eric. "Conclusion: The State as Vampire—Hegemonic Projects, Public Ritual, and Popular Culture in Mexico, 1600–1990." In *Rituals of Rule, Rituals of Resistance: Public Celebrations and the Popular Culture in Mexico*, edited by William H. Beezley, Cheryl English Martin, and William E. French, 343–74. Wilmington, Del.: Scholarly Resources, 1994.

Varey, Simon, Rafael Chabrán, and Dora B. Weiner, eds. *Searching for the Secret of Nature: The Life and World of Dr. Francisco Hernández.* Stanford: Stanford University Press, 2000.

Vázquez Mantecón, María del Carmen. "La *china* mexicana, mejor conocida como *china poblana*." *Anales del Instituto de Investigaciones Estéticas*, 22, núm. 77 (2000): 123–50.

Viajeros europeas del siglo XIX en México. Mexico City: Fomento Cultural Banamex, 1996.

Warner, Marina. *Monuments and Maidens: The Allegory of the Female Form*. New York: Atheneum, 1985.

Warren, Richard. "Elections and Popular Participation in Mexico, 1808–1836." In *Liberals, Politics, and Power: State Formation in Nineteenth Century Latin America*, edited by Vincent C. Peloso and Barbara A. Tenenbaum, 30–58. Athens: University of Georgia Press, 1996.

Weiner, Richard. *Race, Nation and Market: Economic Culture in Porfirian Mexico*. Tucson: University of Arizona Press, 2004.

Welch, Thomas L., and Myrian Figueras. *Travel Accounts and Descriptions of Latin American and the Caribbean, 1800–1920: A Selected Bibliography*. Washington: Organization of American States, 1982.

Wells, Allen, and Gilbert M. Joseph. "Modernizing Visions, 'Chilango' Blueprints, and Provincial Growing Pains: Merida at the Turn of the Century." *Mexican Studies / Estudios Mexicanos* 8 (summer 1992): 167–215.

West, Nancy. "Fantasy, Photography, and the Marketplace: Oliver Wendell Holmes and the Stereoscope." *Nineteenth-Century Contexts* 19 (1996): 231–58.

West, Robert C. "The *Relaciones Geográficas* of Mexico and Central America, 1740–1792." In *Guide to the Ethnohistorical Sources*, edited by Howard F. Cline. Vol. 12: *Handbook of Middle American Indians*. Austin: University of Texas Press, 1964–76.

Wey Gómez, Nicolás. *The Tropics of Empire: Why Columbus Sailed South to the Indies*. Cambridge: MIT Press, 2008.

Widdifield, Stacie. "El impulso de Humboldt y la mirada extranjera sobre México." *Hacia otra historia del arte en México: De la estructuración colonial a la exigencia nacional (1780–1860)*, edited by Esther Acevedo. Tomo 1: 257–72. Mexico City: Consejo nacional para la Cultura y las Artes, 2001.

———. "El indio re-tratado." In *Hacia otra historia del arte en México: De la estructuración colonial a la exigencia nacional (1780–1860)*, edited by Esther Acevedo. Tomo 1:241–56. Mexico City: Consejo Nacional para la Cultura y las Artes, 2001.

———. *The Embodiment of the National in Late Nineteenth-Century Mexican Painting*. Tucson: University of Arizona Press, 1996.

———. "Manuel Tolsá's Equestrian Portrait of Charles IV: Art History, Patrimony, and the City." *Journal X* 8, no. 1 (2003): 61–83.

———. "Modernizando el pasado: La recuperación de arte y su historia, 1860–1920." In *Hacia otra historia del arte en México*. Tomo 2: *Amplitud del modernismo y la modernidad (1861–1920)*, edited by Stacie Widdifield. 2: 69–98. Mexico City: Consejo Nacional para la Cultura y las Artes, 2004.

Widdifield, Stacie, ed. *Hacia otra historia del arte en México*: Tomo 2: *Amplitud del modernismo y la modernidad (1861–1920)*. Mexico City: Consejo Nacional para la Cultura y las Artes, 2001.

Wierich, Jochen. "Struggling through History: Emanuel Leutze, Hegel and Empire." *American Art* 15, no. 2 (summer 2001): 52–71.
Wilcox, Scott. "El panorama de Leicester Square." In *Viajeros europeas del siglo XIX en México*, 126–35. Mexico City: Fomento Cultural Banamex, 1996.
Winichakul, Thongchai. *Siam Mapped: A History of the Geo-Body of a Nation.* Honolulu: University of Hawai'i Press, 1994.
Winship, George Parker. "Character de la Conquista Espanola en America y en Mexico Segun los Textos Historicos de los Historiadores Primitivos [typographical errors in original]." Review in *American Historical Review* 7, no. 4 (July 1902): 757–58.
Withers, Charles W. J. *Placing the Enlightenment: Thinking Geographically about the Age of Reason.* Chicago: University of Chicago Press, 2007.
Wolcott, Roger, ed. *The Correspondence of William Hickling Prescott 1833–1847.* New York: Houghton Mifflin Company, 1925,
Wolff, Hans, ed. *America: Early Maps of the New World.* Munich: Prestel, 1992.
———, ed. "Newly Discovered Islands, Regions and People: The Letters of Christopher Columbus, Amerigo Vespucci and Hernán Cortés." In *America: Early Maps of the New World*, edited by Hans Wolff, 103–8. Munich: Prestel, 1992.
Wolter, John A., and Ronald E. Grim. *Images of the World: The Atlas through History.* Washington: Library of Congress, 1997.
Wood, Denis, with John Fels. *The Power of Maps.* New York: Guilford Press, 1992.
Wright-Rios, Edward. "Indian Saints and Nation States: Ignacio Manuel Altamirano's Landscapes and Legends." *Mexican Studies / Estudios Mexicanos* 20 (winter 2004): 47–68.
Zavala, Huguette. "América inventada: Fiestas y espectáculos en la Europa de los siglos XVI al XX." In XVII Coloquio Internacional de Historia del Arte: *Arte, historia e identidad en América: Visiones comparativas*, edited by Gustavo Curiel and Renato González Mello y Juana Gutiérrez Haces. Tomo 1:33–50. Mexico City: Universidad Nacional Autónoma de México, Instituto de Investigaciones Estéticas, 1994.

INDEX

MAGALI M. CARRERA is Chancellor Professor of Art History at the University of Massachusetts, Dartmouth. She is the author of *Imagining Identity in New Spain: Race, Lineage, and the Colonial Body in Portraiture and Casta Paintings* (2003).

Library of Congress Cataloging-in-Publication Data

Carrera, Magali Marie
Traveling from New Spain to Mexico : mapping practices
of nineteenth-century Mexico / Magali M. Carrera.
p. cm.
Includes bibliographical references and index.
ISBN 978-0-8223-4976-1 (cloth : alk. paper)
ISBN 978-0-8223-4991-4 (pbk. : alk. paper)
1. García Cubas, Antonio, 1832–1912.
2. Cartography—Mexico—History—19th century.
3. Mexico—Maps. I. Title.
GA483.7.A1C377 2011
912.7209′034—dc22
2010049645